A Photographic Atlas for the MICROBIOLOGY LABORATORY

2nd EDITION

Michael J. Leboffe
San Diego City College

Burton E. Pierce
San Diego City College

Morton Publishing Company
925 W. Kenyon Ave., Unit 12
Englewood, Colorado 80110
http://www.morton-pub.com

Book Team

Publisher: Douglas N. Morton
Biology Editor: Chris Rogers
Production Manager: Joanne Saliger
Typography: Ash Street Typecrafters, Inc.
Cover Design: Bob Schram, Bookends, Inc.

ISBN 0-89582-461-2

10 9 8 7 6 5 4 3 2 1

Printed in the United States of America

Preface

This new edition of *A Photographic Atlas for the Microbiology Laboratory* has been revised and expanded. It still is intended to act as a supplement to introductory microbiology laboratory manuals, but now it may also be used by professional microbiologists in training as an inexpensive alternative to more extensive diagnostic microbiology texts. It is not designed to replace textbooks or lab manuals, nor is it intended to replace actual performance of the techniques. Rather, the photographs and text are supplied to help in interpretation of results and provide the theory behind the tests.

The *Atlas* is divided into thirteen sections, each devoted to a particular component of a typical introductory microbiology course. Topics within the first eight sections include the following components.

PURPOSE The purpose describes why the procedure is done. If used to detect the presence of a particular enzyme, it is listed in this section. Also included in the purpose is any special medical application of the procedure.

PRINCIPLE The principle comprises several features. First, each procedure is used to probe some aspect of bacterial physiology, morphology or biochemistry. These are addressed and the procedure is put into the context of the living microbe. Chemical structures of relevant metabolic intermediates are also shown. Second, the theory behind the procedure itself is discussed. Chemical structures of reagents are usually included. Third, while instructions on *how* to perform each procedure are omitted (since these may be obtained in your laboratory manual or the companion volume, *Exercises for the Microbiology Laboratory, by Pierce and Leboffe*), helpful hints to avoid common pitfalls are provided.

Most of what was in the first edition remains in the second, but the following changes have been made:

- Sections 1 through 4, 6 and 7 have had new photographs added and some old ones replaced with better ones. Some of the artwork was improved, too.
- Many new differential tests were added to Section 5. In this section, tests are listed alphabetically by the ***test* name, with the medium listed parenthetically.** Multiple test media (such as the *Enterotube®*) are listed by the product name.
- The former Section 8 (Host Defenses, Immunology and Serology) has been split into two sections. The host immune cells and organs are now by themselves in Section 8. Section 9, Molecular Techniques, Immunology and Serology, contains some new serological tests and useful procedures like electrophoresis.
- Section 10 covers viruses. It now includes photographs of viral cytopathic effects on various cell cultures.
- Section 11 is entirely new with photographs of major bacterial pathogens along with a short synopsis about the pathogen and the diseases it causes. This replaces the Medical Applications component of the First Edition and consolidates the information.
- Section 12, the Fungi, has had a few pathogens added to it.
- Section 13, Parasitology, has been greatly expanded with the addition of photographs depicting diagnostic stages of helminths and the inclusion of more protozoan parasites.
- Appendices A, B, and C provide detail on glycolysis, Krebs Cycle and fermentation. Appendix D provides a listing of selected references for students who wish to expand on the information provided herein.

We believe good preparation is essential for success in the Microbiology laboratory. We hope this edition of the *Atlas* helps make your microbiological experience a positive one.

Michael J. Leboffe
Burton E. Pierce

Acknowledgments

As with the First Edition, we were again overwhelmed at the willingness of so many people to assist us in producing this work. Our colleagues graciously honored our requests for help and we remain grateful to you all. We also appreciate the adopters (and nonadopters) who offered suggestions for improving the *Atlas*. Without you, there would have been no Second Edition.

Thanks to the following individuals for their contributions to one or both editions:

Dianne Anderson — San Diego City College
David Brady — San Diego City College
Joyce Costello — San Diego City College
Deborah Durand — UCSD Department of Medicine
Jeff Foti — Ward's Natural Science Establishment, Inc.
Susan Garrison — Carolina Biological Supply
Deborah Gelfand — San Diego City College
Mark Kamps — UCSD School of Medicine, Department of Pathology
Randall Kottel — San Diego Mesa College
Shannon McGrath — UCSD School of Medicine, Department of Pathology
Bill Morse — Ward's Natural Science Establishment, Inc.
Darla Newman — Mesa College
Robert Olisa, III, Lt. U.S. Navy — NEPMU 5, Naval Station, San Diego
Nicole O'Brien — San Diego State University
Michele Pierce — San Diego City College
Brett Rustin — San Diego City College
Rachel D. Schrier — UCSD Department of Pathology
David Singer — San Diego City College
Students of the Summer 1995 General Microbiology Class, San Diego City College
Students of the Spring 1998 Morning and Evening General Microbiology Classes, San Diego City College
Joy Sussman — Becton Dickinson Company
Rodrigo Villar — Centers for Disease Control
Robert-Eli Anthony Wheeler and Margaret Wheeler
Clayton Wiley — University of Pittsburgh
Gary Wisehart — San Diego City College
Alan Zeglarski — Instrument Service Co.

Special thanks are warranted for certain individuals. Bob Waddell, Tom Lee, Alan Kuritzky and William McClellan of Scientific Instrument Company, Temecula, CA made an Olympus Ultraviolet Photomicroscope with automatic exposure available to us. Without their assistance, the photomicrographs would have been impossible. We are also grateful to Jerome Hunter, President, Carol Dexheimer, Business Manager, and Marianne Tortorici, Dean of the School of Natural Sciences for their assistance in arranging use of San Diego Community College District facilities through the Civic Center Program.

Newly-formed connections with two establishments expanded our resources and enabled us to provide more thorough coverage of introductory microbiology. First, Dr. Barbara Hemmingsen, Dr. Ron Monroe, Marlene De Mers, and Bruce Wingerd of the San Diego State University Biology Department provided assistance by supplying media, organisms, advice, and access to microscope slides. Second, we developed an invaluable association with the County of San Diego Public Health Laboratory. Our deepest appreciation to Dr. Chris Peter, **Chief Microbiologist** and Dr. Sue Sabet, **Assistant Chief Microbiologist**, and to their entire staff for opening their laboratory and providing specialized media, samples, equipment, and advice not available to us otherwise. Each staff member contributed in some way, and we are indebted to all of you for your help. Special recognition goes to Microbiologists Let Negado and Jane Zackary. They were cooperative beyond belief, fitting our requests around their busy schedules, often on short notice and always without complaint. Your contributions and friendships are highly valued.

The support of Doug Morton and Christine Morton of Morton Publishers has been greatly appreciated through both editions. Their continued confidence and enthusiasm in this project made it possible for us to focus on creativity rather than on business matters. We are also indebted to Chris D. Rodgers, Morton Publishers' Biology Editor, for his advice, publishing insights, and useful surveys. Finally, thanks to Joanne Saliger of Ash Street Typecrafters, Inc. for her creative design of both editions of the *Atlas*. You are truly the "Wizard of Englewood"!

Lastly, thanks and love go to our wives Karen Leboffe and Michele Pierce for their emotional support and understanding that our temporary preoccupation with bacteriological media and photographs does not diminish our love for them. The same goes for the others in the Leboffe clan: Nathan, Alicia, and Eric — Dad's back!

Contents

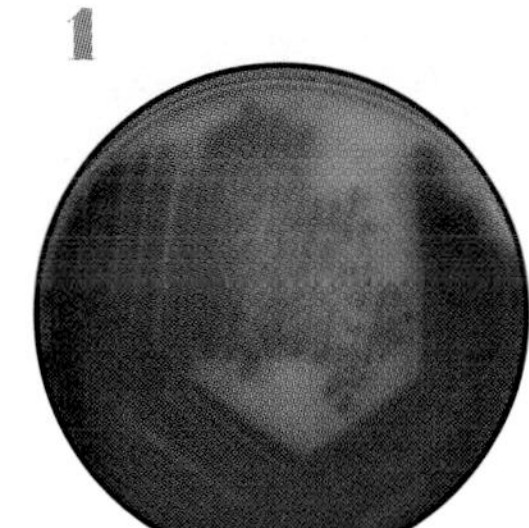

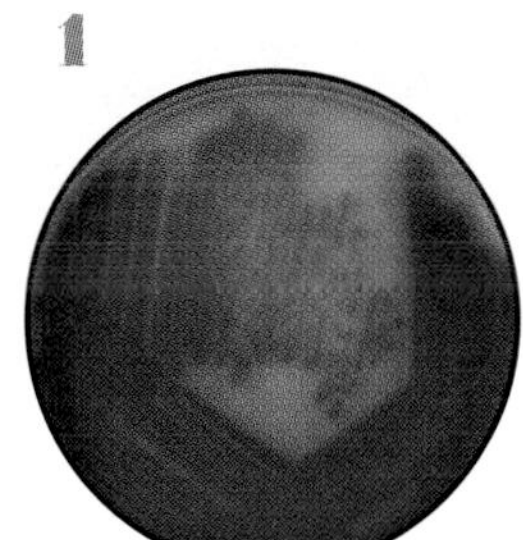

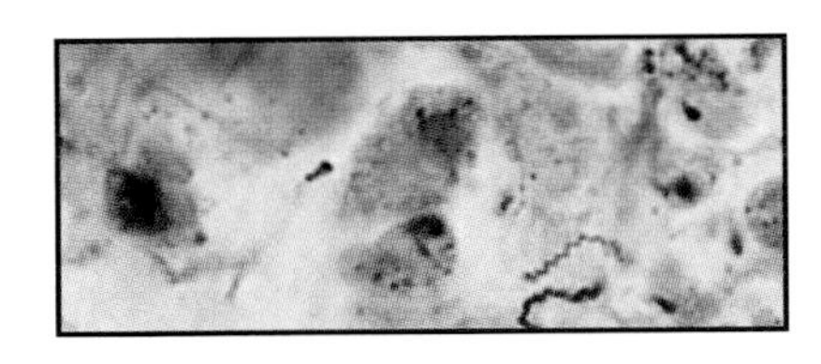

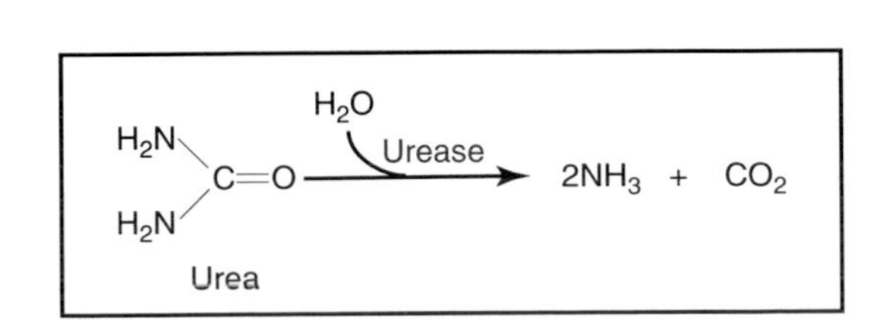

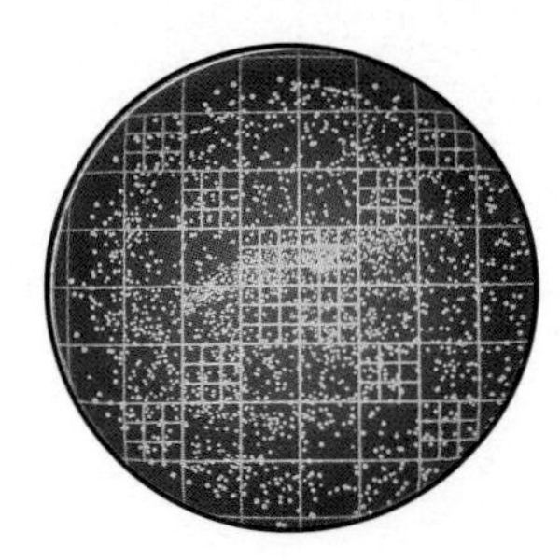

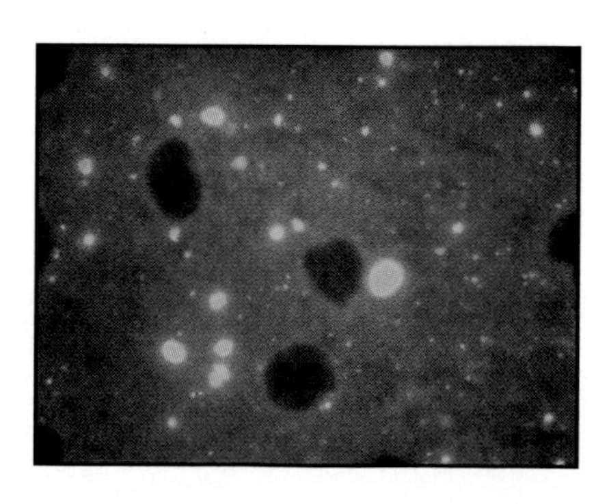

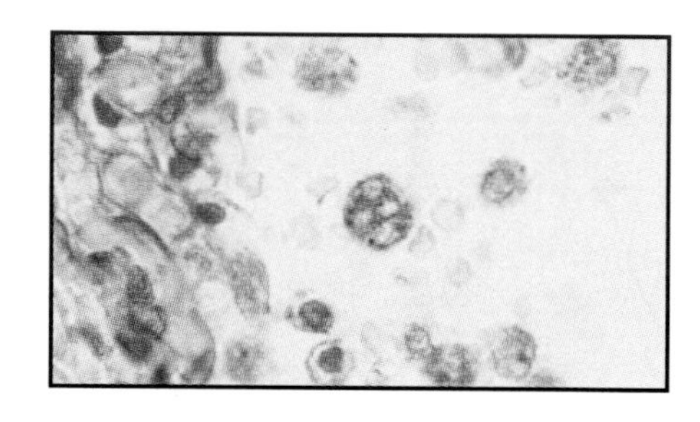

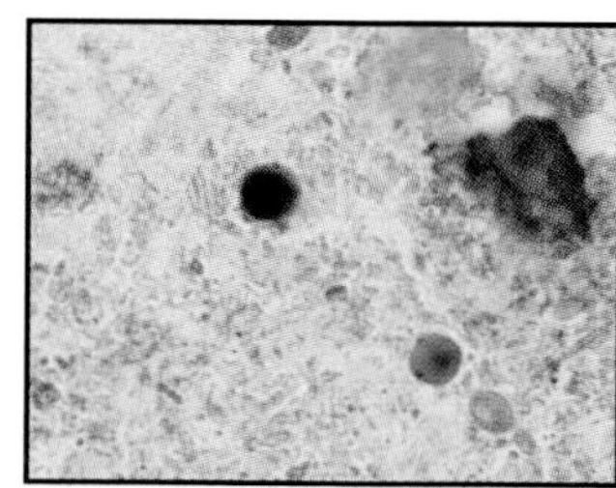

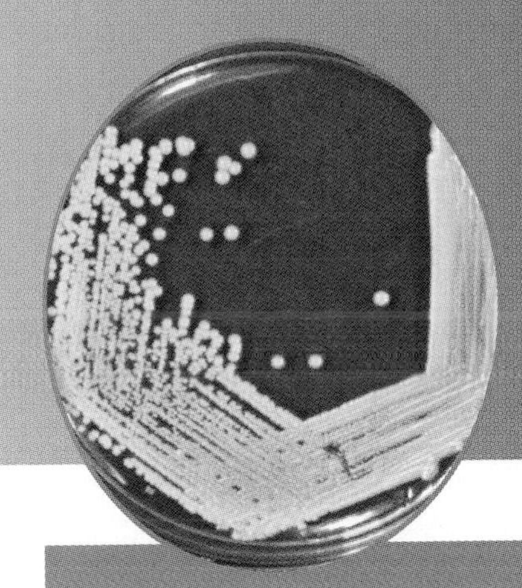

Bacterial Growth Patterns 1

COLONY MORPHOLOGY

PURPOSE Recognizing different bacterial colony morphologies on agar plates is useful for distinguishing between different species in a mixed culture. Once identified as different, cells from a colony may be transferred to a sterile medium to begin a pure culture.

PRINCIPLE When a single bacterial cell is deposited on an agar surface which supplies its nutrient needs, it begins to divide. One cell makes two, two make four, four make eight . . . one billion make two billion, and so on. Eventually, a visible mass of cells is found on the plate where the original cell was deposited. This mass of cells is called a *colony*.

Figures 1-1 through 1-16 show some variety of bacterial colony forms and characteristics. The basic categories include colony shape, margin (edge), elevation, color, and texture. Colony shape may be described as *circular*, *irregular*, or *punctiform* (tiny). The margin may be *entire* (smooth, with no irregularities), *undulate* (wavy), *lobate* (lobed), *filamentous*, or *rhizoid* (branched like roots). Colony

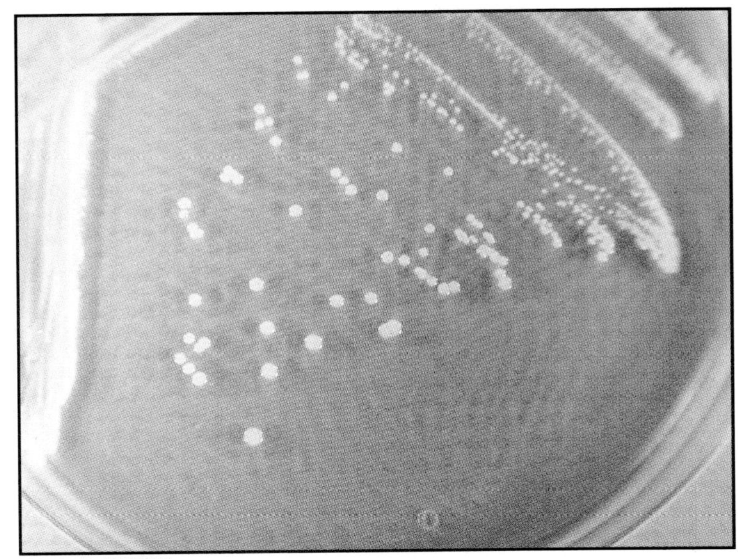

FIGURE 1-2 White, raised, circular and entire colonies of *Staphylococcus epidermidis* viewed from above with reflected light.

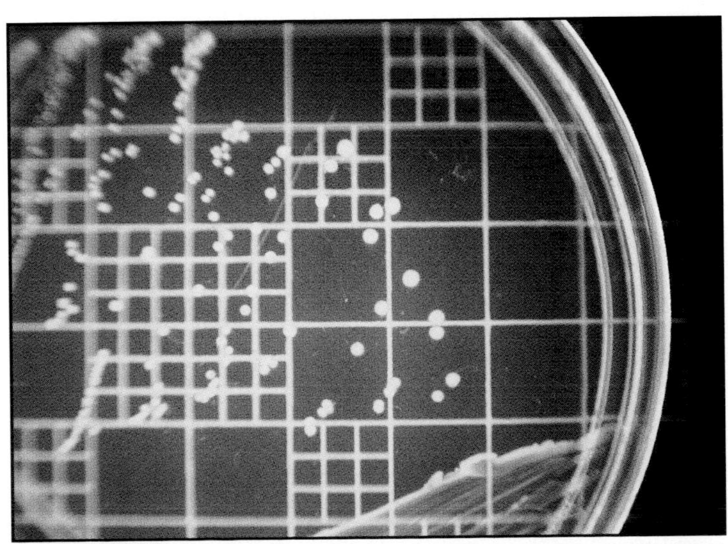

FIGURE 1-1 *Enterococcus faecium* colonies viewed from above with transmitted light are cream colored and circular with an entire margin.

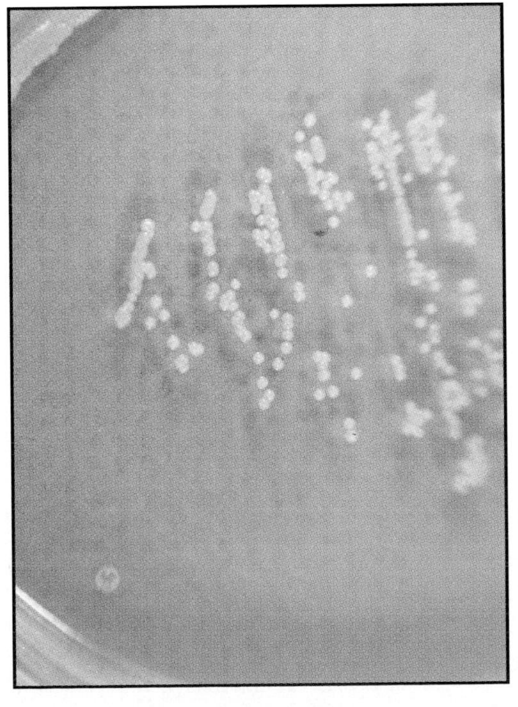

FIGURE 1-3 Convex yellow colonies of *Micrococcus luteus* as seen from above.

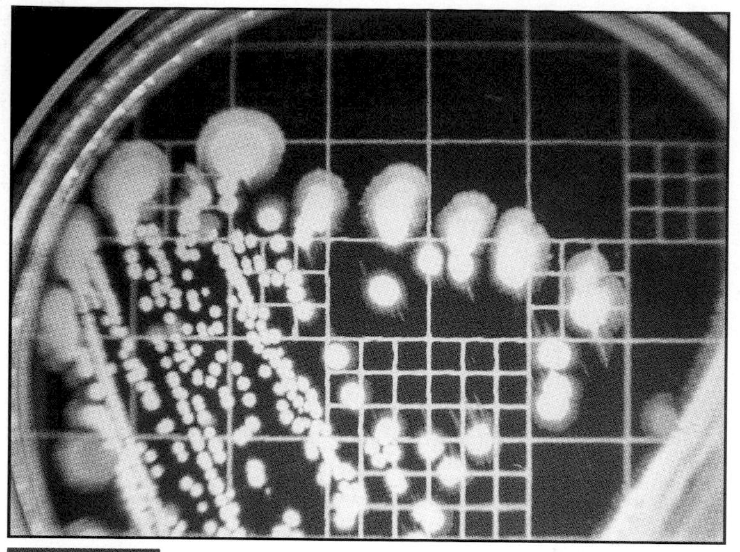

FIGURE 1-4 *Bacillus subtilis* with an opaque center and spreading edge.

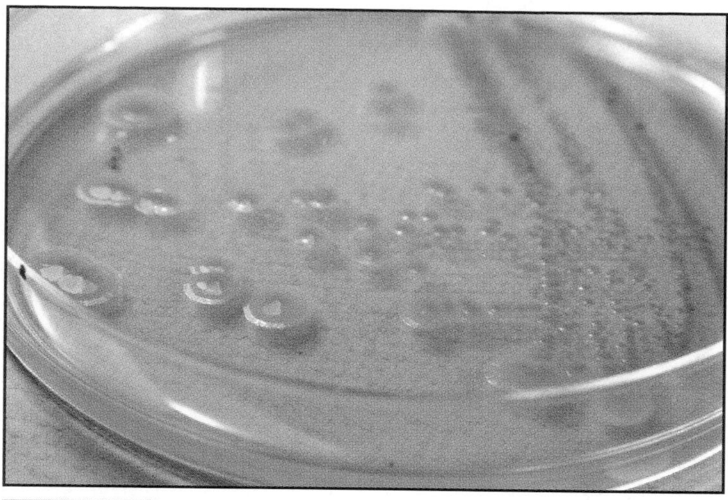

FIGURE 1-5 Shiny, umbonate colonies of *Serratia marcescens*. By viewing the plate at an angle in reflected light, the elevation is easier to determine.

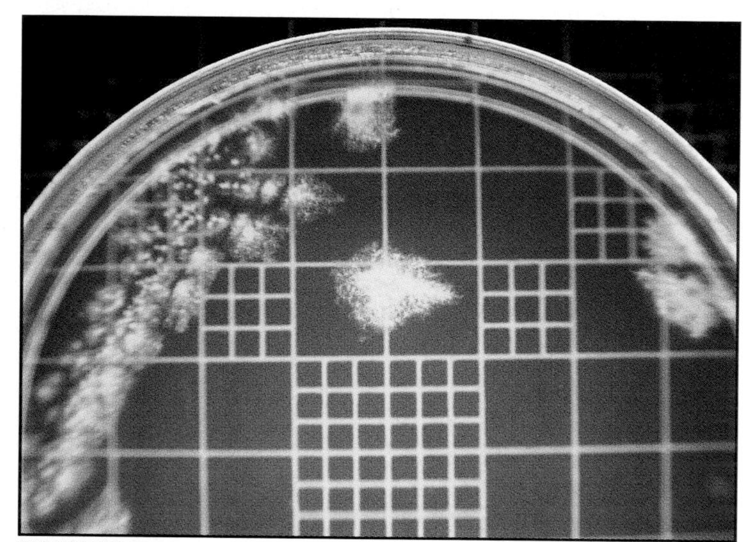

FIGURE 1-6 Irregular, rhizoid growth is demonstrated by these colonies of *Clostridium sporogenes*.

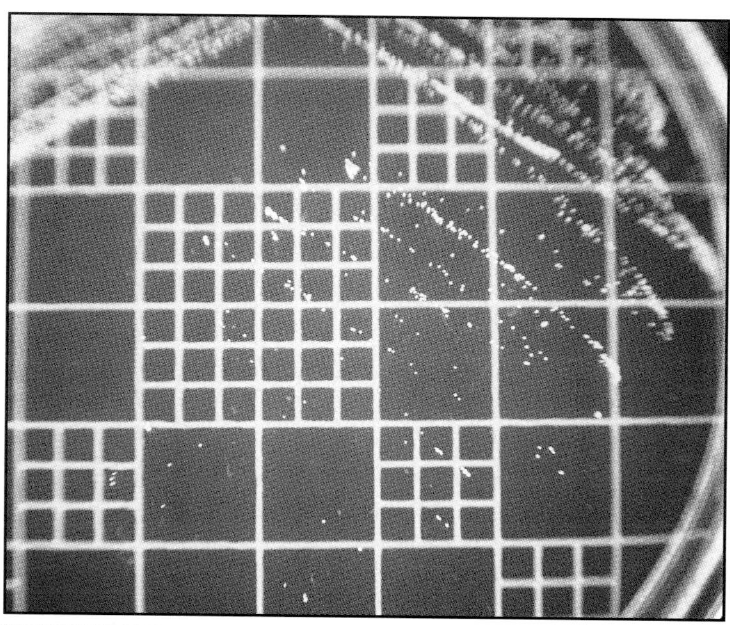

FIGURE 1-7 Punctiform colonies of *Mycobacterium smegmatis*.

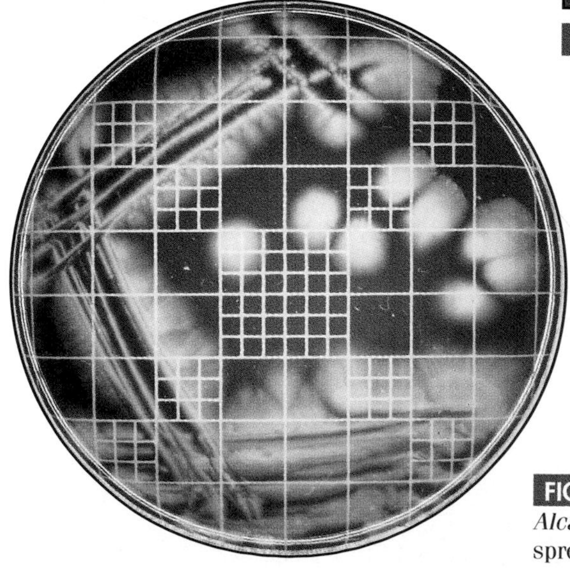

FIGURE 1-8 *Alcaligenes faecalis* demonstrates spreading and translucent growth.

FIGURE 1-9 Raised colonies of *Klebsiella pneumoniae* viewed at an angle in reflected light.

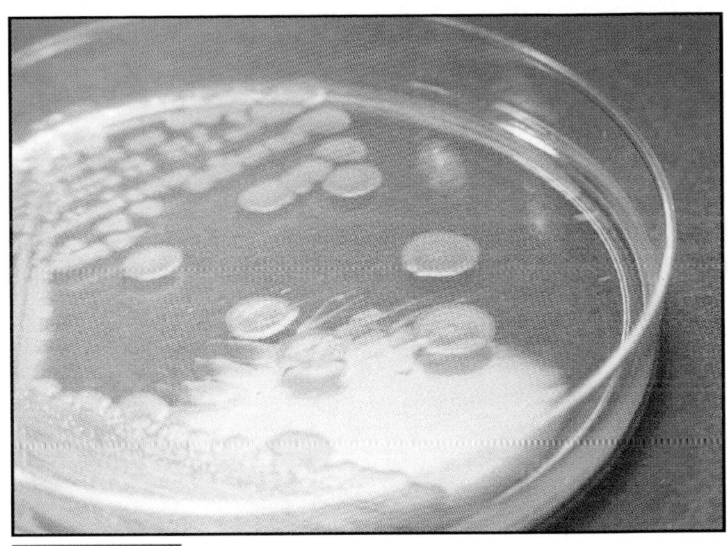

FIGURE 1-10 *Bacillus subtilis* colonies illustrating a raised margin and dull surface.

FIGURE 1-11 Mucoid colonies of *Pseudomonas aeruginosa* grown on Endo agar.

FIGURE 1-12 Members of the genus *Proteus* will swarm at certain intervals and produce a pattern of rings due to their motility. This photograph shows swarming of *Proteus vulgaris* on DNase agar.

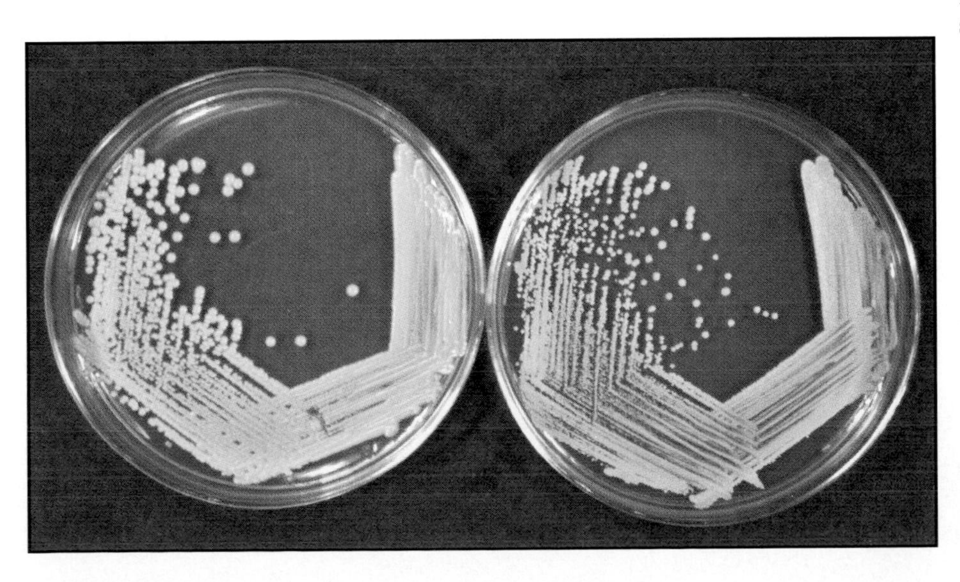

FIGURE 1-13 Closely related species may look very different, as seen in these plates of *Micrococcus luteus* (left) and *M. roseus* (right).

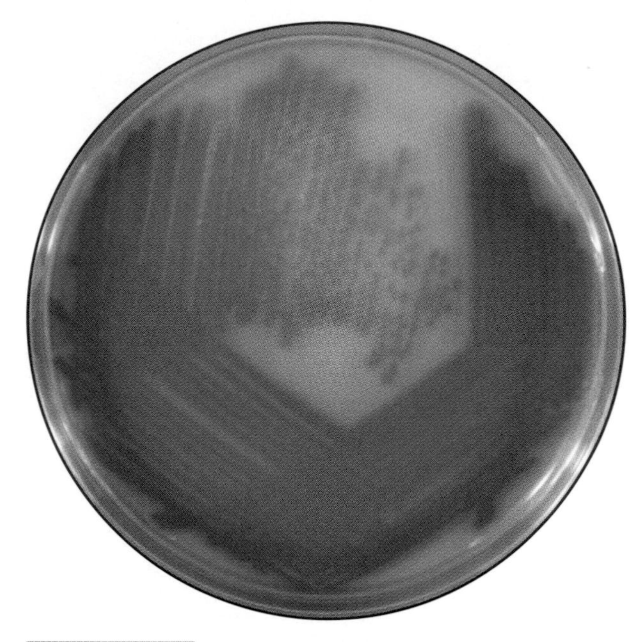

FIGURE 1-14 *Pseudomonas aeruginosa* often produces a characteristic diffusible blue-green pigment.

elevations include *flat, raised, convex, pulvinate* (very convex), and *umbonate* (raised in the center). Colony texture may be *moist, mucoid,* or *dry*. Pigment production is another useful characteristic, but it may be influenced by environmental factors such as temperature and nutrient supply (Figs. 1-15 and 1-16). Colony color may be combined with optical properties such as *opaque, translucent, shiny,* or *dull*.

You should not be overwhelmed by these terms. Most of these terms replace descriptive phrases, so they are intended to make colony morphology description easier, not more difficult. Nor should you feel there is always a single, correct description for each organism on every medium. Rather, use these terms in a way that is meaningful to you and realize it's more important to recognize each when you see it than to memorize all the descriptive terms for a particular species.

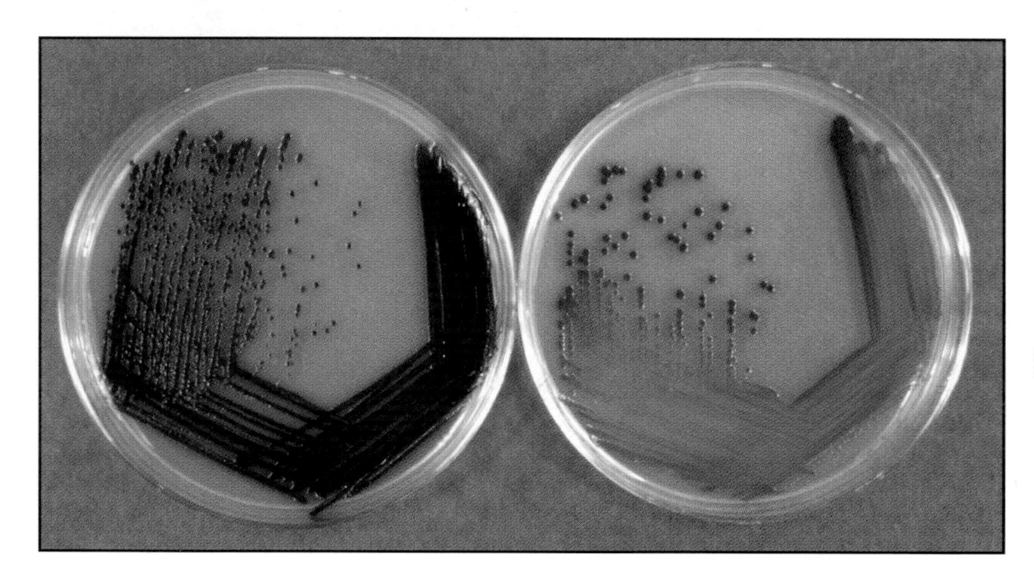

FIGURE 1-15 Pigment production may be influenced by environmental factors such as nutrient availability. *Chromobacterium violaceum* produces a much more intense purple pigment when grown on trypticase soy agar (left) than when grown on nutrient agar (right).

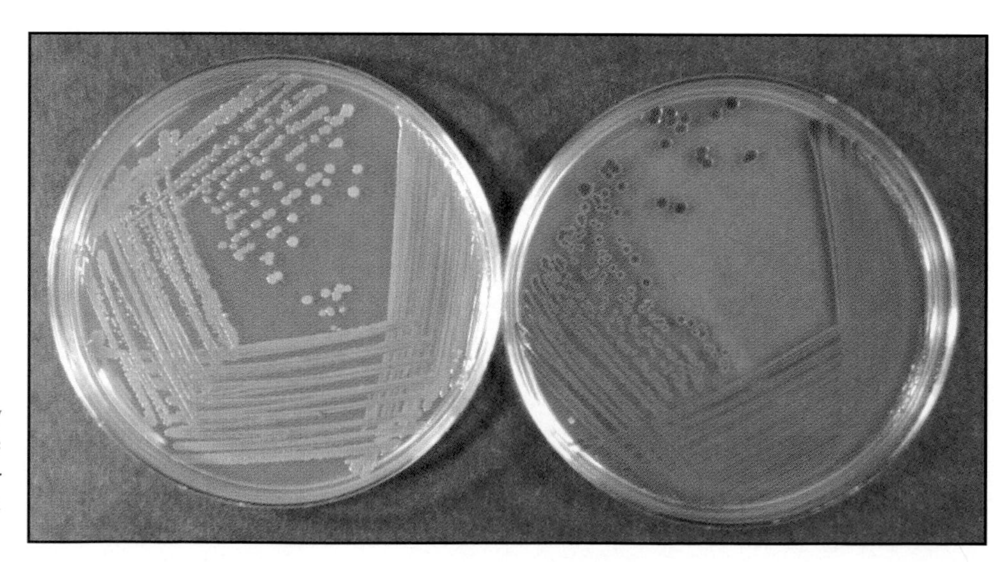

FIGURE 1-16 Pigment production may be influenced by temperature. Less orange pigment is produced by *Serratia marcescens* when grown at 37°C (left) than when grown at 25°C (right).

GROWTH IN BROTH

PURPOSE Characteristic growth patterns in broth may be of use in identifying an unknown bacterium.

PRINCIPLE Growth pattern in broth may be used to distinguish between microbes during identification. Each may be the result of specific cell structures. Common patterns in broth are *uniform fine turbidity* (cloudiness), *flocculent* (clumps of growth), and *sediment*, as shown in Figure 1-17. Figure 1-18 shows two types of surface growth: a *ring* and a *pellicle* (surface membrane).

FIGURE 1-17 Growth patterns in broth. From left to right: *Bacillus cereus* (sediment), *Clostridium sporogenes* (flocculent), and *Pseudomonas aeruginosa* (uniform fine turbidity).

FIGURE 1-18 Broth cultures of *Chromobacterium violaceum* (left) and *Corynebacterium xerosis* (right) illustrating a surface ring and a pellicle, respectively. Both also exhibit fine turbidity and a slight sediment.

AEROTOLERANCE

PURPOSE Recognizing the aerotolerance of a specimen is of use in identifying an unknown bacterium.

PRINCIPLE Bacteria are classified according to their aerotolerance — that is, their ability to grow in the presence of oxygen. Organisms requiring oxygen are *strict (obligate) aerobes*, whereas those that cannot live in the presence of oxygen are *strict (obligate) anaerobes*. Between these two extremes are *facultative anaerobes* that grow both aerobically and anaerobically, and *microaerophiles* that require oxygen, but at a concentration less than is found in the atmosphere.

Agar deep tubes may be used to determine aerotolerance. A stab inoculation is made to the bottom of the agar in a test tube (Fig. 1-19). With such a small surface area for oxygen diffusion into the agar, the aerobic zone extends to a depth of only about 1 cm from the surface. After incubation, the portion(s) of the tube supporting growth allows aerotolerance determination. Growth only at the top indicates an obligate aerobe. Growth only in the lower portion indicates an obligate anaerobe. Facultative anaerobes grow well in both aerobic and anaerobic zones. Microaerophiles grow in the aerobic zone, but below the surface.

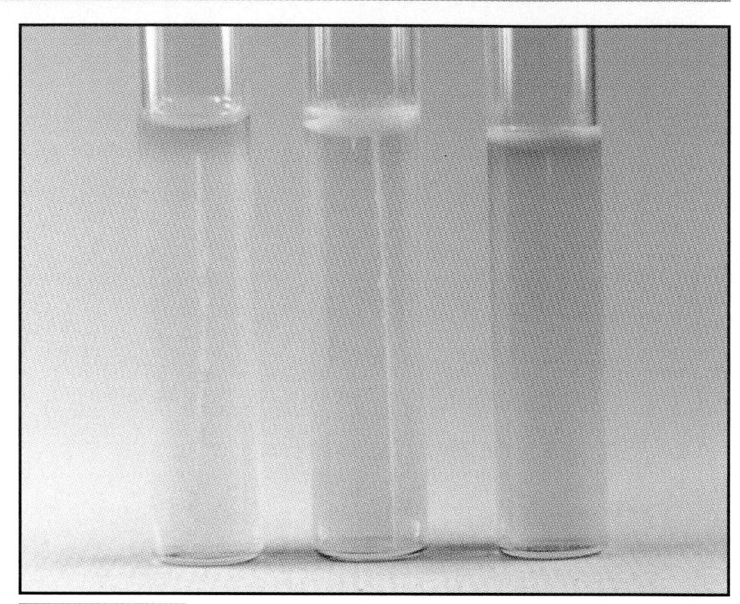

FIGURE 1-19 Agar deep stab tubes showing microbial aerotolerance. From left to right: *Clostridium sporogenes* (strict anaerobe), *Staphylococcus aureus* (facultative anaerobe), and *Pseudomonas aeruginosa* (strict aerobe).

ANAEROBIC CULTURE METHODS

PURPOSE These procedures allow growth of anaerobes under anaerobic conditions.

PRINCIPLE Bacteria typically are grown under aerobic conditions which support growth of all aerotolerance categories except for strict anaerobes. Creating conditions for anaerobic growth requires extra effort, and many systems have been developed. Three anaerobic culture methods are discussed in this section.

One anaerobic culture medium is *thioglycolate broth*. It contains sodium thioglycolate, a reducing agent that reduces free O_2 to water, thus making the broth anaerobic. A dye, such as methylene blue or resazurin, may be included to indicate where oxygen remains in the medium. Resazurin dye is pink if oxidized and colorless if reduced; methylene blue is blue if oxidized, colorless if reduced. Depending on which dye is used, thioglycolate tubes will have a pink or blue band near the surface where oxygen diffuses in. In addition to allowing growth of anaerobes, thioglycolate may also be used to determine aerotolerance. Interpretation of aerotolerance is the same as for agar deep tubes (Fig. 1-20).

A second method of growing anaerobes is the *anaerobic jar* (Fig. 1-21). Inoculated plates to be grown anaerobically are placed inside. Then a gas generator packet is opened, water is added, and the lid is immediately clamped on the jar. Sodium borohydride and sodium bicarbonate in the packet react with the water and produce H_2 and CO_2 gases. Palladium then catalyzes the conversion of H_2 and O_2 to water, as shown in Figure 1-22. Removal of free oxygen produces anaerobic conditions within the jar. A methylene blue strip is included to indicate if anaerobic conditions are actually created. Methylene blue is blue if oxidized and white if reduced. Therefore, a functional anaerobic jar should have a white strip within an hour or so of the addition of water.

Aerotolerance may be determined by inoculating plates with the same organisms and incubating one aerobically and the other anaerobically in the jar (Figs. 1-23 and 1-24).

The anaerobic jar may also be used to grow microaerophilic organisms, especially members of the genus *Campylobacter*. The gas generator packet works under the same principle as the anaerobic packet but produces less hydrogen. Therefore, not all of the oxygen in the jar is reduced to water. The resulting atmosphere is one which still contains oxygen, but at less than atmospheric concentration.

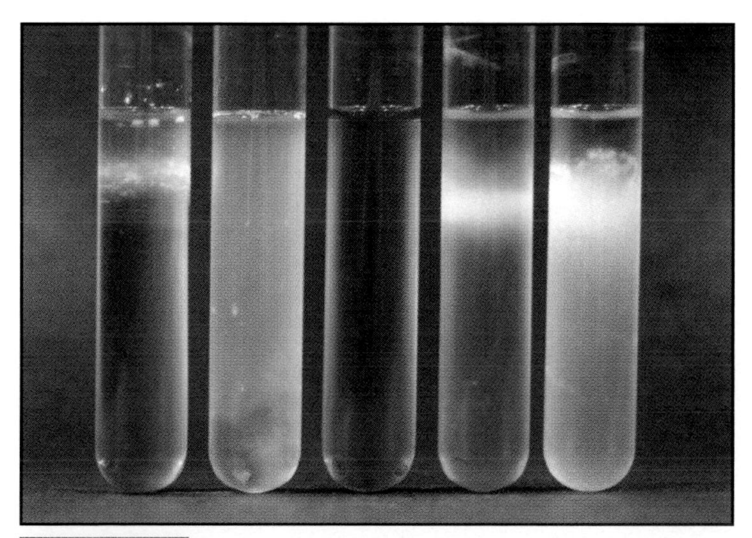

FIGURE 1-20 Thioglycolate tubes demonstrating aerotolerance. Pictured are, from left to right: *Moraxella catarrhalis* (obligate aerobe), *Escherichia. coli* (facultative anaerobe), uninoculated control, *Clostridium sporogenes* (microaerophile), and *Clostridium sporogenes* (anaerobe). Note the pink band in the control tube due to O_2 oxidizing the resazurin dye.

FIGURE 1-21 The anaerobic jar. Note the white methylene blue strip and the open packet which has discharged H_2 and CO_2 gases. The palladium, contained in the packet, catalyzes the conversion of H_2 and O_2 to water as shown in Figure 1-22.

$$2H_2 + O_2 \xrightarrow{Pd} 2H_2O$$

FIGURE 1-22 The conversion of H_2 and O_2 to water is catalyzed by palladium.

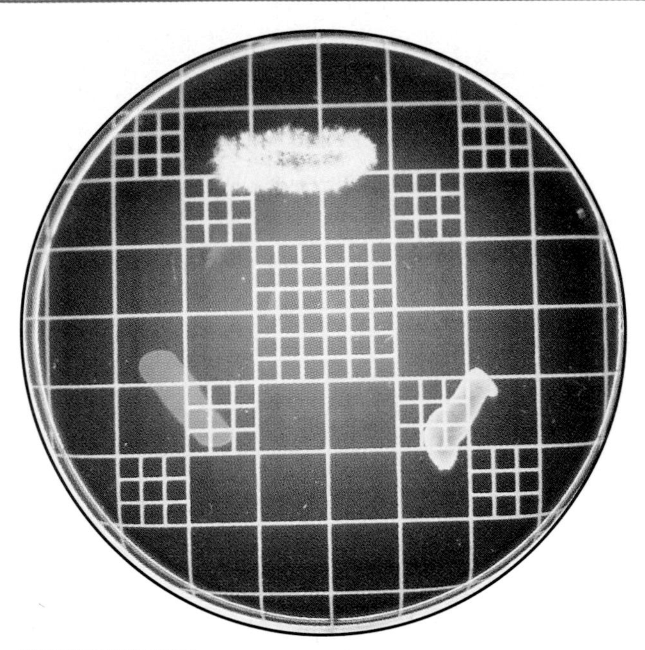

FIGURE 1-23 A nutrient agar plate inoculated with *Pseudomonas aeruginosa* (left), *Clostridium sporogenes* (top) and *Staphylococcus aureus* (right) and incubated in an anaerobic jar. Compare the amount of growth with the plate in Figure 1-24.

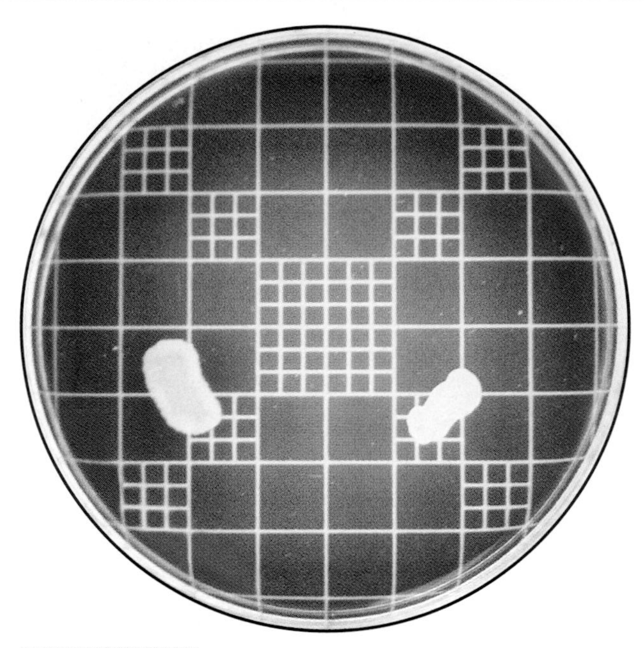

FIGURE 1-24 A nutrient agar plate inoculated with *Pseudomonas aeruginosa* (left), *Clostridium sporogenes* (top) and *Staphylococcus aureus* (right) and incubated aerobically. Compare the amount of growth with the plate in Figure 1-23.

Isolation Techniques and Selective Media 2

STREAK PLATE METHOD OF ISOLATION

PURPOSE Most tests designed to identify bacterial species require the use of *pure cultures*. Streaking a plate produces isolated growth of the microbial species in a mixed culture sample. Once isolated, pure cultures may be started by transferring a portion of each colony to a sterile medium.

PRINCIPLE Bacteria from a mixed culture are streaked over the agar surface in a pattern that deposits them progressively farther apart (Figs. 2-1 and 2-2). Toward the end of the pattern, the resulting colonies should be separate from all others and may be used (*picked*) to start pure cultures.

While it is generally safe to assume a pure culture has been started from picking an isolated colony, such is not always the case. Contaminants may be slow-growing and not yet visible on the plate. Contaminants may also be trapped in slime or among chains of bacteria. Lastly, if a selective medium has been streaked for isolation, a contaminant may be alive but not dividing (a *bacteriostatic* state) until transferred to the nonselective medium used for pure culture. Since the consequence of not producing a pure culture is almost certainly misidentification of the microbe, culture purity should be confirmed by periodic streak plates and Gram stains.

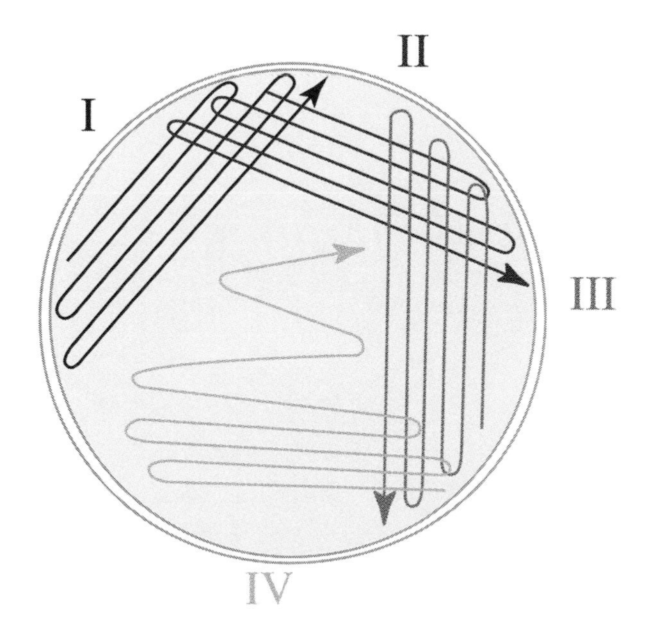

FIGURE 2-1 The quadrant method of streaking a plate for isolation. The agar surface is streaked as in I. After *flaming the loop*, the plate is rotated almost 90° and streaked as in II. The process is repeated for streaks III and IV.

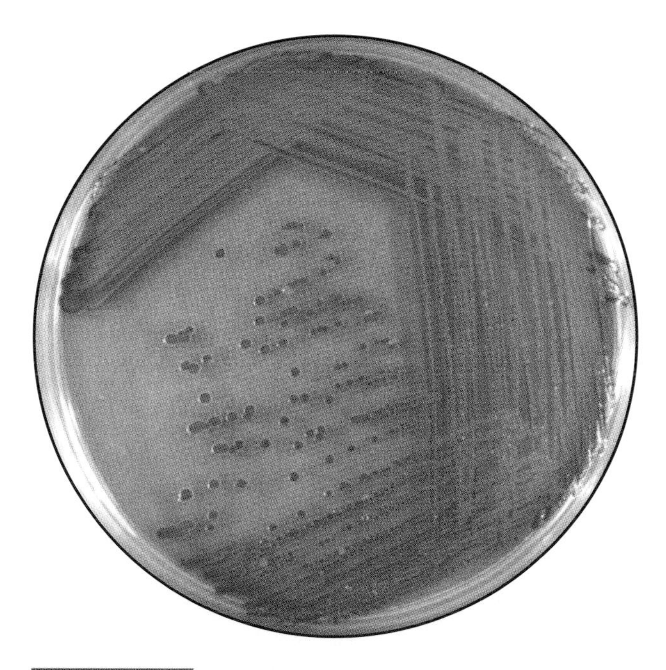

FIGURE 2-2 A streak plate of *Serratia marcescens* after incubation. Note the decreasing density of growth in the four streak patterns. On this plate, isolation is first obtained in the fourth streak. Cells from individual colonies may be transferred to sterile media to start pure cultures of each.

DESOXYCHOLATE (DOC) AGAR

PURPOSE Desoxycholate agar is an undefined, selective, and differential medium used for isolation and differentiation among the *Enterobacteriaceae* — the enteric (gut) bacteria. Enteric bacteria are facultatively anaerobic Gram-negative rods. They may be divided into those that produce acid from lactose fermentation (the *coliforms*) and those that do not. Coliforms include the usually nonpathogenic *Escherichia coli* and *Enterobacter aerogenes*. The lactose nonfermenter group includes such pathogens as *Salmonella typhi* and *Shigella dysenteriae*. Desoxycholate agar allows a quick preliminary indication of whether a specimen contains enteric pathogens.

PRINCIPLE Desoxycholate Agar contains nutrients, including lactose, sodium desoxycholate, citrate, and neutral red. Desoxycholate is a component of bile and inhibits growth of Gram-positive organisms. Citrate is included to increase the action of desoxycholate. Acid end products from lactose fermentation lower the pH (Fig. 2-3). When the pH gets below 6.8, the colorless neutral red turns a reddish color. Thus, Gram-negative lactose fermenters will appear some shade of red, whereas Gram-negative lactose nonfermenters will remain colorless (Fig. 2-4).

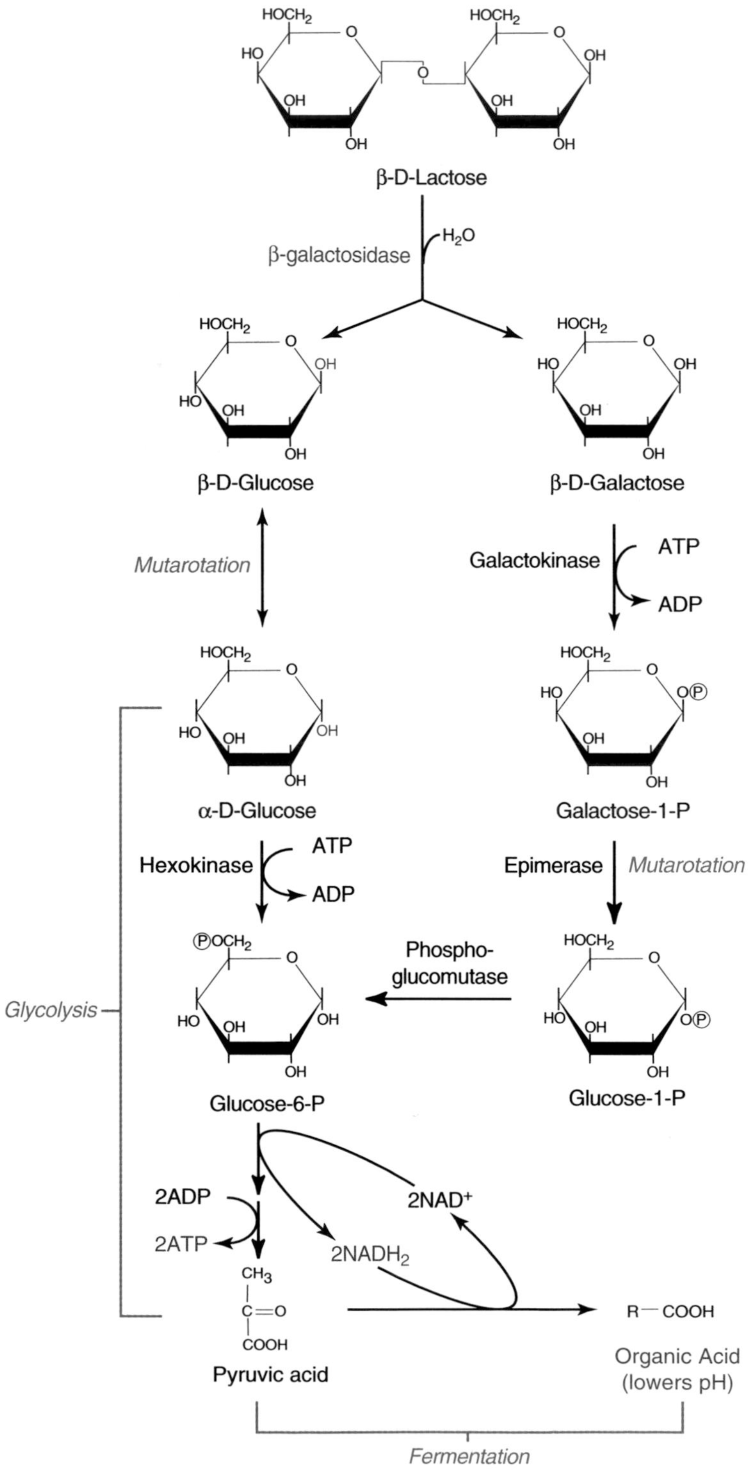

FIGURE 2-3 Lactose fermentation with acid end products.

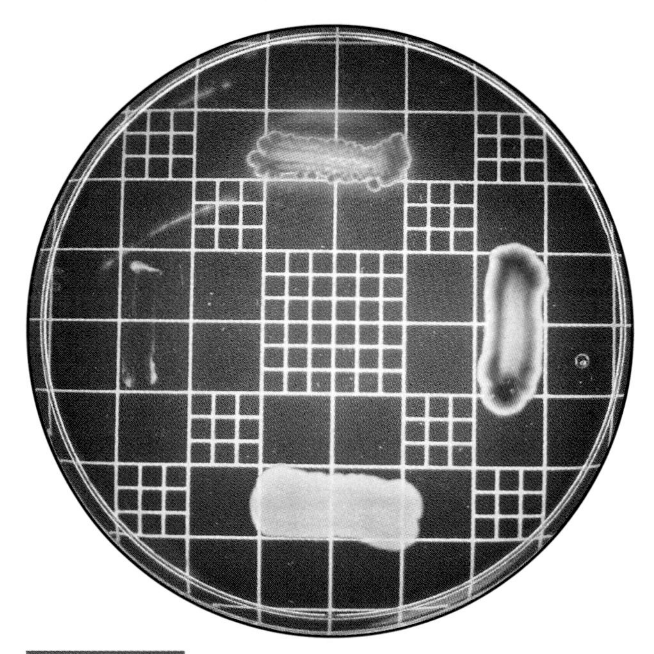

FIGURE 2-4 DOC medium inoculated with (clockwise from top): *Escherichia coli, Enterobacter aerogenes, Proteus vulgaris,* and *Micrococcus luteus*. Note the inhibition of the Gram-positive *M. luteus*. Note also the red coloring of the lactose fermenting *E. coli* and *E. aerogenes*. In practice, DOC would be streaked for isolation.

ENDO AGAR

PURPOSE Endo agar is a selective and differential medium used to determine the presence of enteric bacteria and differentiate the coliforms from the noncoliforms in testing water potability, wastewater, and dairy products. Enteric bacteria are facultatively anaerobic Gram-negative rods. They may be divided into those that produce acid from lactose fermentation (the *coliforms*) and those that do not. Coliforms include the usually nonpathogenic *Escherichia coli* and *Enterobacter aerogenes*. The lactose nonfermenter group includes such pathogens as *Salmonella typhi* and *Shigella dysenteriae.*

PRINCIPLE Endo agar contains sodium sulfite and basic fuchsin to inhibit growth of Gram-positive organisms, making the medium selective. Digested animal tissue and lactose are the nutrient sources. Lactose fermentation by the coliforms turns the growth red or pink due to the reaction of sodium sulfite with the fermentation intermediate acetaldehyde (Fig. 2-5). Heavy lactose fermenters (such as *Escherichia coli*) may also produce a metallic sheen (Fig. 2-6). Lactose nonfermenters produce colorless to slightly pink growth.

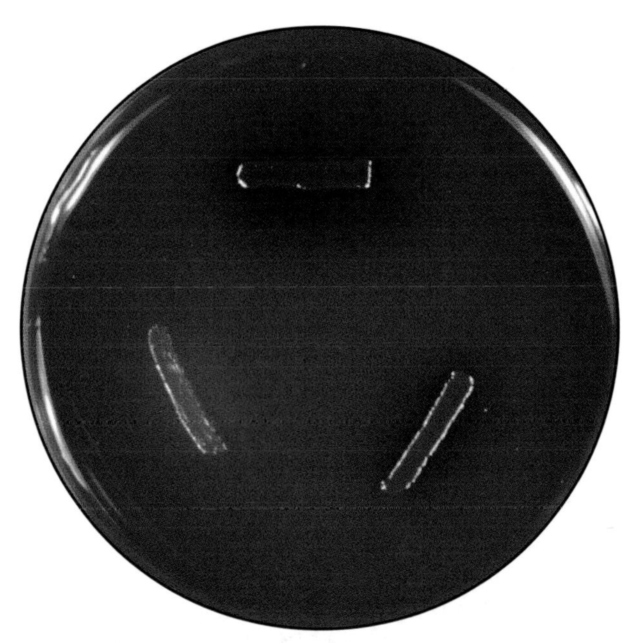

FIGURE 2-5 Endo Agar inoculated with *Escherichia coli* (top), *Enterobacter aerogenes* (lower right) and *Shigella sonnei* (lower left). Notice the difference in the intensity of the pink between *E. aerogenes* (a coliform) and *S. sonnei* (not a coliform).

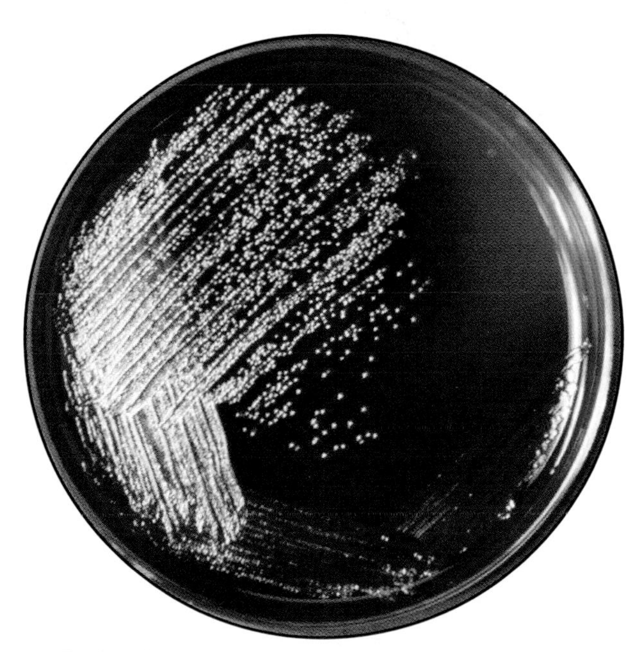

FIGURE 2-6 Endo Agar streaked with *E. coli* to illustrate the metallic sheen.

EOSIN METHYLENE BLUE (EMB) AGAR

PURPOSE Eosin methylene blue agar is a selective and differential medium used for isolation and differentiation among members of the *Enterobacteriaceae* — the enteric (gut) bacteria. Enteric bacteria are facultatively anaerobic Gram-negative rods. They may be divided into those that produce acid from lactose fermentation (the *coliforms*) and those that do not. Coliforms include the usually non-pathogenic *Escherichia coli* and *Enterobacter aerogenes*. The lactose nonfermenter group includes such pathogens as *Salmonella typhi* and *Shigella dysenteriae*. Eosin methylene blue agar allows a quick preliminary indication of whether a specimen contains enteric pathogens. EMB may be streaked for isolation or used in procedures such as the membrane filter technique.

PRINCIPLE Eosin methylene blue agar is an undefined and selective medium that contains the aniline dyes methylene blue and eosin which inhibit Gram-positive bacteria, thus favoring growth of Gram-negative enterics. In addition to the dyes, the medium also contains lactose. Lactose makes EMB a differential medium in that it allows distinction between lactose fermenters and lactose nonfermenters. Large amounts of acid from lactose fermentation (Fig. 2-3) cause the dyes to precipitate on the colony surface, producing a characteristic green metallic sheen (some strains do not produce the green sheen and appear almost black). Smaller amounts of acid production result in a pink coloration of the growth. Nonfermenting enterics do not produce acid so their colonies remain colorless or take on the coloration of the medium (Fig. 2-7).

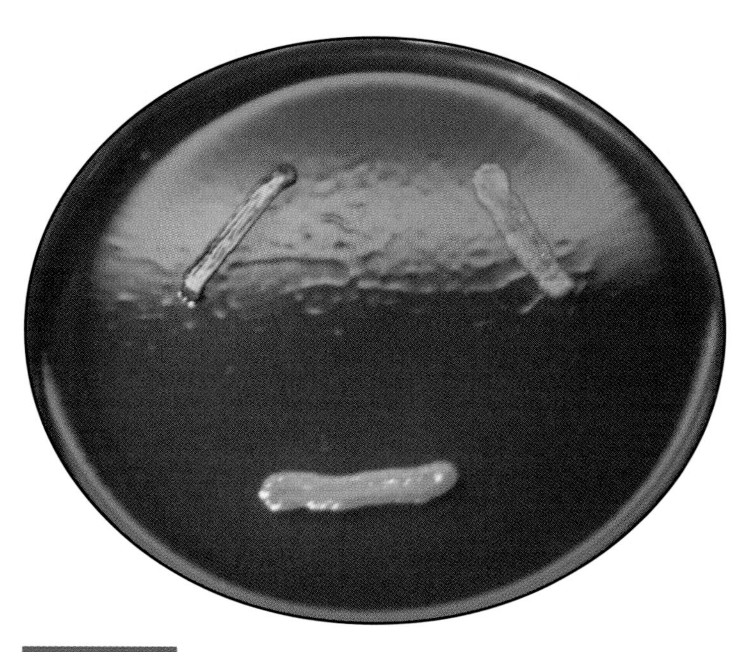

FIGURE 2-7 *Escherichia coli* (top left), *Shigella flexneri* (top right) and *Enterobacter aerogenes* (bottom) on EMB. Note the characteristic green metallic sheen of *E. coli* and the pink coloration of *E. aerogenes*. The difference in color is due to degree of acid production. Some *E. coli* do not produce the green sheen and appear black.

HEKTOEN ENTERIC (HE) AGAR

PURPOSE Various media have been formulated to isolate enteric pathogens. Hektoen enteric agar is used to isolate and differentiate *Salmonella* and especially *Shigella* species from other Gram-negative enteric pathogens in clinical samples.

PRINCIPLE Hektoen agar is an undefined medium with yeast extract and peptone supplying the nutrients. A comparatively high concentration of bile salts inhibits Gram-positive and some Gram-negative organisms, but not *Salmonella* and *Shigella* species. Three carbohydrates — lactose, sucrose and salicin — and the dyes bromthymol blue and acid fuchsin allow differentiation between enterics due to the colony and medium colors produced. Bromthymol blue is yellow when acidic, blue when alkaline. Gram-negatives that ferment lactose with acid end products (*i.e.*, coliforms) will produce yellow to salmon-pink colonies. *Salmonella* and *Shigella* are lactose nonfermenters. *Salmonella* produces blue-green colonies and *Shigella* produces raised, green, moist colonies. Further differentiation is obtained by the ability or inability of the organisms to reduce sulfur. Reduction of thiosulfate to H_2S by the organism produces a black coloration due to the combination of H_2S with ferric ion in the medium (Fig. 2-8).

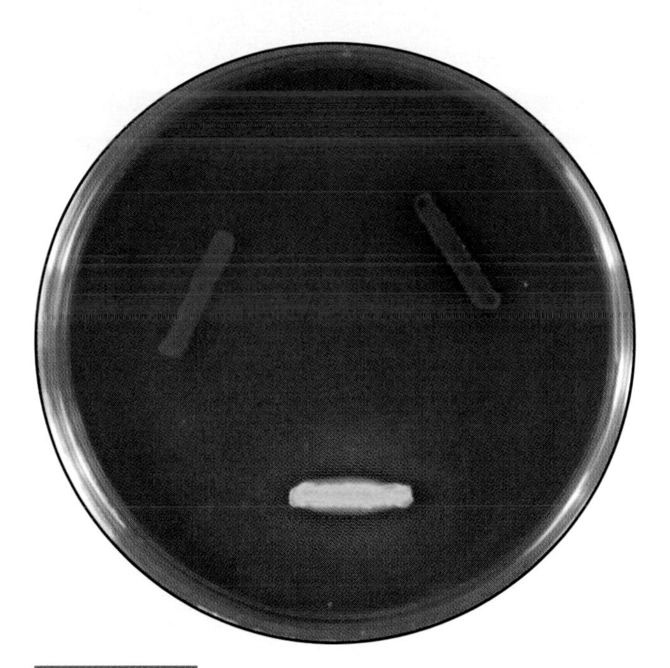

FIGURE 2-8 Hektoen enteric agar inoculated with (clockwise from the upper left), *Shigella flexneri, Proteus mirabilis* and *Escherichia coli. E. coli* is salmon-colored because it ferments the lactose. *Shigella* does not ferment lactose and is blue-green. *P. mirabilis* is also a nonfermenter, but does have some black at either end indicating an ability to reduce sulfur. In clinical use, Hektoen agar is streaked for isolation.

MacConkey Agar

PURPOSE MacConkey agar is a selective and differential medium used to isolate and differentiate members of the *Enterobacteriaceae* — the enteric (gut) bacteria — as well as some *Staphylococcus* and *Enterococcus* species. Enteric bacteria are facultatively anaerobic Gram-negative rods. They may be divided into those that produce acid from lactose fermentation (the *coliforms*) and those that don't. Coliforms include the usually nonpathogenic *Escherichia coli* and *Enterobacter aerogenes*. The lactose nonfermenter group includes such pathogens as *Salmonella typhi* and *Shigella dysenteriae*. MacConkey Agar allows a quick preliminary indication of whether a specimen contains enteric pathogens.

PRINCIPLE MacConkey agar is an undefined medium containing nutrients, including lactose, as well as bile salts, neutral red and crystal violet. Bile salts and crystal violet inhibit growth of Gram-positive bacteria, making MacConkey agar a selective medium. (In a variation, more bile salts are added and crystal violet is left out, making the medium less selective and capable of isolating Gram-positive *Enterococcus* and some *Staphylococcus* species.) Neutral red is a pH indicator that is colorless above a pH of 6.8 and red at a pH less than 6.8. Acid accumulating from lactose fermentation (Fig. 2-3) turns the colorless neutral red a red color. Coliforms, therefore, turn a shade of red on MacConkey agar whereas lactose nonfermenters remain colorless (Fig. 2-9).

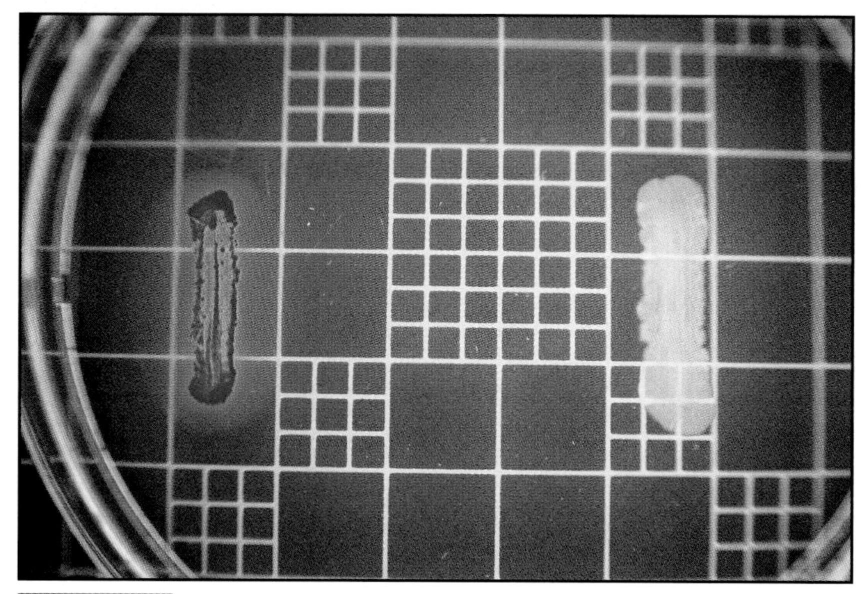

FIGURE 2-9 *Escherichia coli* (left) is red, indicating acid products from lactose fermentation. As a result of lowered pH, the bile salts may also precipitate around the growth, as seen here. *Providencia stuartii* (right) is a lactose nonfermenter and remains colorless on MacConkey agar. If crystal violet had been left out of the medium, fecal species of *Enterococcus* and some species of *Staphylococcus* would appear red or pink. In practice, MacConkey agar would be streaked for isolation.

MANNITOL SALT AGAR (MSA)

PURPOSE Mannitol salt agar is both selective and differential. It favors organisms capable of tolerating high sodium chloride concentration and also distinguishes bacteria based on their ability to ferment mannitol. It is used to differentiate pathogenic *Staphylococcus* species, which ferment mannitol, from the less pathogenic members of the genus *Micrococcus*, which do not.

PRINCIPLE Mannitol salt agar is formulated with 7.5% NaCl. This makes it highly selective since most bacteria cannot tolerate this level of salinity. Organisms not suited for this environment will show stunted growth or no growth at all.

Phenol red is the pH indicator included in the medium. Phenol red is yellow below pH 6.8, red at pH 7.4 to 8.4, and pink at 8.4 and above. The development of yellow halos around the bacterial growth is an indication that mannitol has been fermented with the production of acid (Fig. 2-10). No color change or formation of pink color is a negative result (Fig. 2-11).

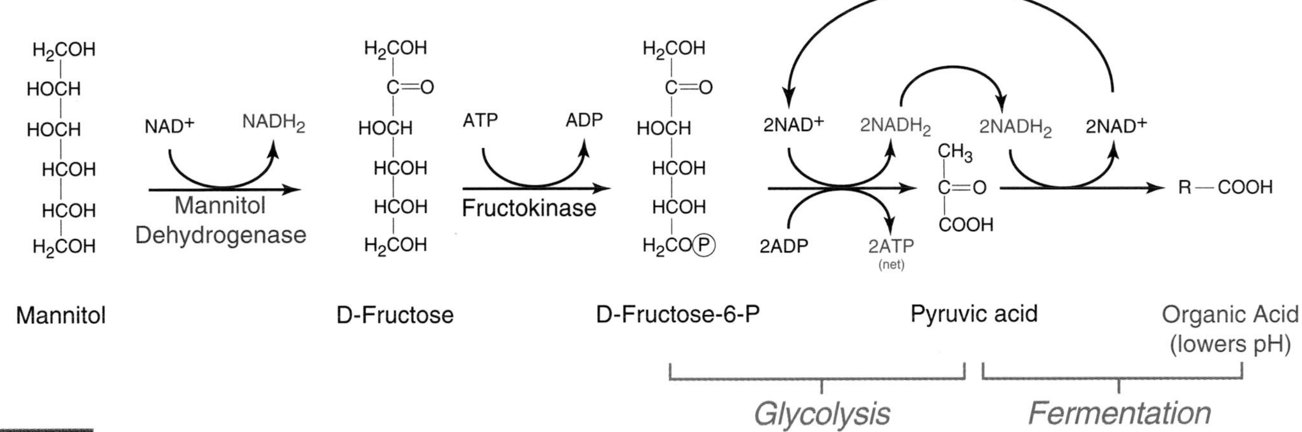

FIGURE 2-10 Fermentation of mannitol with acid end products.

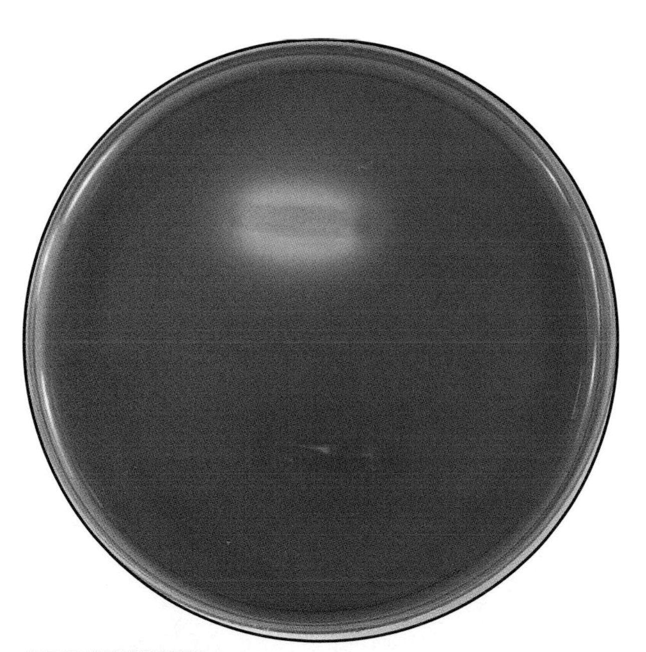

FIGURE 2-11 Mannitol salt agar inoculated with *Staphylococcus aureus* (above) and *S. epidermidis* (below). (Some strains of *S. epidermidis* are inhibited by this medium.) The yellow halo around *S. aureus* is due to mannitol fermentation with acid end products. For clinical applications, MSA would be streaked for isolation.

PHENYLETHYL ALCOHOL (PEA) AGAR

PURPOSE This selective medium is used on mixed bacterial cultures to isolate Gram-positive cocci from any Gram-negative organisms that may be present.

PRINCIPLE Phenylethyl alcohol agar is an undefined, selective medium which inhibits or prevents growth of most Gram-negative organisms by interfering with DNA synthesis. The mechanism and effectiveness, however, differ slightly from species to species.

Phenylethyl alcohol agar contains nutrients, sodium chloride, and 0.025% phenylethyl alcohol (PEA). When a mixture of Gram-negative and Gram-positive organisms is streaked onto a plate of PEA, the Gram-positive organisms will grow and the Gram-negative organisms will show either stunted growth or no growth at all. The low concentration of PEA is used because high concentrations can be toxic to Gram-positive as well as Gram-negative bacteria.

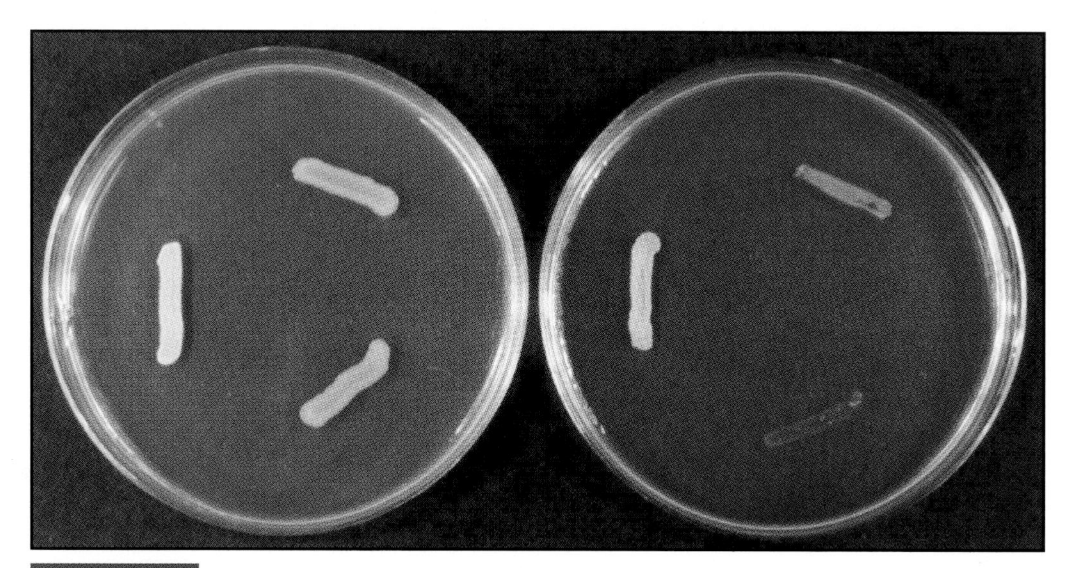

FIGURE 2-12 Growth of *Staphylococcus aureus* (left), *Klebsiella pneumoniae* (upper right), and *Citrobacter diversus* (lower right) on nutrient agar (left plate) and on PEA (right plate). The growth of Gram-positive *S. aureus* is about the same on both plates, but the growth of the two Gram-negatives is clearly inhibited by the PEA, though to different degrees. In practice, the medium would be streaked for isolation.

SALMONELLA-SHIGELLA (SS) AGAR

PURPOSE *Salmonella-Shigella* agar is a selective medium originally used for the isolation of *Salmonella* and many *Shigella* species (*i.e.*, lactose nonfermenting enteric bacteria) from the lactose fermenting enterics (the coliforms). It is no longer recommended for isolation of *Shigella*, since Hektoen and XLD agars are more effective, but is still of use in isolating *Salmonella* species.

PRINCIPLE *Salmonella-Shigella* agar is an undefined, differential, and selective medium with bile salts and brilliant green dye acting as the selective agents against Gram-positives and many Gram-negatives. Lactose is the fermentable carbohydrate and neutral red is the pH indicator. Sodium thiosulfate is the sulfur source. Coliform bacteria that are able to grow will produce reddish colonies since the neutral red changes from colorless to red in conditions of low pH. *Salmonella* and *Shigella* species will be colorless due to their inability to ferment lactose. *Salmonella* and *Proteus* (a slow lactose-fermenting enteric) species may reduce thiosulfate to H_2S which then reacts with ferric ion in the medium to produce a black center in the colonies (Fig. 2-13).

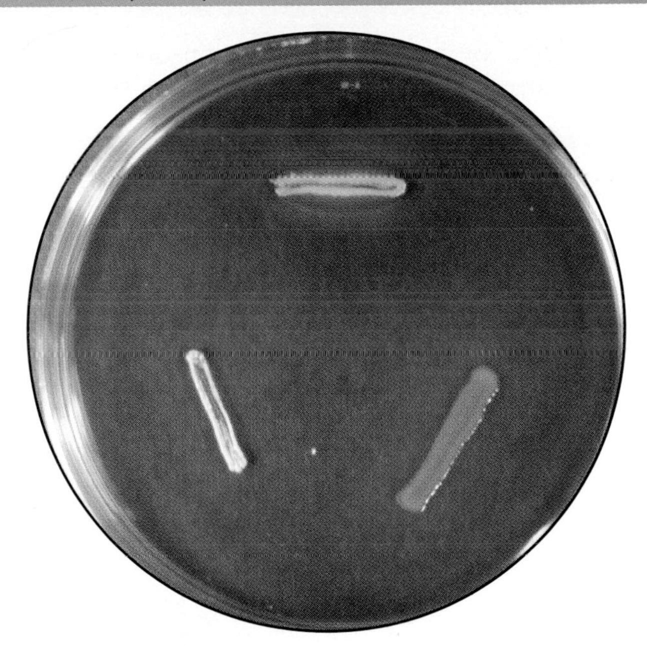

FIGURE 2-13 *Escherichia coli* (top), *Shigella flexneri* (lower right), and *Salmonella typhimurium* (lower left) grown on SS agar. Coliform colonies turn red due to the acid produced during lactose fermentation; noncoliforms remain colorless. Sulfur reduction, if it occurs, is detected by a black color in the center of the colonies. In practice, this medium would be streaked for isolation using the clinical or environmental specimen.

TELLURITE GLYCINE AGAR

PURPOSE Tellurite glycine agar is an undefined, selective, and differential medium used for the isolation of coagulase-positive staphylococci from various sources. The most common and clinically important coagulase-positive staphylococcus is *Staphylococcus aureus.*

PRINCIPLE Tellurite glycine agar is an undefined, selective, and differential medium. Potassium tellurite and lithium chloride select against coagulase-negative staphylococci and Gram-negative bacteria. Mannitol favors the growth of coagulase-positive staphylococci since they are able to ferment it. Coagulase-positive staphylococci are differentiated by their ability to reduce tellurite and produce black growth (Fig. 2-14).

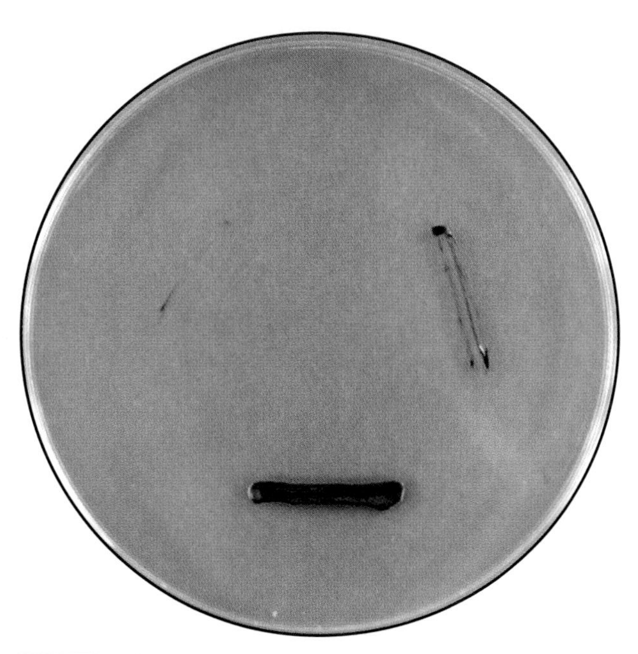

FIGURE 2-14 Clockwise from the top right: Gram-positive, coagulase-negative *Staphylococcus epidermidis,* Gram-positive, coagulase-positive *Staphylococcus aureus,* and Gram-negative *Escherichia coli,* grown on tellurite glycine agar. Notice that *E. coli* is inhibited, and that the two staphylococci may be differentiated by their ability to reduce tellurite and turn the growth black.

THIOSULFATE CITRATE BILE SALTS SUCROSE (TCBS) AGAR

PURPOSE Thiosulfate citrate bile salts sucrose agar is an undefined, selective, and differential medium used for the primary isolation of *Vibrio* species. Specimens suspected of fecal contamination are streaked on the plate in an effort to recover *Vibrio cholerae*, the most important pathogen of the genus.

PRINCIPLE TCBS agar is an undefined, differential, and selective medium. Its alkaline pH promotes growth of *Vibrio cholerae*, while bile salts (in the form of oxgall and sodium cholate) act as selective agents and inhibit the growth of Gram-positive bacteria. Sucrose is the fermentable carbohydrate and bromthymol blue is the pH indicator. Bacteria fermenting with acid end products (such as *Vibrio cholerae*) produce yellow colonies; those that don't ferment sucrose are blue (Figs. 2-15 and 2-16). Species able to reduce thiosulfate to H_2S produce black colonies due to the reaction of H_2S with ferric ion in the medium.

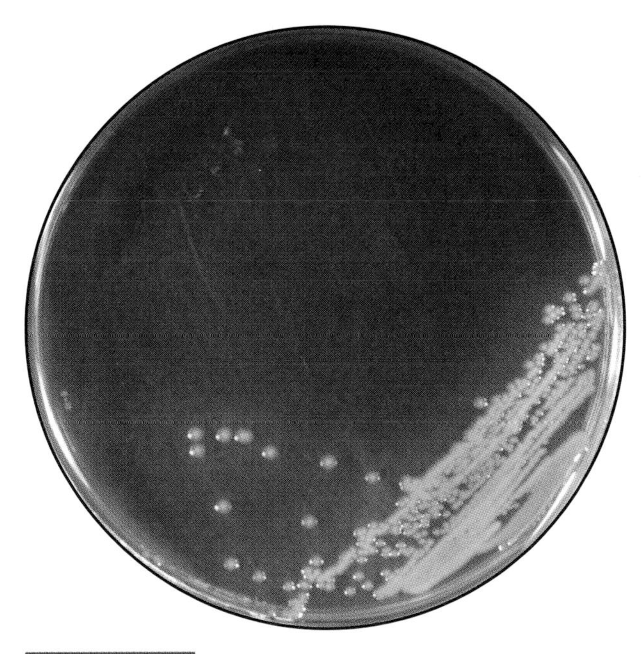

FIGURE 2-15 *Vibrio cholerae* streaked on TCBS agar. The large, yellow colonies are indicative of *V. cholerae*.

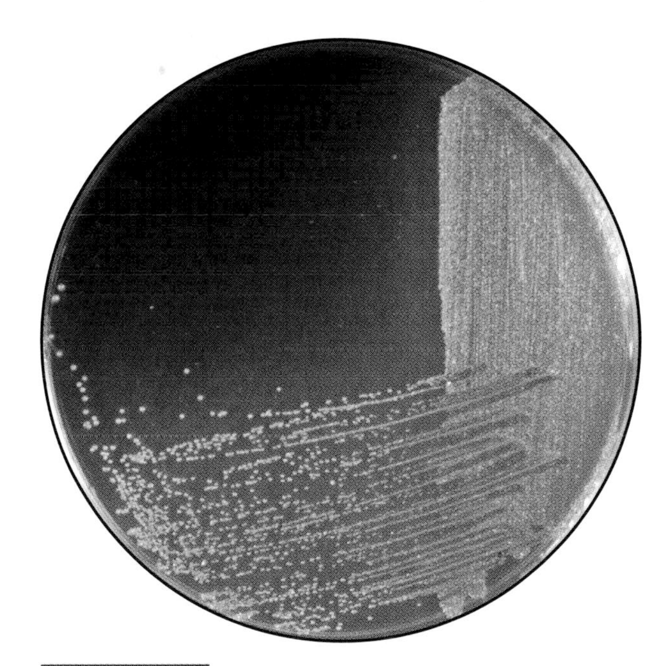

FIGURE 2-16 *Enterococcus faecalis* streaked on TCBS agar. This enteric streptococcus may also be recovered from fecally contaminated samples, however, its yellow colonies are much smaller than those of *V. cholerae*. Compare the *E. faecalis* colonies with those in Figure 2-15.

XYLOSE LYSINE DESOXYCHOLATE (XLD) AGAR

PURPOSE Xylose lysine desoxycholate agar is a selective and differential medium used to isolate and identify enteric pathogens, especially *Shigella*, from fecal samples suspected of fecal contamination.

PRINCIPLE Xylose lysine desoxycholate agar is a selective *and* differential medium. The selective agent sodium desoxycholate inhibits Gram-positive organisms. Xylose, L-lysine, and ferric ammonium citrate allow detection of fermentation, decarboxylation, and the ability to reduce sulfur, respectively (Fig. 2-17). Phenol red, the pH indicator, is yellow below pH 6.8 and red above pH 7.4. A yellow color in the medium is evidence of fermentation with acid production, while a red color indicates decarboxylation of lysine with alkaline end products. Sodium thiosulfate is the sulfur source. Its reduction to H_2S is detected by ferric ion in the medium which produces the black precipitate FeS.

Shigellae, which do not ferment xylose but decarboxylate lysine, appear red on the medium. Salmonellae, which ferment xylose but then decarboxylate the lysine also appear as red colonies but with black centers due to the reduction of sulfur to H_2S. Other enterics are prevented from decarboxylating the lysine by the inclusion of large amounts of lactose and sucrose which the organisms preferentially ferment. These organisms appear yellow on the medium.

FIGURE 2-17 XLD agar inoculated with (clockwise from top): *Proteus mirabilis* (positive for sulfur reduction), *Salmonella typhimurium* (atypically negative for sulfur reduction), and *Escherichia coli* (positive for lactose fermentation).

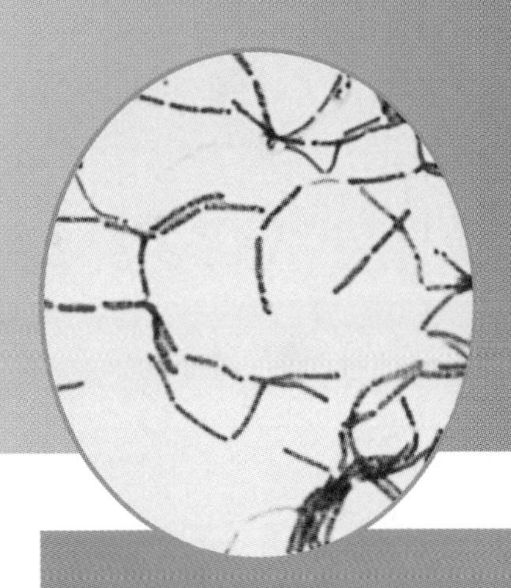

Bacterial Cellular Morphology and Simple Stains

3

SIMPLE STAINS

PURPOSE Since cytoplasm is transparent, cells are usually stained with a colored dye to make them more visible under the microscope. Cell morphology, size, and arrangement may then be determined.

PRINCIPLE Stains are solutions consisting of a solvent and a colored solute called the *chromophore*. Since bacterial cells typically have a negative charge on their surface, they are most easily colored by basic stains with a positively charged chromophore (Fig. 3-1). (A basic chromophore gives up a hydroxide ion [OH^-] or picks up a hydrogen ion [H^+], either of which leaves it with a positive charge.) Common basic stains include methylene blue, crystal violet and safranin. Examples of these may be seen in Figure 3-2 and the Gallery of Bacterial Cell Diversity section.

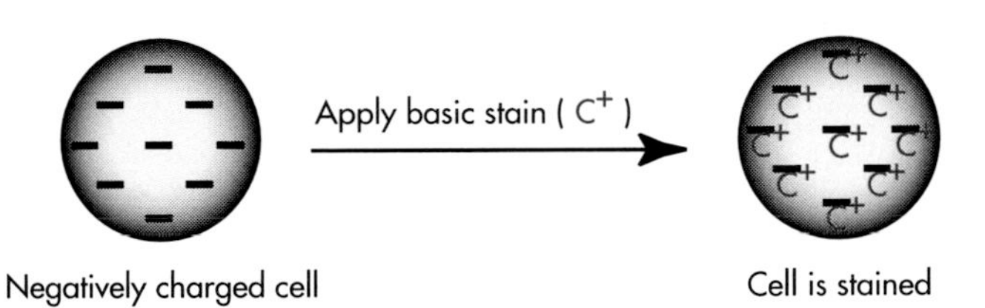

FIGURE 3-1 Basic stains have a positively charged chromophore (C^+) which forms an ionic bond with the negatively charged bacterial cell, thus colorizing the cell.

Basic stains are applied to bacterial smears that have been heat-fixed. Heat-fixing kills the bacteria, makes them adhere to the slide, and coagulates cytoplasmic proteins to make them more visible.

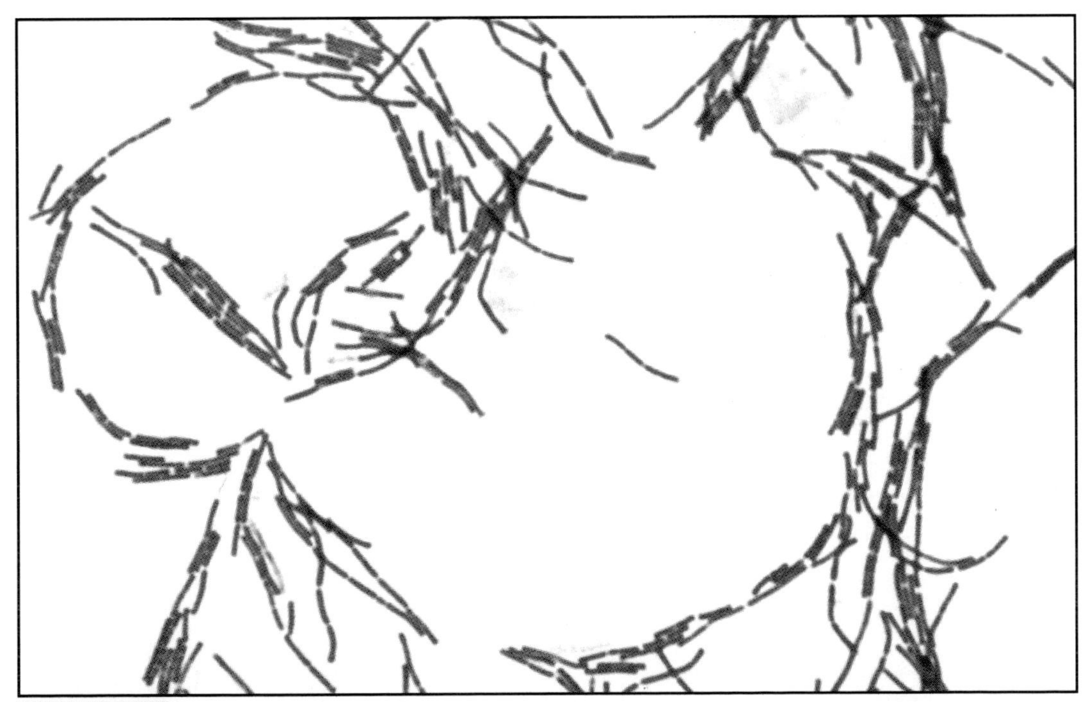

FIGURE 3-2 A simple stain using safranin, a basic stain. The organism is *Bacillus subtilis* (X1520)

NEGATIVE STAIN

PURPOSE A negative stain may be used to determine morphology and cellular arrangement in bacteria that are too delicate to withstand heat-fixing. A primary example is the spirochete *Treponema*, which is distorted by the heat-fixing of other staining techniques. Where accurate size determination is crucial, a negative stain may be used since it produces minimal cell shrinkage.

PRINCIPLE Stains are solutions consisting of a *solvent*, usually water or an alcohol, and a colored solute called the *chromophore*. Stains with an acidic chromophore are used for the negative staining technique. (An acidic chromophore gives up a hydrogen ion [H^+] which leaves it with a negative charge.) The negative charge on the bacterial surface repels the acidic chromophore, so the cell remains unstained against a colored background (Figs. 3-3 and 3-4). Examples of acidic stains are eosin and nigrosin.

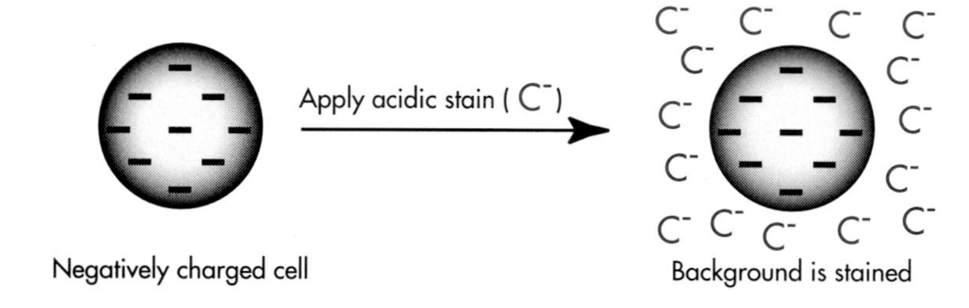

FIGURE 3-3 Acidic stains have a negatively charged chromophore (C^-) that is repelled by negatively charged cells. Thus, the background is colored and the cell remains transparent.

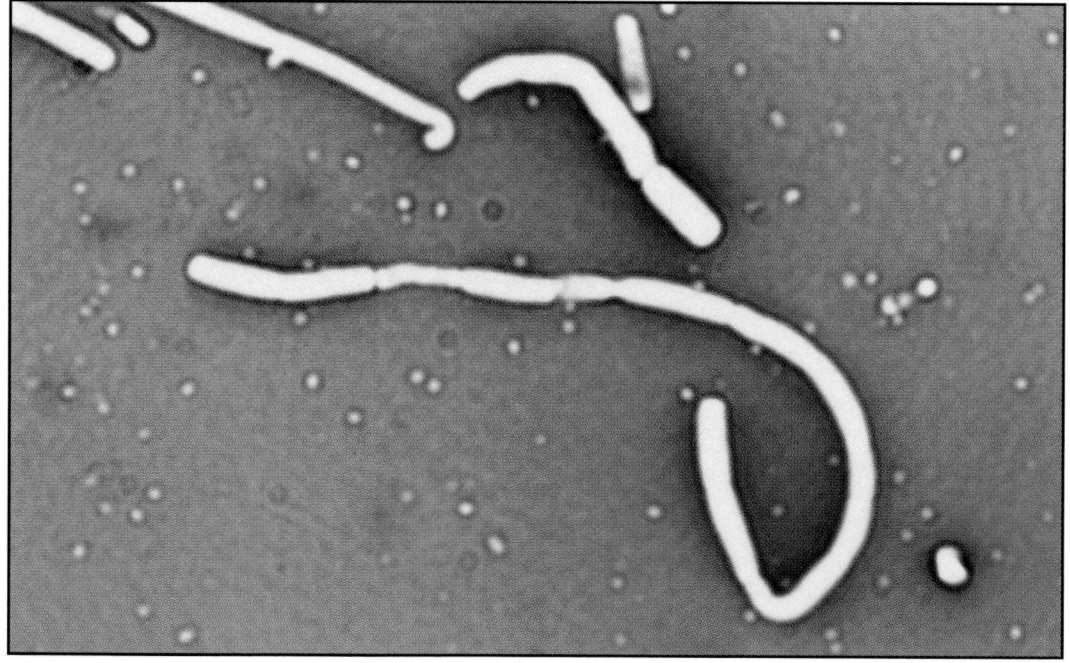

FIGURE 3-4 *Bacillus cereus* negatively stained with nigrosin (X1320).

A GALLERY OF BACTERIAL CELL DIVERSITY

Bacterial Cell Shapes

Bacterial cells are much smaller than eukaryotic cells (Fig. 3-5) and come in a variety of shapes and arrangements. Determining cell shape is an important first step in identifying a bacterial species.

Cells may be spheres (*cocci*, sing. *coccus*), rods (*bacilli*, sing. *bacillus*) or spirals (*spirilla*, sing. *spirillum*). Variations on these shapes include slightly curved rods (*vibrios*) and flexible spirals (*spirochetes*). Examples of cell shapes are shown in Figures 3-6 through 3-12.

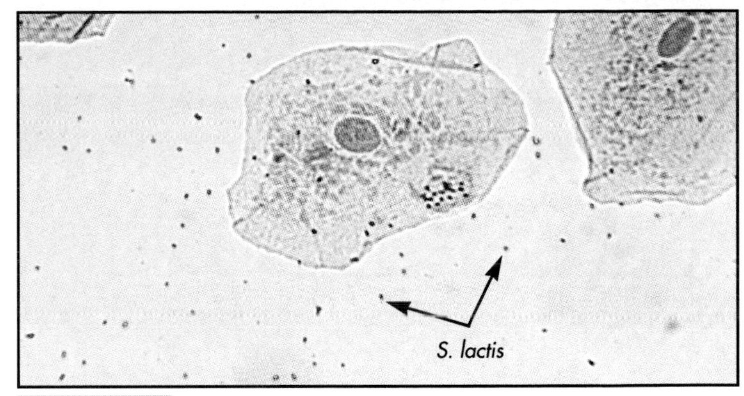

FIGURE 3-5 Relative sizes of eukaryotic and prokaryotic cells (X396). A human cheek cell (stained with safranin) and *Streptococcus lactis* (stained with crystal violet).

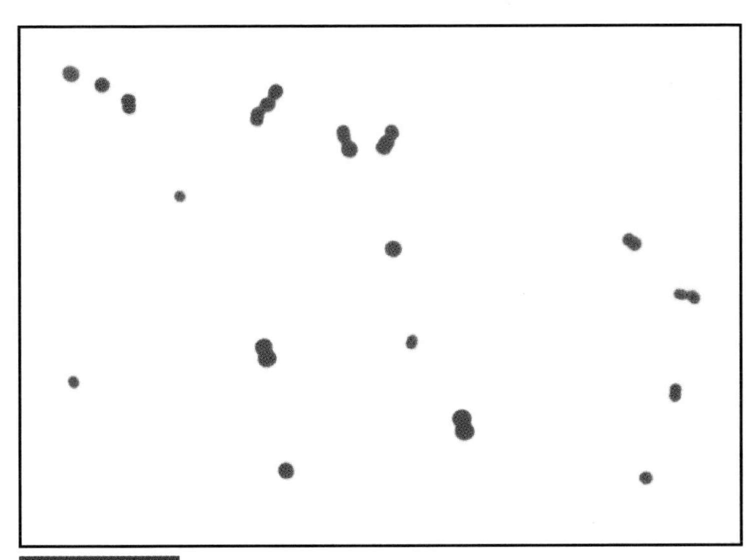

FIGURE 3-6 *Moraxella catarrhalis*, a coccus, stained with crystal violet. *M. catarrhalis* is an inhabitant of the human upper respiratory tract, especially the nasal cavity, and is rarely pathogenic (X1000).

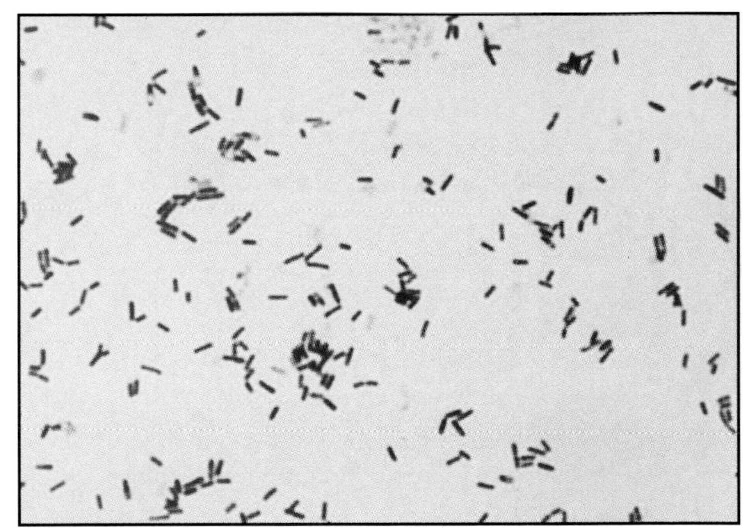

FIGURE 3-7 *Bacillus subtilis*, a bacillus found in soil, stained with crystal violet (X2112).

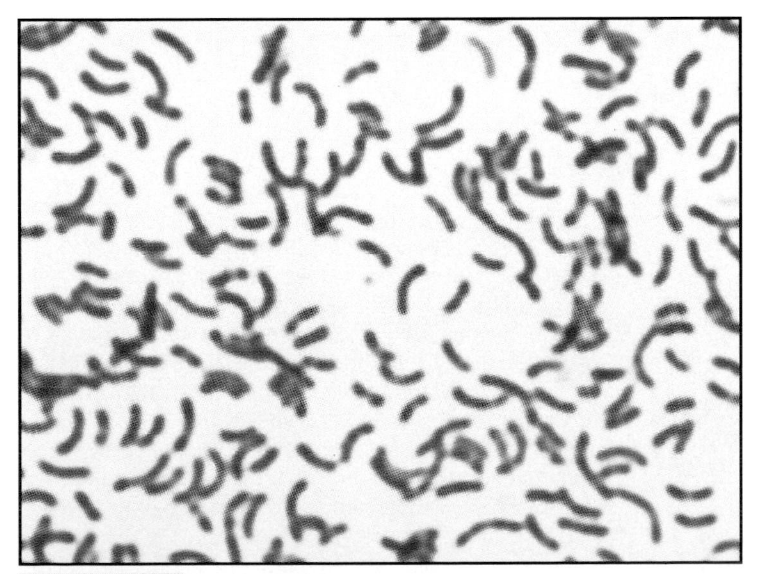

FIGURE 3-8 *Rhodospirillum rubrum*, a spirillum, grown on an agar slant and stained with carbolfuchsin. Compare its size and shape with Fig. 3-9 (X2640).

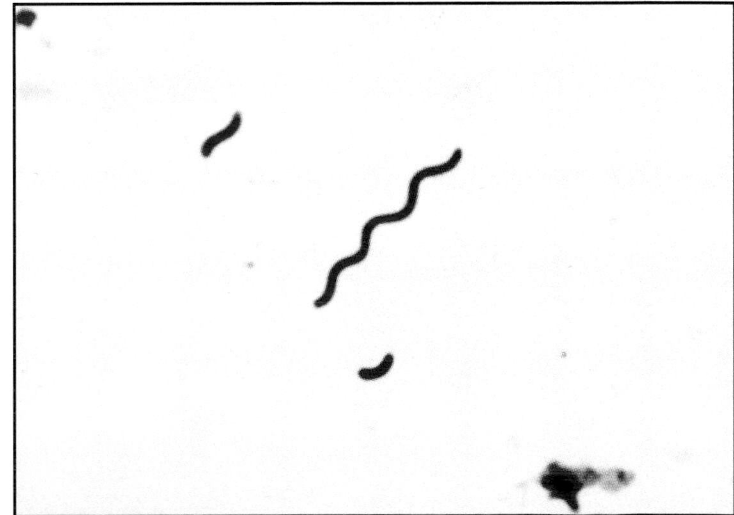

FIGURE 3-9 *Rhodospirillum rubrum* grown in nutrient broth and stained with crystal violet. Compare its size and shape with Fig. 3-8 (X2640).

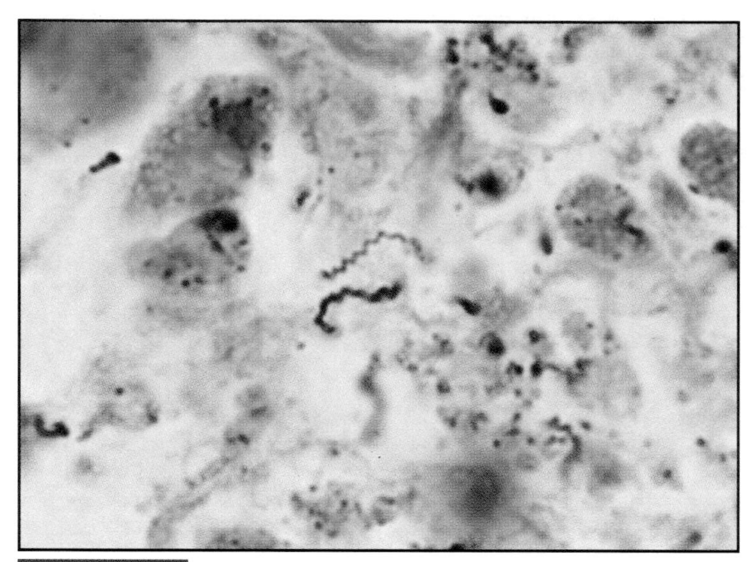

FIGURE 3-10 *Treponema pallidum*, a spirochete, is the causative agent of syphilis in humans (X2640).

FIGURE 3-11 *Vibrio vulnificus* is a vibrio found in shell fish and marine habitats (X1000).

Bacterial Cell Arrangements

Cell arrangement, determined by the number of planes in which cell division occurs and whether the cells separate after division, is also useful in bacterial identification. Cell arrangements and the prefix or root word used to indicate each are listed below.

diplo-	pairs of cells
strepto-	chains of cells
sarcina	group of eight cells
staphylo-	irregular cluster of cells

Cocci may exhibit any of these arrangements, depending on their division planes as shown in Figure 3-12. Bacilli are found as single cells or in chains. Spirilla are rarely seen as anything other than single cells. Figures 3-13 through 3-19 illustrate common cell arrangements.

Exceptions to the more common categories occur. In Figure 3-20, *Corynebacterium diphtheriae* illustrates *pleomorphism*. That is, there is a variety of cell shapes — slender, ellipsoidal or ovoid rods — in a given sample. Second, snapping division produces either a *palisade* or *angular* arrangement of cells. Figure 3-21 illustrates a phenomenon characteristic of some mycobacteria in which they grow in parallel chains called *cords*.

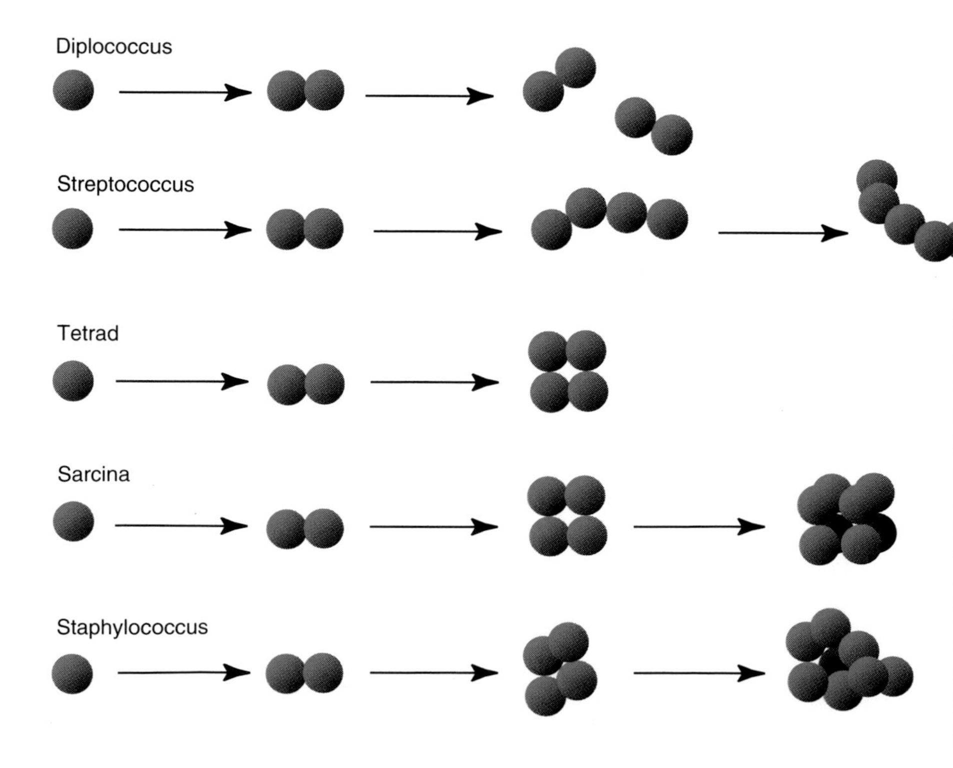

FIGURE 3-12 Division patterns among cocci. In diplococci, there is a single division plane and cells are generally found in pairs. Streptococci also have a single division plane, but the cells remain attached to form chains of variable length. If there are two division planes (X and Y), the cells form tetrads. Sarcinae divide in three planes (X, Y and Z) to produce a regular cuboidal arrangement of cells. Staphylococci divide in more than three planes (X, Y, Z and oblique) to produce a characteristic grapelike cluster of cells. NOTE: Rarely will a sample be composed of just one arrangement. The more complex the arrangement, the more likely scattered examples of simpler arrangements will be found. Look for the most common arrangement.

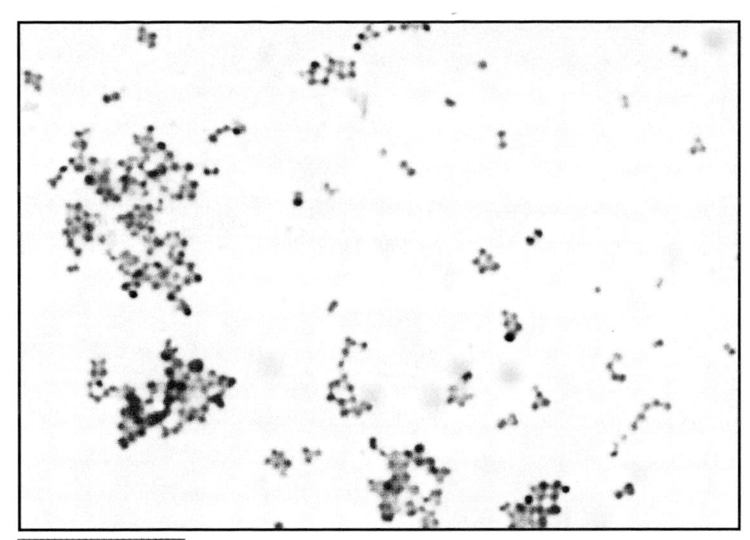

FIGURE 3-13 *Streptococcus pneumoniae* is a diplococcus (X1000). The stain used in this preparation was crystal violet.

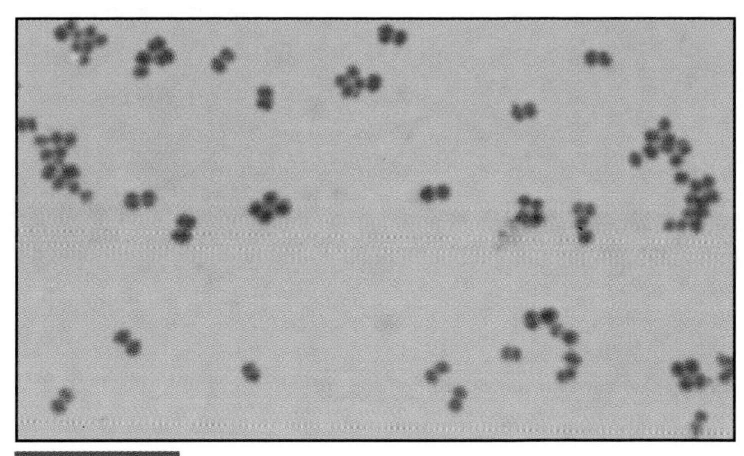

FIGURE 3-14 *Neisseria gonorrhoeae* is a diplococcus that causes gonorrhea in humans. Members of this genus produce diplococci with adjacent sides flattened (X2640).

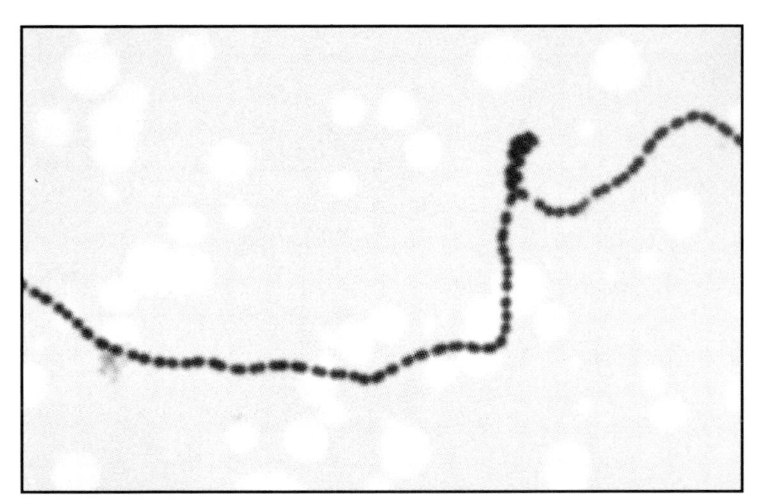

FIGURE 3-15 *Enterococcus faecium* is a streptococcus. This specimen is from a broth culture and was stained with crystal violet (X1000).

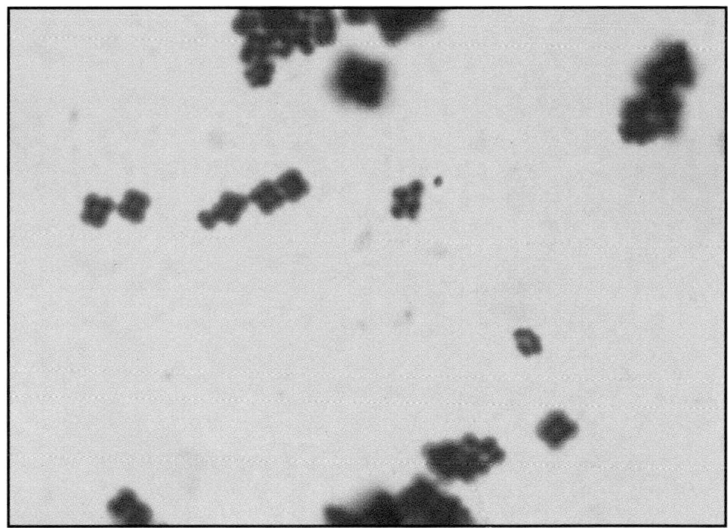

FIGURE 3-16 Tetrads of *Micrococcus roseus* stained with carbolfuchsin (X2640).

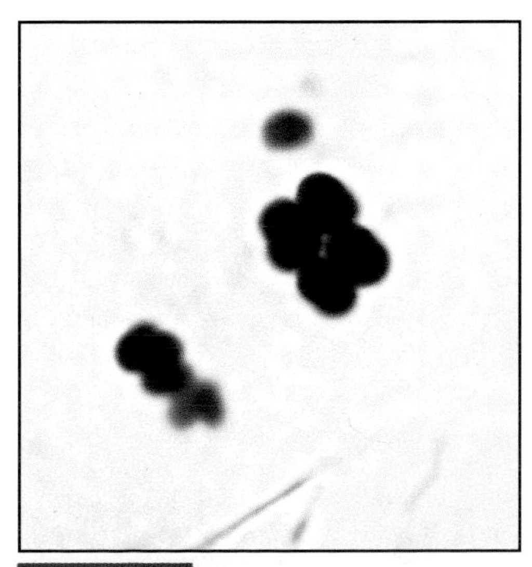

FIGURE 3-17 *Sarcina maxima* exhibits the sarcina organization (X1716).

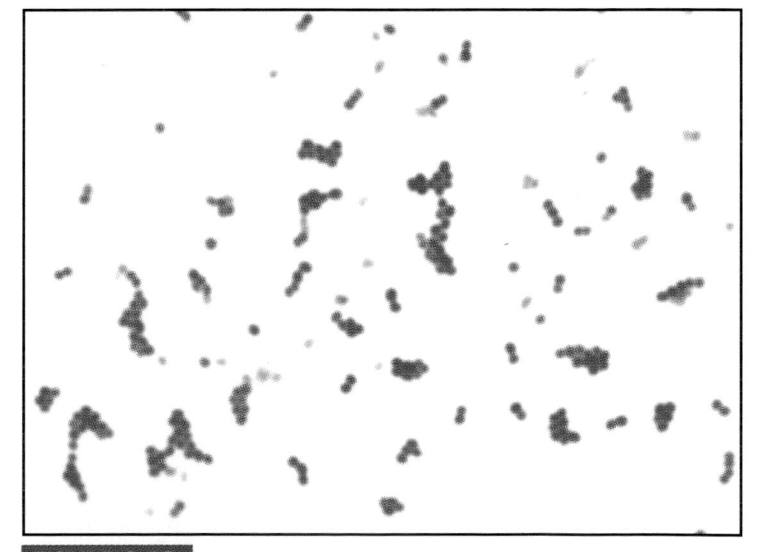

FIGURE 3-18 *Staphylococcus aureus* from broth culture showing the staphylococcus arrangement of cells. *S. aureus* is a common opportunistic pathogen of humans. The stain was crystal violet (X1000).

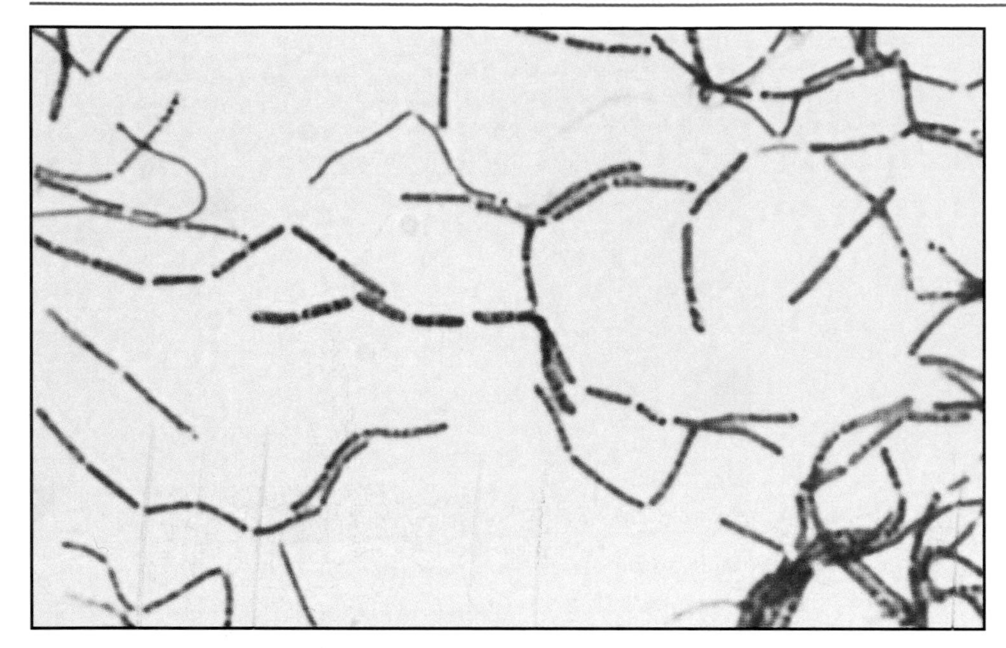

FIGURE 3-19 *Bacillus megaterium* is a streptobacillus and was stained with crystal violet (X2400).

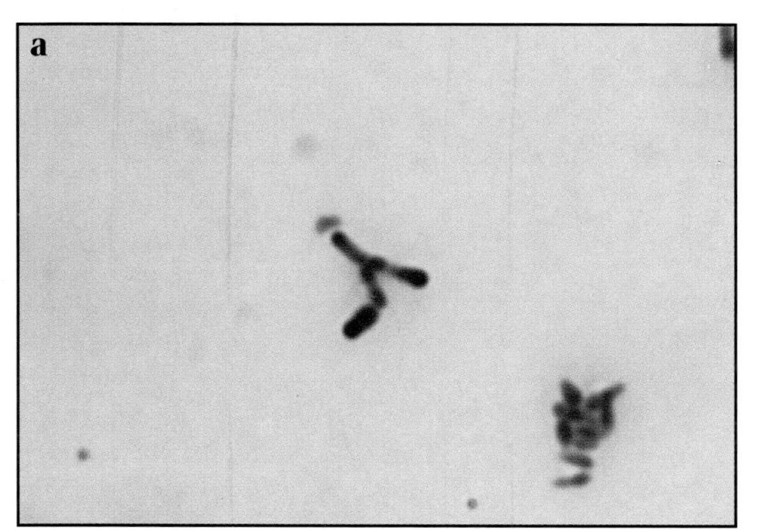

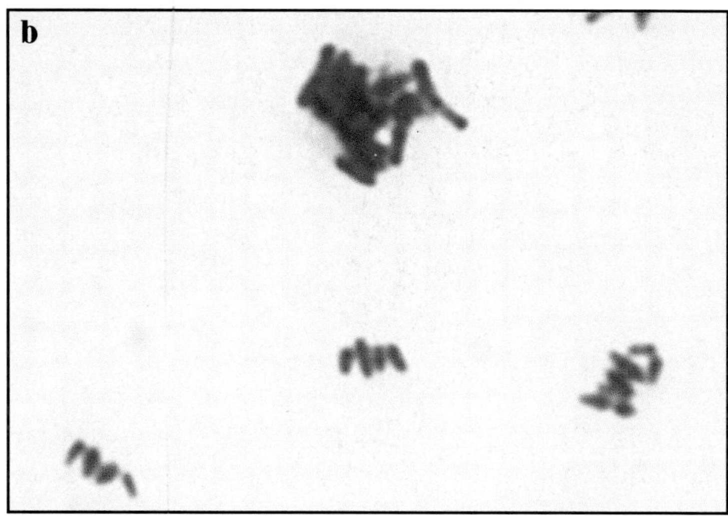

FIGURE 3-20 *(a) Corynebacterium diphtheriae* exhibits an angular arrangement of cells produced by snapping division typical of the genus (X3432). (*b*) *Corynebacterium diphtheriae* illustrating characteristic palisade arrangement of cells (X3432).

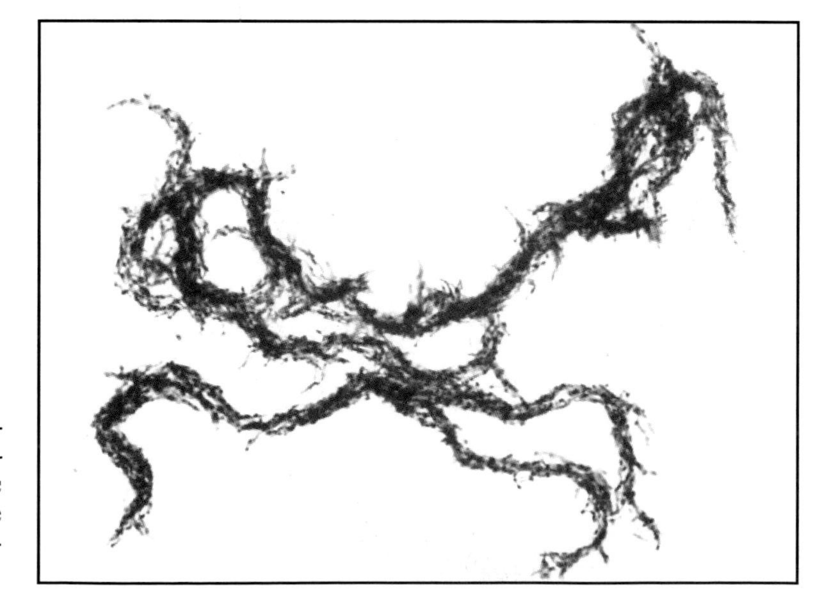

FIGURE 3-21 Cording of *Mycobacterium tuberculosis. M. tuberculosis* aggregates in characteristic cords due to the adhesion of the waxy, acid-fast cell walls of the organisms. (Acid-fast stain, X600).

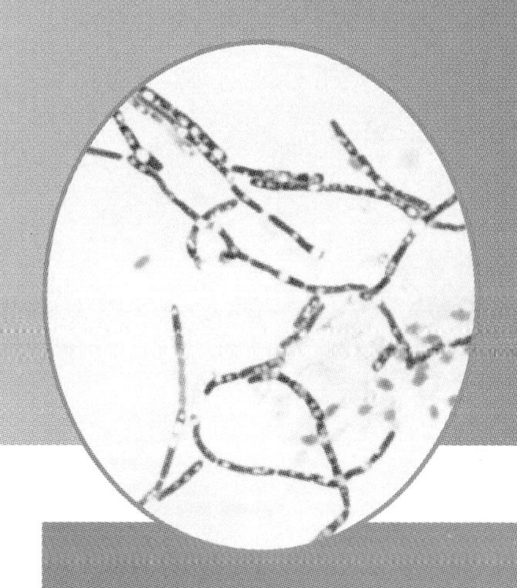

Bacterial Cellular Structures and Differential Stains 4

GRAM STAIN

PURPOSE The Gram stain is used to distinguish between Gram-positive and Gram-negative cells. It is probably the most important and widely used microbiological differential stain. The Gram stain is typically the first differential test run on a specimen brought into the laboratory for identification.

PRINCIPLE The Gram stain is a differential stain in which a decolorization step occurs between the application of two basic stains. As shown in Figure 4-1, the primary stain is crystal violet. Iodine is added as a *mordant* to enhance crystal violet staining by forming a crystal violet-iodine complex. Alcohol decolorization follows and is the most critical step in the procedure. Gram-negative cells are decolorized by the alcohol (95% ethanol) whereas Gram-positive cells are not. Gram-negative cells are thus able to be colorized by the counterstain safranin. Upon successful completion of a Gram stain, Gram-positive cells appear purple and Gram-negative cells appear red (Fig. 4-2).

Electron microscopy and other evidence have allowed microbiologists to determine that the ability to resist alcohol decolorization or not is based on the different wall constructions of Gram-positive and Gram-negative cells. Gram-negative cell walls have a higher lipid content than Gram-positive cell walls. It is thought that the alcohol extracts the lipid, making the Gram-negative wall more porous and incapable of retaining the crystal violet-iodine complex, thus decolorizing it. Alcohol dehydration of Gram-positive cell walls also makes them *less* porous, thereby trapping the crystal violet-iodine complex inside.

While some organisms give Gram-variable results, novice microbiologists should not rely on this too much as an explanation of difficulties they are having with the technique. The decolorization step is the most crucial and most likely source of Gram-stain inconsistency. It is possible to *over-decolorize* by leaving the alcohol on too long and get red Gram-*positive* cells. It also is possible to *under-decolorize* and produce purple Gram-*negative* cells. Neither of these situations changes the actual Gram reaction for the organism being stained. Rather, they are false results due to poor technique. Until correct results are consistently obtained, it is recommended that control smears of Gram-positive and Gram-negative organisms be stained along with the organism in question (Fig. 4-3).

Potassium hydroxide provides a nonstain test to confirm Gram reaction for particularly difficult species. Part of a colony is emulsified in a drop of KOH for one minute, then the loop is slowly withdrawn. Release of chromosomal material by Gram-negative cells makes the suspension viscous, stringy, and adhesive. Gram-positives are unchanged and the emulsion remains watery (Fig. 4-4).

FIGURE 4-1 The Gram stain. Gram-positive cells stain violet; Gram-negative cells stain red.

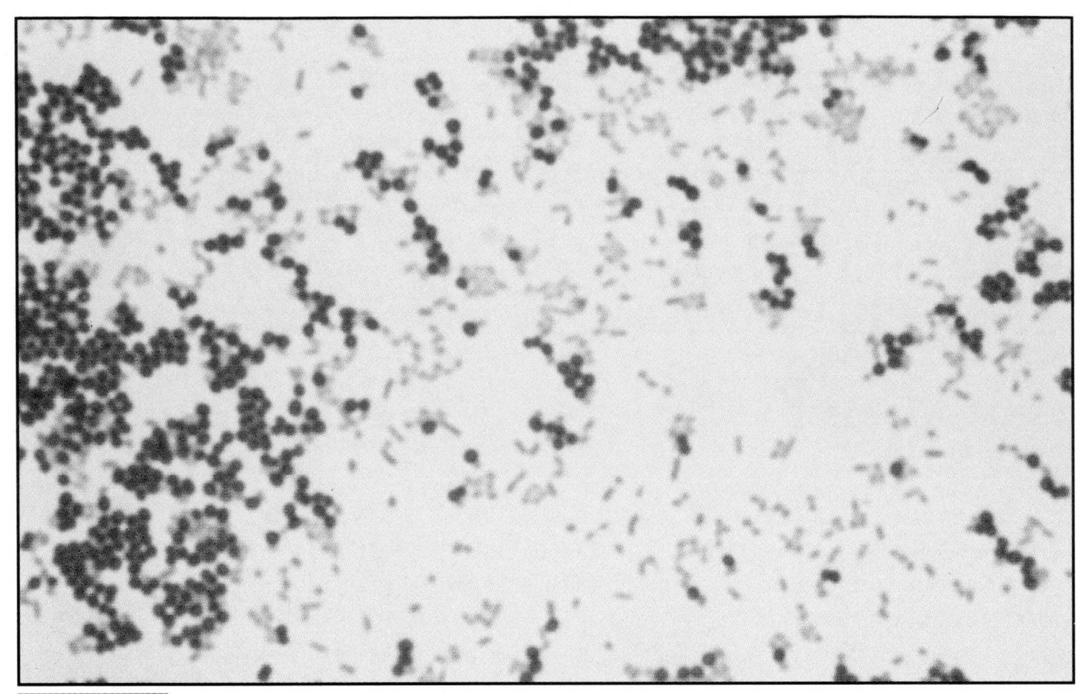

FIGURE 4-2 Gram stain of *Staphylococcus epidermidis* (+) and *Citrobacter diversus* (–) (X2640).

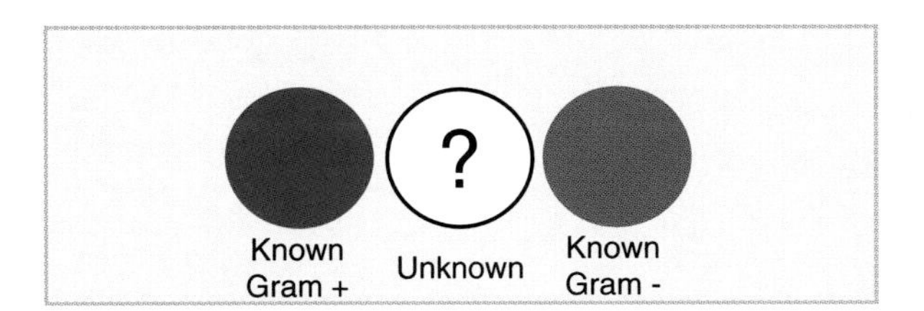

FIGURE 4-3 Until consistent results are obtained, run Gram stains with known Gram-positive and Gram-negative organisms as controls.

FIGURE 4-4 The KOH test for Gram reaction. *Escherichia coli*, a Gram-negative organism, after emulsification in KOH for one minute. The solution has become viscous and stringy due to the release of chromosomal material from the cells.

ACID-FAST STAIN

PURPOSE The acid-fast stain is a differential stain used to detect cells capable of retaining a primary stain when treated with an acid alcohol. The acid-fast stain is an important differential stain for identifying bacteria in the genus *Mycobacterium,* some of which are pathogens (*e.g., M. leprae* and *M. tuberculosis*). Members of the actinomycete genus *Nocardia* (*N. brasiliensis* and *N. asteroides* are opportunistic pathogens) are partially acid-fast.

PRINCIPLE Acid-fast cells contain a large amount of the waxy material *mycolic acid* in their walls. The wax makes it difficult for cells to adhere to a glass slide, so the smear may be prepared using a drop of serum rather than water. Following heat-fixing, it may be stained in one of two ways: the Ziehl-Neelsen (ZN) method, which uses heat as part of the staining procedure, and the Kinyoun method, which is a "cold" stain.

In the ZN method (Fig. 4-5), the phenolic compound carbolfuchsin is used as the primary stain because it is lipid soluble and penetrates the waxy cell wall. Typical aqueous stains are repelled by the waxy wall. Staining by carbolfuchsin is further enhanced by steam heating the preparation to drive the stain into the cell. Acid alcohol is used to decolorize nonacid-fast cells; acid-fast cells resist this decolorization. A counterstain of methylene blue is then applied. Acid-fast cells stain reddish-purple; nonacid-fast cells stain blue (Fig. 4-6).

The Kinyoun method (Fig. 4-7) uses a slightly more lipid soluble and concentrated carbolfuchsin as the primary stain. These properties allow it to penetrate the acid-fast walls without the use of heat. Decolorization with acid alcohol is followed by a contrasting counterstain, such as brilliant green (Fig. 4-8).

Fluorescent dyes, such as auramine or rhodamine, are used in some laboratories, and are actually preferable to traditional carbolfuchsin stains because of their higher sensitivity. The *fluorochrome* combines specifically with mycolic acid. Acid alcohol is used for decolorization and potassium permanganate is the counterstain. When observed under the microscope with UV illumination, acid-fast cells are yellow and nonacid-fast cells are not seen (Fig. 4-9).

	Acid-Fast	Nonacid-Fast
Cells prior to staining are transparent.		
After staining with carbolfuchsin, cells are reddish-purple. Steam heat enhances the entry of carbolfuchsin into cells.		
Decolorization with acid alcohol removes stain from acid-fast negative cells.		
Methylene blue is used to counterstain acid-fast negative cells.		

FIGURE 4-5 The Ziehl-Neelsen acid-fast stain. Acid-fast cells stain reddish-purple; nonacid-fast cells stain blue.

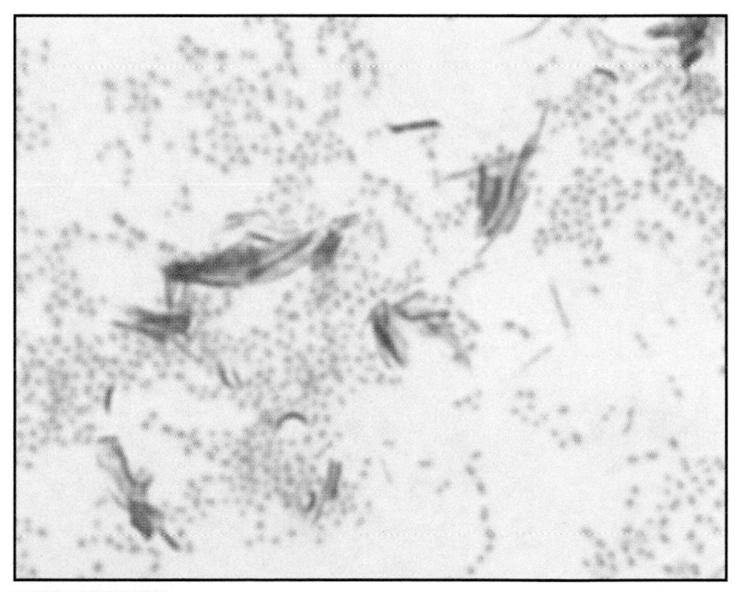

FIGURE 4-6 Acid-fast stain of *Mycobacterium smegmatis* (+) and *Enterococcus faecium* (–) by the Ziehl-Neelsen method (X2640).

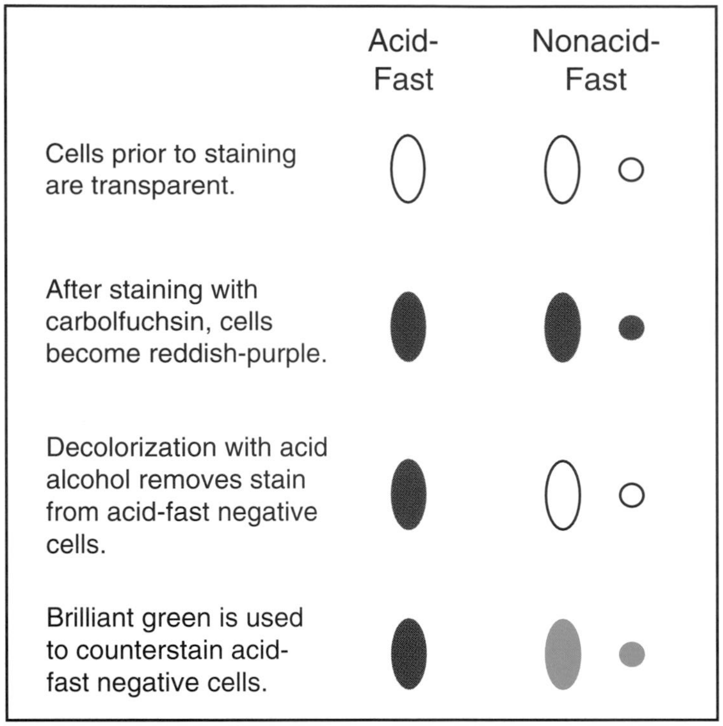

Acid-Fast | Nonacid-Fast

Cells prior to staining are transparent.

After staining with carbolfuchsin, cells become reddish-purple.

Decolorization with acid alcohol removes stain from acid-fast negative cells.

Brilliant green is used to counterstain acid-fast negative cells.

FIGURE 4-7 The Kinyoun acid-fast stain. Acid-fast cells stain reddish-purple; nonacid-fast cells stain green.

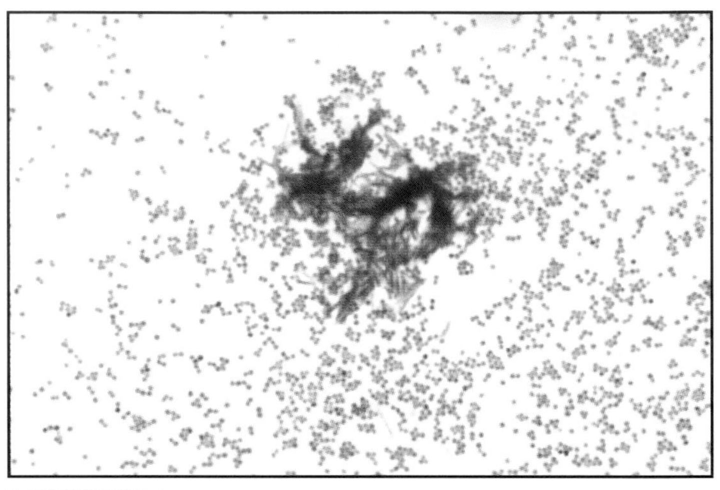

FIGURE 4-8 Acid-fast stain *of Mycobacterium smegmatis* (+) and *Staphylococcus epidermidis* (−) by the Kinyoun method (1000X).

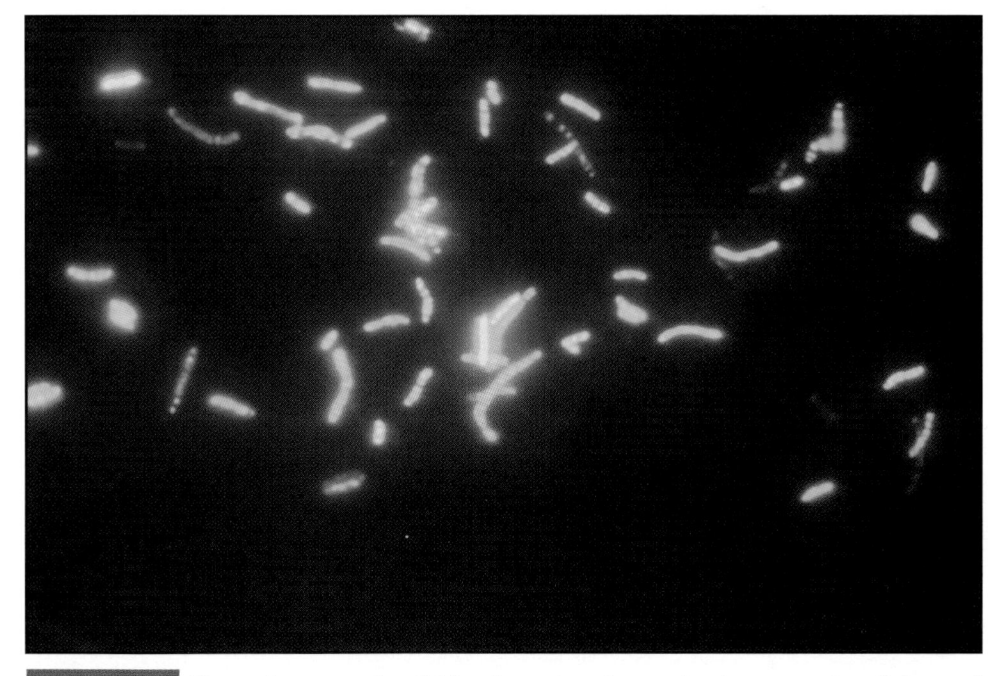

FIGURE 4-9 Fluorochrome stain of *Mycobacterium kansasii* using auramine O dye and observed under ultraviolet light (1000X).

CAPSULE STAIN

PURPOSE The capsule stain is a differential stain used to detect cells capable of producing an extracellular capsule. Capsule production increases virulence in some microbes (such as *Bacillus anthracis* and *Streptococcus pneumoniae*) by making them less vulnerable to phagocytosis. *Streptococcus mutans* produces an insoluble capsule that enables it to adhere to tooth enamel. Other bacteria become trapped in the capsule and form the plaque brushing and flossing removes. Acid production from these bacteria erodes the enamel and causes dental caries (cavities).

PRINCIPLE Capsules are composed of mucoid polysaccharides or polypeptides, the chemical structure of which makes staining difficult. The capsule stain technique takes advantage of this phenomenon by simply staining *around* the cells. Typically, an acidic stain, such as congo red, which stains the background, and a basic stain which colorizes the cell proper, are used. The capsule remains unstained and appears as a white halo between the cells and the colored background (Fig. 4-10).

Since this technique begins as a negative stain, cells are spread in a film with the acidic stain and are not heat-fixed. Heat-fixing causes shrinkage of the cells, leaving an artifactual white halo around them that might be interpreted as a capsule. As an alternative to heat-fixing, cells may be emulsified in a drop of serum to promote adherence to the glass slide.

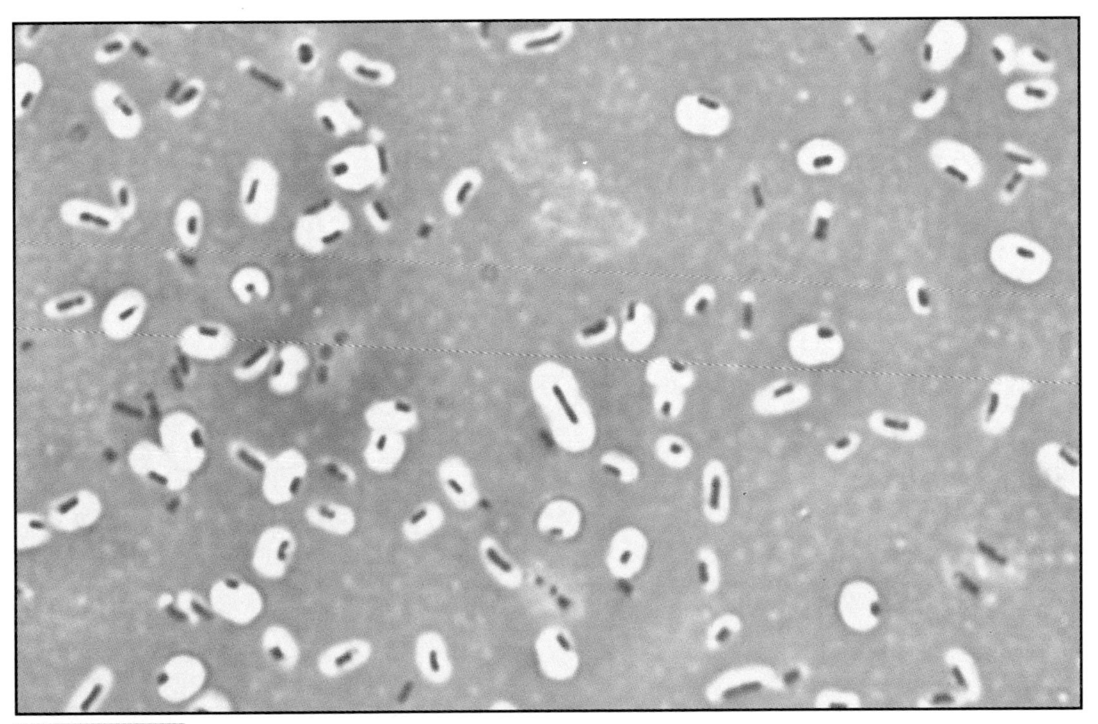

FIGURE 4-10 Capsule stain of *Klebsiella pneumoniae*. The acidic stain colorizes the background while the basic stain colorizes the cell, leaving the capsules as unstained, white clearings around the cells (X2640).

SPORE STAIN

PURPOSE The spore stain is a differential stain used to detect the presence and location of spores in bacterial cells. Only a few genera produce spores. Among them are the genera *Bacillus* and *Clostridium*. Most members of *Bacillus* are soil, freshwater, or marine *saprophytes*, but a few are pathogens, such as *B. anthracis*. Most members of *Clostridium* are soil or aquatic saprophytes, or inhabitants of human intestines, but three pathogens are fairly well known: *C. tetani*, *C. botulinum*, and *C. perfringens*.

PRINCIPLE A spore is a dormant form of the bacterium that allows it to survive lean environmental conditions. Spores have a tough outer covering made of the protein *keratin* and are resistant to heat and chemicals. The keratin also resists staining, so extreme measures must be taken to stain the spore. In the Schaeffer-Fulton method (Fig. 4-11), a primary stain of malachite green is forced into the spore by steaming the bacterial emulsion. Malachite green is water soluble and has a low affinity for cellular material, so vegetative cells and spore mother cells may be decolorized with water and counterstained with safranin (Fig. 4-12).

Spores may be located in the middle of the cell (central), at the end of the cell (terminal), or between the end and middle of the cell (subterminal). These are shown in Figures 4-13 through 4-15. Spore shape may also be of diagnostic use. Spores may be spherical or elliptical (oval).

Members of the genus *Corynebacterium* may exhibit club-shaped swellings (Fig. 4-16) that might be confused with spores. A spore stain will distinguish between true spores and these structures.

	Spore producer	Spore nonproducer
Cells and spores prior to staining are transparent.		
After staining with Malachite green, cells and spores are green. Heat is used to force the stain into spores, if present.		
Decolorization with water removes stain from cells, but not spores.		
Safranin is used to counterstain cells.		

FIGURE 4-11 The Schaeffer-Fulton spore stain. Upon completion, spores are green and vegetative cells are red.

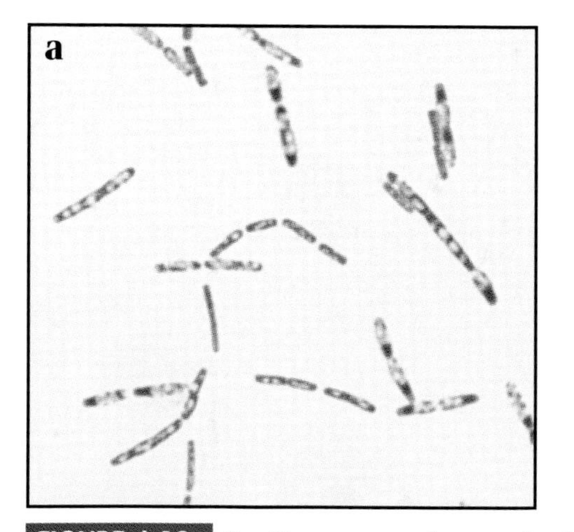

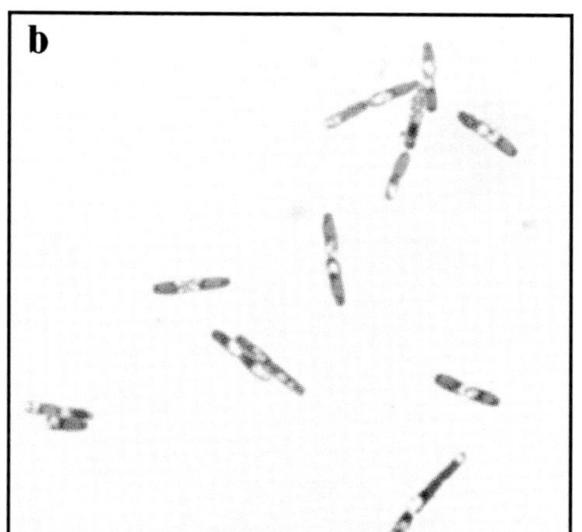

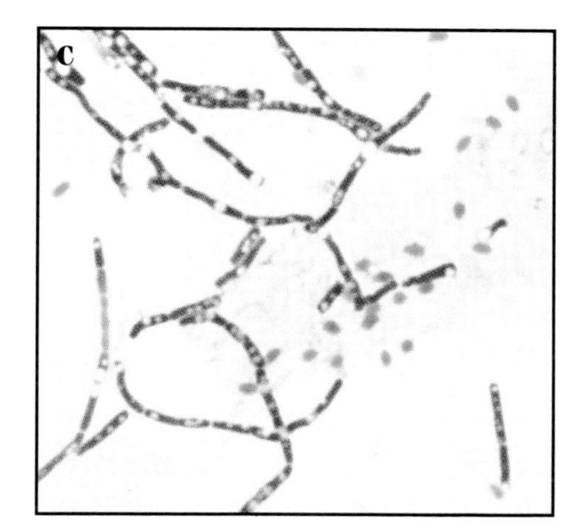

FIGURE 4-12 *Bacillus cereus* culture stained at various stages of growth by the Schaeffer-Fulton technique (1200X). *(a)* Vegetative cells of *B. cereus* prior to sporulation (24 hours after inoculation). *(b)* Central elliptical spores of *B. cereus* (48 hours old). *(c) B. cereus* free spores (72 hours old).

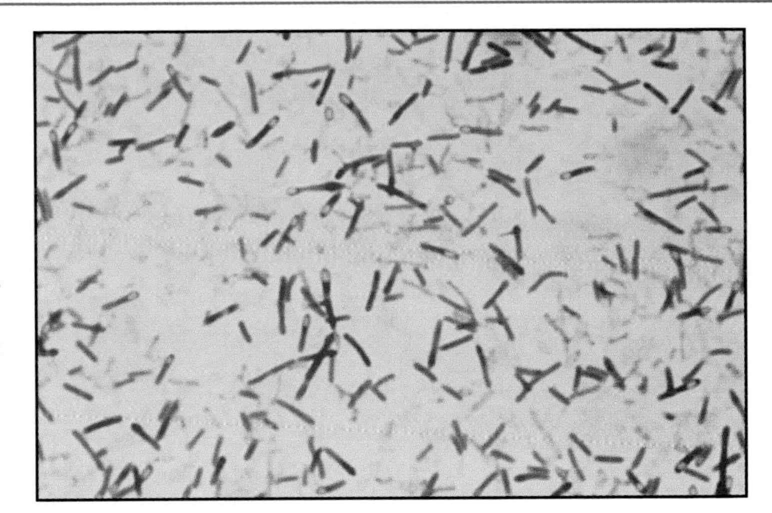

FIGURE 4-13 The subterminal spores of *Clostridium botulinum* are evident as unstained, white ovals in this preparation using a simple stain (X1000).

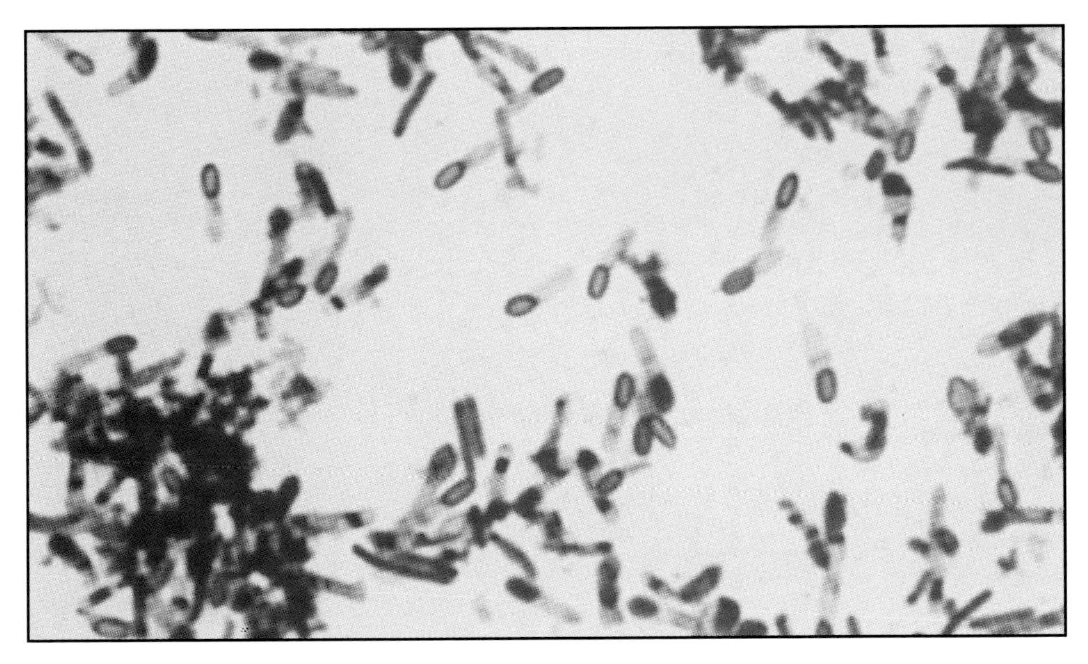

FIGURE 4-14 Elliptical terminal spores of *Clostridium tetani* stained with an alternative spore stain (X2640).

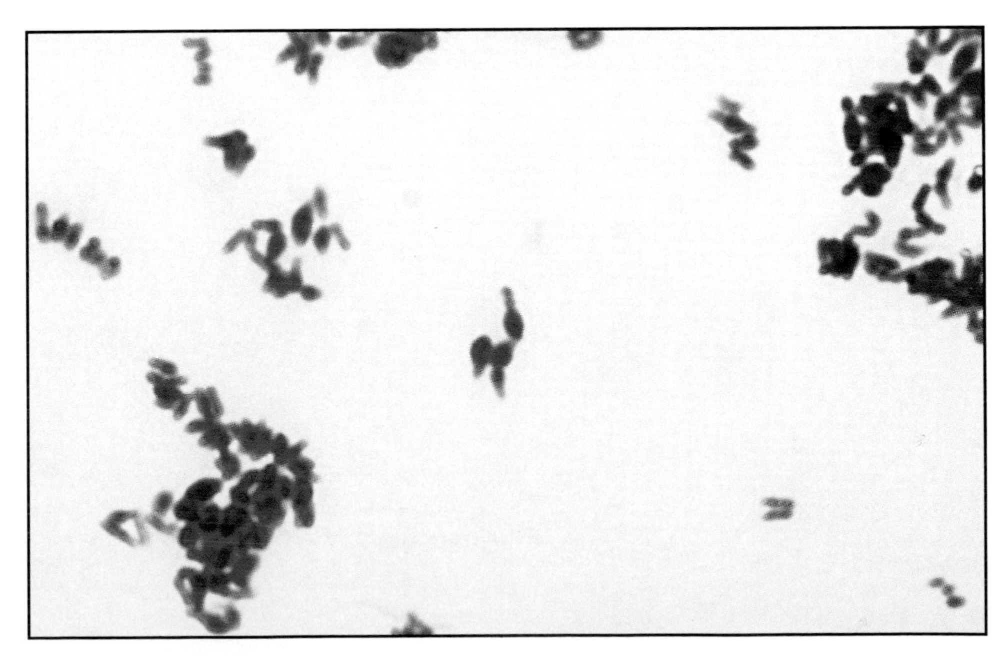

FIGURE 4-15 Club-shaped swellings of *Corynebacterium diphtheriae* should not be confused with spores (X2640).

FLAGELLA STAIN

PURPOSE The flagella stain allows the direct observation of flagella. Presence and arrangement of flagella may be useful in identifying bacterial species. Important pathogens, such as *Bordetella pertussis, Vibrio cholerae, Listeria monocytogenes, Pseudomonas aeruginosa, Salmonella typhi*, and *Proteus vulgaris* are motile. Nonmotile pathogens include *Shigella dysenteriae, Yersinia pestis, Neisseria gonorrhoeae, N. meningitidis,* and *Staphylococcus aureus.*

PRINCIPLE Bacterial flagella typically are too thin to be observed with the light microscope and ordinary stains. Various special flagella stains have been developed which use a mordant to assist in encrusting flagella with stain to a visible thickness. Most require experience and advanced techniques, and are typically not performed in beginning microbiology classes. A variety of techniques is illustrated in the accompanying figures.

The number and arrangement of flagella may be observed with a flagella stain. A single flagellum is said to be *polar* and the cell has a *monotrichous* arrangement (Fig. 4-16). Other arrangements (shown in Figs. 4-17 through 4-19) include *amphitrichous*, with flagella at both ends of the cell; *lophotrichous*, with tufts of flagella at the end of the cell; and *peritrichous*, with flagella emerging from the entire cell surface.

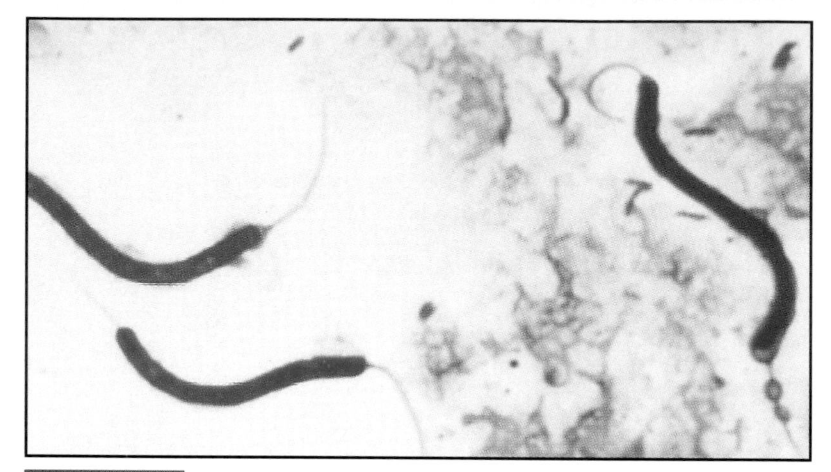

FIGURE 4-17 Amphitrichous flagella of *Spirillum volutans* (X3432).

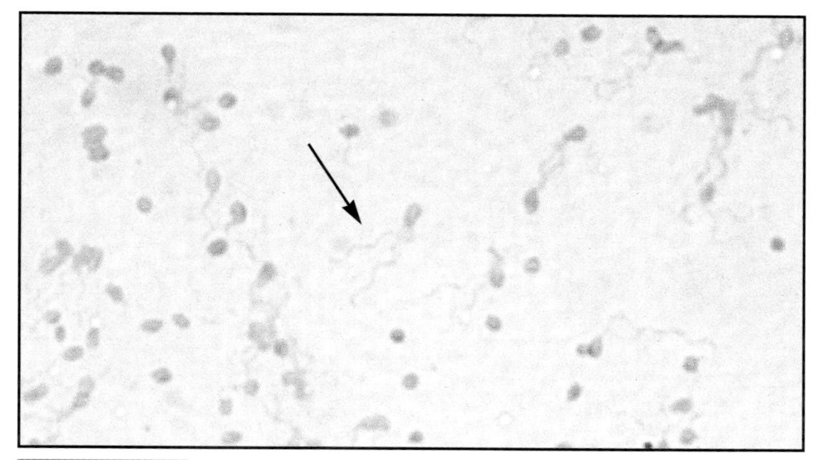

FIGURE 4-18 *"Pseudomonas reptilivora"* illustrates lophotrichous flagella (X2000). Flagella are not seen on all cells because they are extremely delicate and are often destroyed in the staining process.

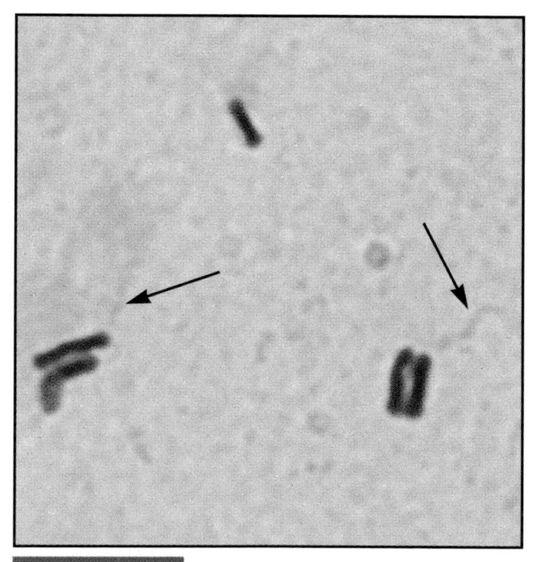

FIGURE 4-16 Polar flagella of *Pseudomonas aeruginosa* (X3432). *P. aeruginosa* is often suggested as a positive control for flagellar stains.

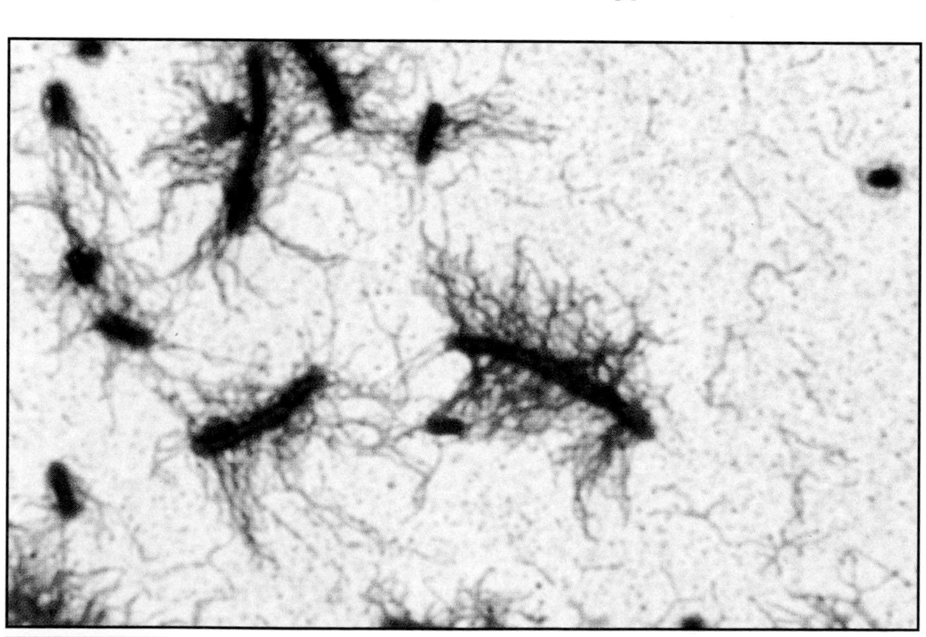

FIGURE 4-19 Peritrichous flagella of *Proteus vulgaris* (X2640).

WET MOUNT AND HANGING DROP PREPARATIONS

PURPOSE Most bacterial microscopic preparations kill the organisms. Simple wet mounts and the hanging drop technique allow observation of living cells to determine motility. They also are used to see natural cell size, arrangement and shape. All of these may be useful characteristics in the identification of a microbe.

PRINCIPLE A wet mount preparation is made by placing bacteria in a drop of water on a microscope slide and covering it with a cover glass. Since no stain is used and most cells are transparent, viewing is best done with as little illumination as possible (Fig. 4-20). Motility may often be observed at medium or high dry magnification, but viewing must be done quickly due to drying of the preparation. As the water recedes, bacteria will appear to be herded across the field. This is not motility. You should look for independent motion of the cells.

A hanging drop preparation allows longer observation of the specimen since it doesn't dry out as quickly. A thin ring of petroleum jelly is applied to the four edges on one side of a cover glass. A drop of water is then placed in the center of the cover glass, and living microbes are transferred into it. A depression microscope slide is carefully placed over the cover glass in such a way that the drop is received into the depression and is undisturbed. The petroleum jelly causes the cover glass to stick to the slide, so the preparation may be picked up, inverted so the cover glass is on top, and placed under the microscope for examination. As with the wet mount, viewing is best done with as little illumination as possible. The petroleum jelly forms an air-tight seal that prevents drying of the drop, allowing a long period for observation of cell size, shape, binary fission and motility.

If these techniques are done to determine motility, the observer must be careful to distinguish between true motility and *Brownian motion* due to collisions with water molecules. In the latter, cells will appear to vibrate in place. With true motility, cells will exhibit independent movement over greater distances.

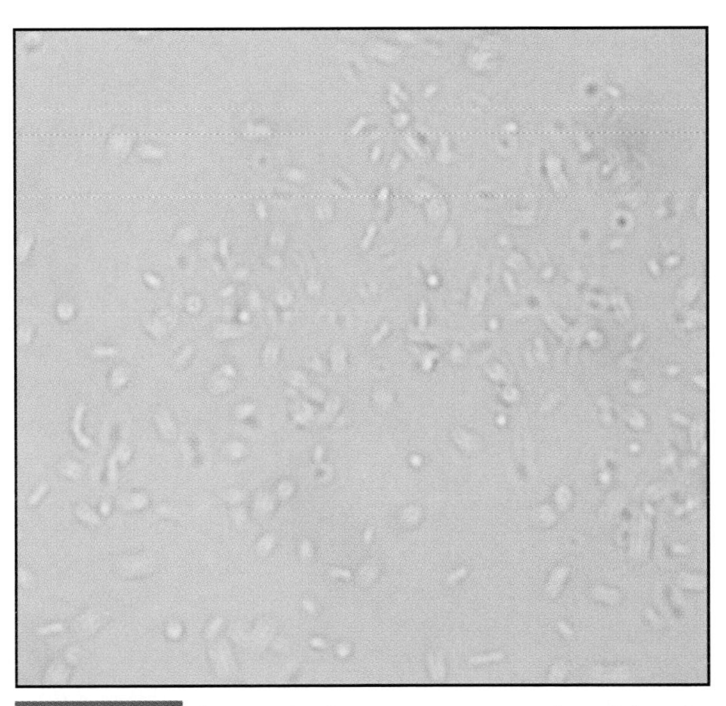

FIGURE 4-20 An unstained wet mount preparation of *Pseudomonas fluorescens,* a motile Gram-negative rod (X960).

MISCELLANEOUS STRUCTURES

FIGURE 4-21 *Bacillus cereus* stained to show the nucleoplasm (X2640).

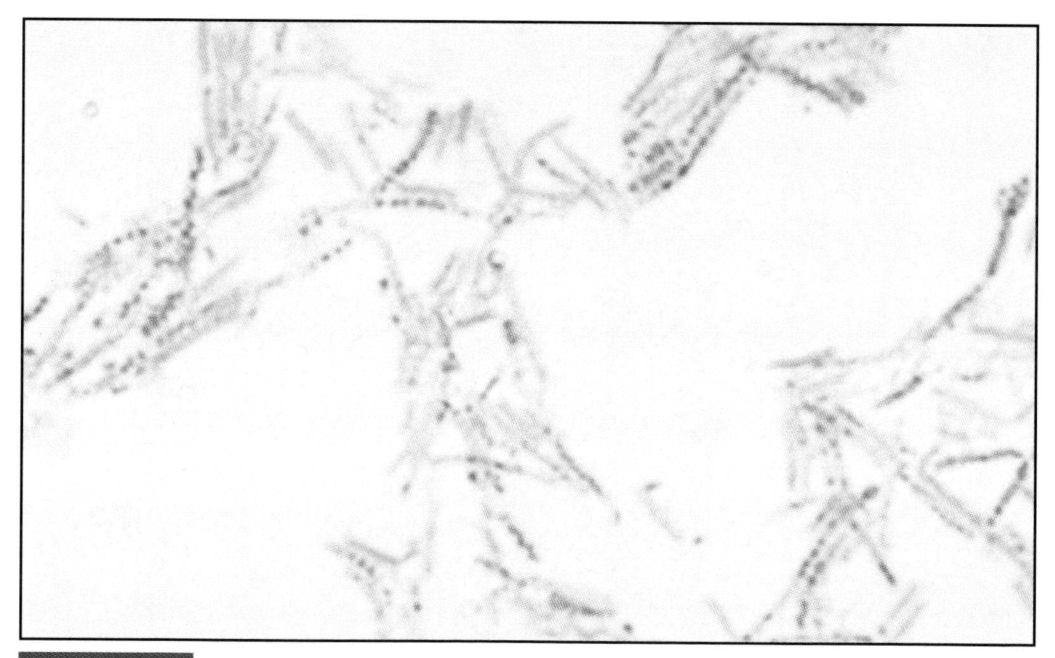

FIGURE 4-22 Dark poly-β-hydroxybutyrate (PHB) granules serve as a carbon and energy reserve (Sudan black B stain, X2640).

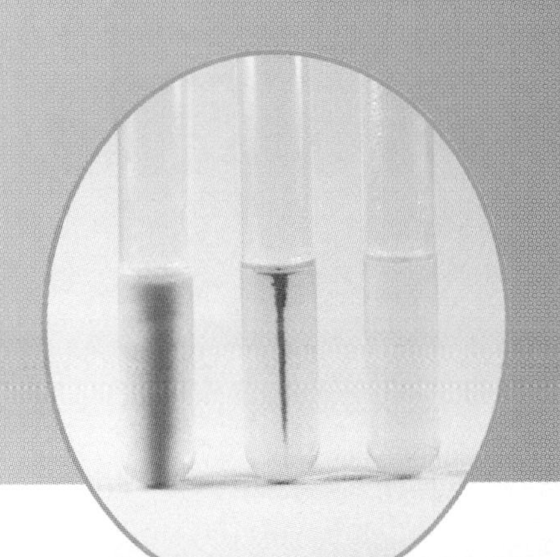

Differential Tests

5

API 20 E® FOR IDENTIFICATION OF ENTEROBACTERIACEAE AND OTHER GRAM-NEGATIVE BACTERIA

PURPOSE This multitest system permits rapid identification of fast-growing, nonfastidious *Enterobacteriaceae* and other Gram-negative bacteria using 23 standard biochemical tests.

PRINCIPLE The API 20 E® system (Fig. 5-1) consists of a strip with 20 tubes containing dehydrated substrates and indicators (where applicable). There is also a cupule above each tube. A bacterial suspension is prepared according to strict standards and is then transferred to the tubes. Inoculation of the tubes must follow rigid guidelines: some tests (ADH, LDC, ODC, H2S (H_2S) and URE) work best if the tube is underfilled, some (CIT, VP and GEL) require that the cupule be filled, and others (ADH, LDC, ODC, H2S (H_2S) and URE) require a layering of mineral oil above the bacterial suspension prior to incubation. In all cases, tube inoculation reconstitutes the dehydrated medium. The entire strip is then incubated for 18 to 24 hours. A nonselective medium (*e.g.*, nutrient agar or trypticase soy agar) is also inoculated with the suspension to provide growth for the oxidase test and additional tests or serology, if necessary.

Results of tests not requiring additional reagents are read first; tests that do require reagents are then performed and read in a specific order. Test results may be compared to the differential charts in the Analytical Profile Index to determine the best match for identification. Alternatively, an identification value may be determined. The tests are clustered into groups, and each positive result within a group is assigned a number — either 4, 2 or 1. The numbers within each group are added together, their sums being used to make a seven digit number. Thus, the combination of test results gives a unique ID value for each organism. The ID value is then located in the API Profile Recognition System database and identification is made.

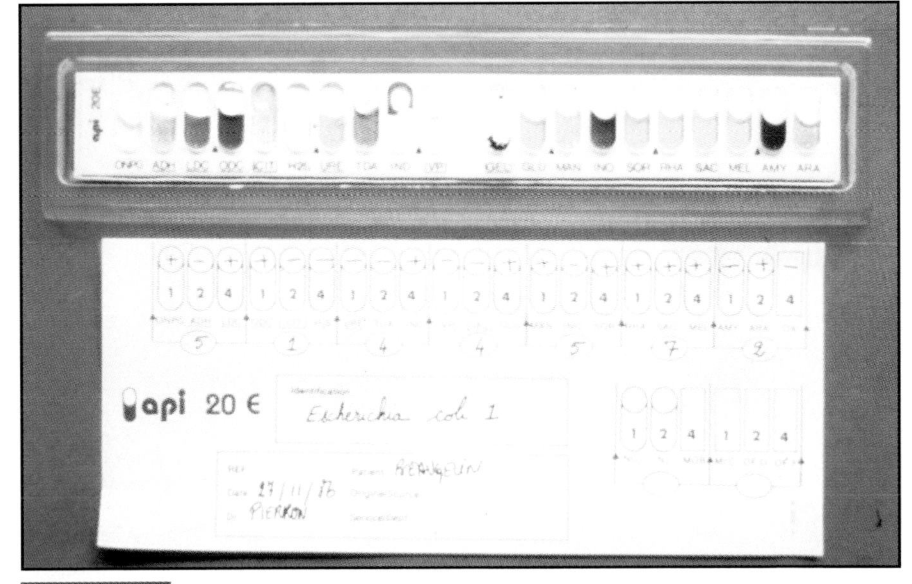

FIGURE 5-1 An inoculated API 20 E® test strip and score sheet. The tests shown are: ONPG = o-Nitrophenyl-β-D-galactopyranoside (yellow is positive), ADH = Arginine dihydrolase (red is positive), LDC = Lysine decarboxylase (red is positive), ODC = Ornithine decarboxylase (red is positive), CIT = Citrate utilization (blue is positive), H_2S = Sulfur reduction (black is positive), URE = Urease (red is positive), TDA = Tryptophan deaminase (brownish-red is positive), IND = Indole (red is positive), VP = Voges-Proskauer (red is positive), and GEL = Gelatin liquefaction (black is positive). The various carbohydrate substrates (GLU = Glucose, MAN = Mannitol, INO = Inositol, SOR = Sorbitol, RHA = Rhamnose, SAC = Sucrose, MEL = Melibiose, AMY = Amygdalin, and ARA = (L+) Arabinose) all yield acid if used and turn the medium yellow. Nitrate reduction is read in the glucose tube and catalase is read in the MAN, INO or SOR tubes, both with the addition of appropriate reagents. The combination of positive and negative results is matched to a table of known results and identification is made.

BACITRACIN SUSCEPTIBILITY TEST

PURPOSE This test is used to differentiate coagulase-negative staphylococci (resistant) from *Micrococcus* and *Stomatococcus*. It may also be used for presumptive identification of group A β-hemolytic streptococci. These bacteria are susceptible, as are some members of the group C, F and G streptococci. To differentiate between the streptococci, the test is often run with sulfamethoxazole-trimethoprim (SXT) disks. Group A streptococci are SXT resistant, groups C, F and G are SXT sensitive (see the SXT Susceptibility Test for more information).

PRINCIPLE Bacitracin is a polypeptide antibiotic produced by *Bacillus subtilis* that inhibits bacterial cell wall synthesis and disrupts membrane structure. Disks impregnated with bacitracin (typically 0.04 units/disk) may be placed on plates inoculated to produce a bacterial lawn. After incubation, any zone of clearing around the disk is interpreted as bacitracin susceptibility (Fig. 5-2). The bacitracin disk alone is enough to distinguish between the resistant *Staphylococcus aureus* and the susceptible micrococci.

FIGURE 5-2 Bacitracin susceptibility on a sheep blood agar plate. *Staphylococcus aureus* (R) is on top and *Micrococcus luteus* (S) is below.

β-LACTAMASE TEST

PURPOSE β-lactamase is group of bacterial enzymes that inactivate penicillins and cephalosporins. This test is used to quickly identify if patient isolates are resistant to these antibiotics. It is especially useful in identifying resistant strains of *Neisseria gonorrhoeae*, *Staphylococcus spp.*, and enterococci.

PRINCIPLE Penicillins and cephalosporins are antibiotics which have a similar molecular structure — they each have a β-lactam ring (Fig. 5-3). Both families of antibiotics interfere with peptidoglycan synthesis by competing with peptidoglycan subunits for a site on a membrane-bound transpeptidase involved in cross-linking the polymers. By causing a defective wall to form, these β-lactam antibiotics are bactericidal. However, increasing numbers of bacteria have developed resistance to these antibiotics (as well as others) through the production of one of many (at least 170) β-lactamases. These enzymes hydrolyze β-lactam antibiotics and make them ineffective.

This procedure is one of many to identify β-lactam antibiotic resistance. In this test, paper discs containing nitrocefin, a cephalosporin with a susceptibility to most β-lactamases (including those that hydrolyze penicillins) are used. A smear of the organism is made on the disc. If the organism produces β-lactamase, it will hydrolyze the nitrocefin at the amide bond in the β-lactam ring and produce a pink color (Fig. 5-4).

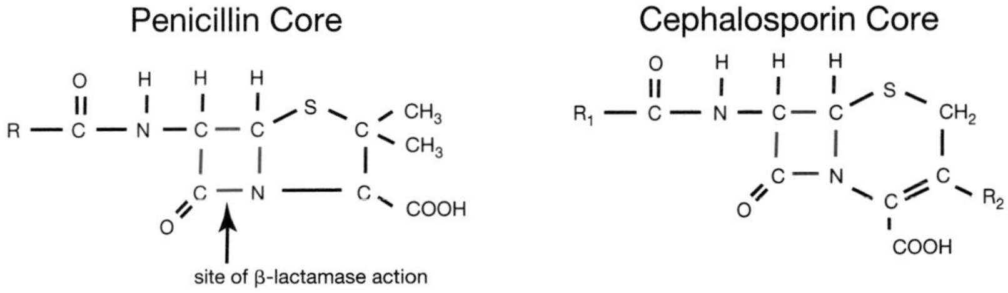

Penicillin Core

site of β-lactamase action

Cephalosporin Core

FIGURE 5-3 Although penicillins and cephalosporins have unique structures, they share a β-lactam ring that makes them bactericidal in the same way — they interfere with peptidoglycan cross-linking. The arrow indicates the site of β-lactamase activity.

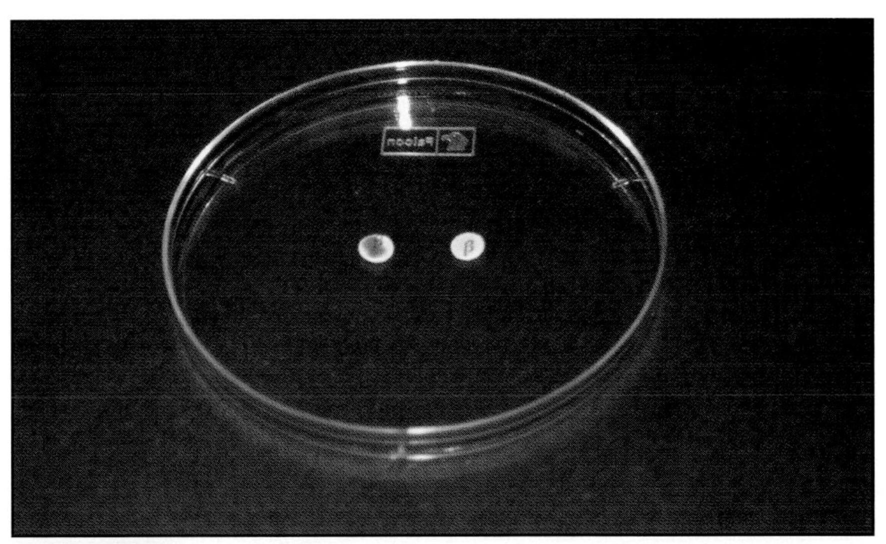

FIGURE 5-4 Cefinase® discs (Becton Dickinson Microbiology Systems, Sparks, MD) for identification of bacteria resistant to β-lactam antibiotics due to β-lactamase production. A β-lactam resistant strain is on the left and a susceptible strain is on the right.

BILE ESCULIN TEST

PURPOSE This test is used to isolate and identify bacteria able to hydrolyze esculin in the presence of bile. It is most commonly used for presumptive identification of group D streptococci and enterococci, all of which are positive. Group D streptococci and enterococci include opportunistic pathogens such as *Enterococcus faecalis, E. faecium* and *Streptococcus bovis.*

PRINCIPLE Bile esculin agar is both a selective and differential medium. It contains esculin and peptone for nutrition, bile to inhibit Gram-positives other than group D streptococci and enterococci, and sodium azide to inhibit the Gram-negatives. Ferric citrate is added as a color indicator.

Esculin is a *glycoside* (a sugar molecule bonded by an acetyl linkage to an alcohol) composed of glucose and esculetin. These linkages (also called *glycosidic linkages*) are easily hydrolyzed under acidic conditions (Fig 5-5). Many species of bacteria can hydrolyze esculin, but relatively few can do so in the presence of bile.

Organisms that split the esculin molecules and use the liberated glucose to supply their energy needs release esculetin into the medium. The free esculetin reacts with ferric citrate in the medium to form a phenolic iron complex which turns the agar slant dark brown to black (Fig. 5-6). An agar slant that is more than half darkened after no more than 72 hours incubation is bile esculin-positive. If less than half the slant has darkened, the result is negative (Fig. 5-7).

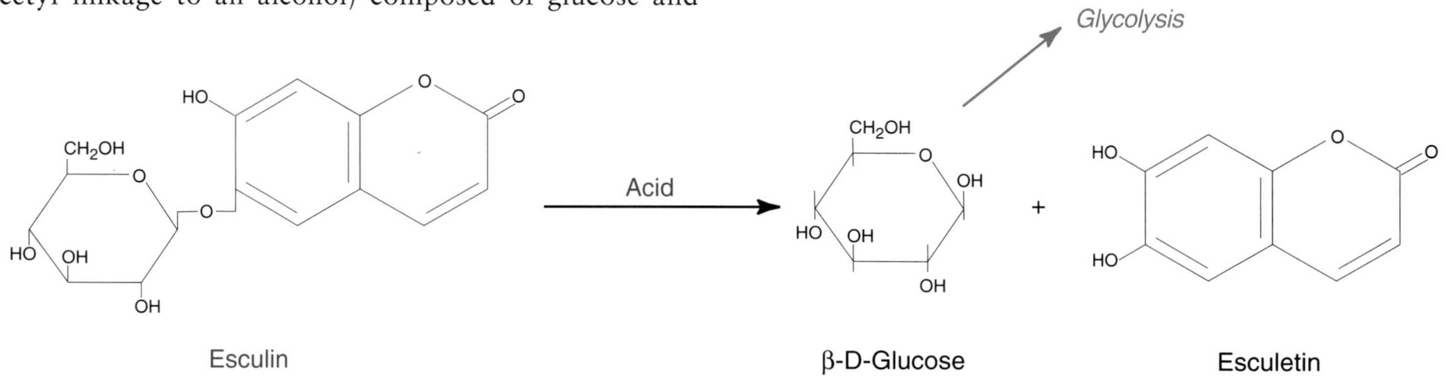

Esculin —Acid→ β-D-Glucose (→ Glycolysis) + Esculetin

FIGURE 5-5 Acid hydrolysis of esculin with the production of esculetin is done by many organisms. However, the group D streptococci and enterococci are unique in their ability to do this in the presence of bile salts.

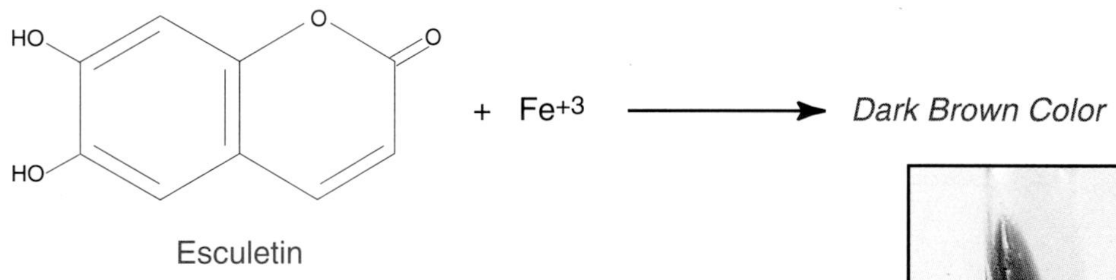

Esculetin + Fe^{+3} → *Dark Brown Color*

FIGURE 5-6 The bile esculin test indicator reaction involves the reaction of esculetin, produced during the hydrolysis of esculin, with Fe^{3+}. The result is a dark brown to black color in the medium.

FIGURE 5-7 The bile esculin test. From left to right: *Enterococcus faecium* (+), *Serratia marcescens* (+), *Citrobacter amalonaticus* (−), and an uninoculated control. A positive result is indicated if more than half the tube turns black within 72 hours.

BLOOD AGAR

PURPOSE Blood agar is commonly used for the cultivation of fastidious organisms requiring specific growth factors not available in other media. Used diagnostically, however, blood agar is an important differential medium used to detect the presence of organisms that produce *hemolysins*. The degree to which these exotoxins hemolyze erythrocytes (RBCs) is a useful diagnostic tool, especially for identification of the genera *Streptococcus*, *Aerococcus*, and *Enterococcus*. Identification of several other species, including members of *Listeria* and *Clostridium* may also require information about their hemolytic characteristics on blood agar.

PRINCIPLE Exotoxins that cause destruction of RBCs (hemolysis) are called hemolysins. There are three categories of hemolysis. β hemolysis is the complete destruction of RBCs and hemoglobin and results in a clearing around the growth on a blood agar plate (Fig. 5-8). Partial destruction of RBCs and hemoglobin (α-hemolysis) produces a greenish discoloration of the blood agar plate (Fig. 5-9). No hemolysis (sometimes called γ-hemolysis) results in no change of the medium (Fig. 5-10).

The hemolysins of *Streptococcus* are called *streptolysins*. Some streptolysins are oxygen-labile, so correct interpretation of hemolysis reaction may require anaerobic

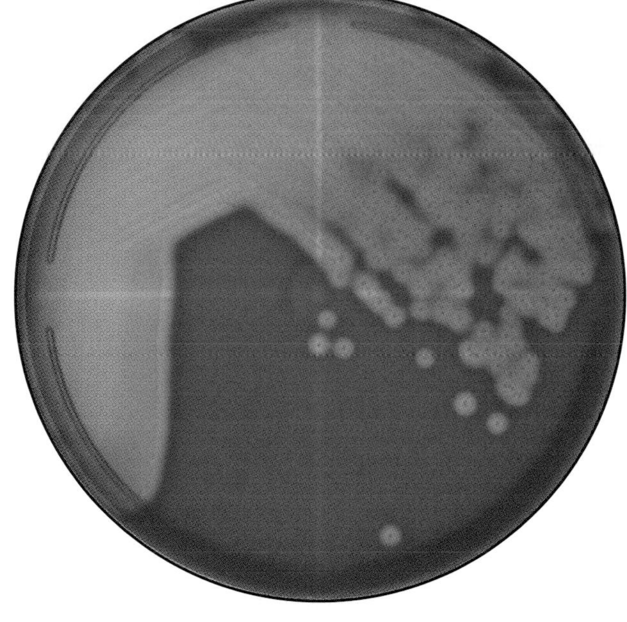

FIGURE 5-8 An unidentified throat culture isolate demonstrates β-hemolysis. The clearing around the individual colonies is due to complete lysis of red blood cells.

FIGURE 5-9 An unidentified throat culture isolate demonstrates α-hemolysis. The greenish zone around the individual colonies is due to incomplete lysis of red blood cells. The two cuts in the agar (arrow) show subsurface β-hemolysis due to an oxygen-labile hemolysin.

FIGURE 5-10 A streak plate of *Enterococcus faecalis* on a sheep blood agar illustrates no hemolysis. This same organism grown on human, bovine, rabbit or equine blood agar may show β-hemolysis.

growth. The simplest method to achieve this is a streak-stab technique in which the blood agar plate is streaked and then stabbed with a loop (Fig. 5-11). The stabs allow subsurface growth in a less aerobic region so oxygen-labile streptolysin activity may be detected.

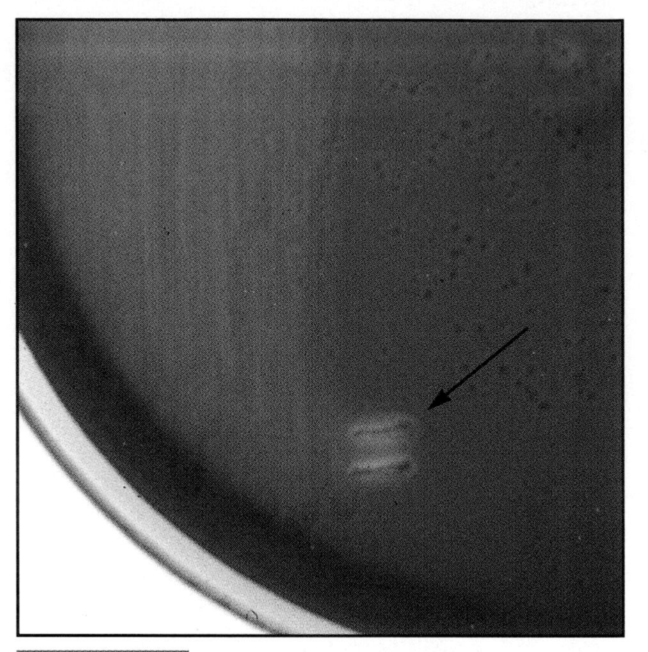

FIGURE 5-11 An unidentified throat culture isolate demonstrates α-hemolysis when growing on the surface, but β-hemolysis around the cuts (arrow).

CASEASE TEST (MILK AGAR)

PURPOSE The casease test is used to identify bacteria capable of hydrolyzing casein with the enzyme *casease*.

PRINCIPLE Milk agar is an undefined medium containing peptone, beef extract and casein. Casein is the protein molecule which gives milk its white color. To be utilized by bacteria these large molecules must be broken down outside the cell to a size small enough to pass through the membrane. Some bacteria secrete the proteolytic exoenzyme casease which hydrolyzes milk protein in the surrounding environment (Fig. 5-12), thus forming smaller, more soluble molecules that can enter the cell. In this test a zone of clearing is created around the colony as casein is digested by casease-positive organisms (Fig. 5-13). No clearing is a negative result.

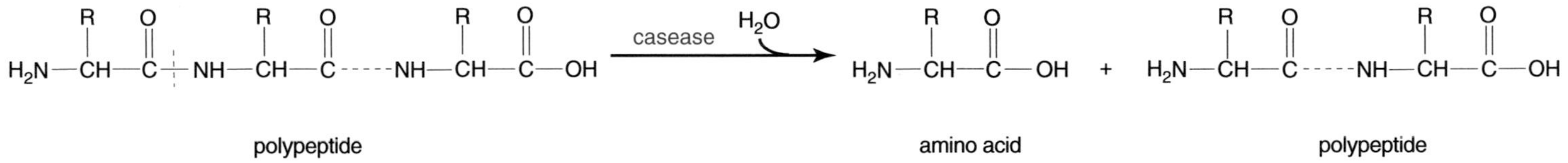

$$\underbrace{H_2N-\overset{R}{\overset{|}{C}H}-\overset{O}{\overset{\|}{C}}-NH-\overset{R}{\overset{|}{C}H}-\overset{O}{\overset{\|}{C}}\cdots NH-\overset{R}{\overset{|}{C}H}-\overset{O}{\overset{\|}{C}}-OH}_{\text{polypeptide}} \xrightarrow[\text{casease}]{H_2O} \underbrace{H_2N-\overset{R}{\overset{|}{C}H}-\overset{O}{\overset{\|}{C}}-OH}_{\text{amino acid}} + \underbrace{H_2N-\overset{R}{\overset{|}{C}H}-\overset{O}{\overset{\|}{C}}\cdots NH-\overset{R}{\overset{|}{C}H}-\overset{O}{\overset{\|}{C}}-OH}_{\text{polypeptide}}$$

FIGURE 5-12 Protein hydrolysis occurs by breaking peptide bonds (one is shown by the red line) between adjacent amino acids to produce short peptides or individual amino acids.

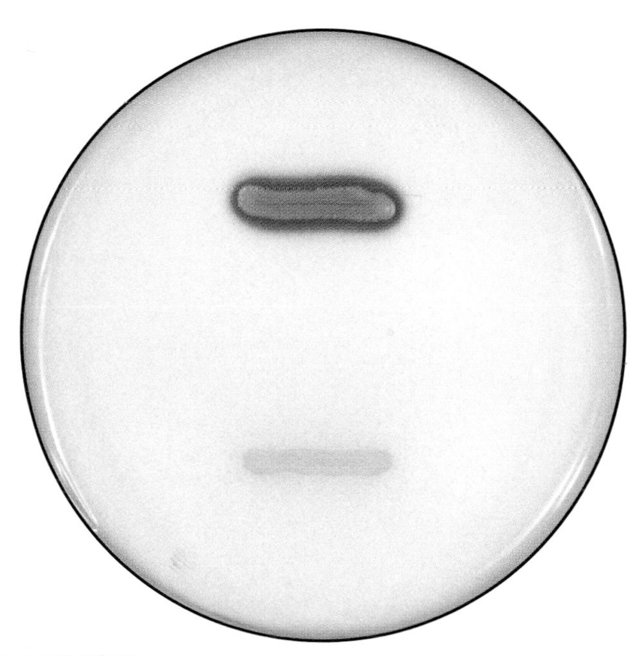

FIGURE 5-13 Milk agar plate with *Bacillus megaterium* (casease-positive) above and *Micrococcus roseus* (casease-negative) below.

CATALASE TEST

PURPOSE This test is used to identify organisms that produce the enzyme catalase. It is most commonly used to differentiate members of the catalase-positive *Micrococcaceae* from the catalase-negative *Streptococcaceae*. Variations on this test may also be used in identification of *Mycobacterium* species.

PRINCIPLE Most aerobic and facultatively anaerobic bacteria produce hydrogen peroxide via the nonenzymatic transfer of electrons from reduced flavoprotein to oxygen, as shown in Figure 5-14. Hydrogen peroxide may also be produced in aerobes and facultative anaerobes enzymatically by superoxide dismutase, as shown in Figure 5-14. Regardless of how it is produced, hydrogen peroxide is a highly reactive molecule that damages cell components. Organisms that produce the enzyme catalase are able to break hydrogen peroxide down into water and oxygen gas, as shown in Figure 5-15.

$$\underset{\text{Reduced Flavoprotein}}{FPH_2} + O_2 \longrightarrow \underset{\text{Oxidized Flavoprotein}}{FP} + \underset{\text{Hydrogen Peroxide}}{H_2O_2}$$

$$2H^+ + \underset{\text{Superoxide Radical}}{2\,O_2^{\bar{\cdot}}} \xrightarrow{\text{Superoxide dismutase}} \underset{\text{Hydrogen Peroxide}}{H_2O_2} + O_2$$

FIGURE 5-14 Hydrogen peroxide may be formed through the transfer of hydrogens from reduced flavoprotein to oxygen or from the action of superoxide dismutase.

$$\underset{\text{Hydrogen Peroxide}}{2H_2O_2} \xrightarrow{\text{Catalase}} 2H_2O + O_2\uparrow$$

FIGURE 5-15 Catalase is an enzyme of aerobes and facultative anaerobes that converts hydrogen peroxide to water and oxygen gas.

The catalase test may be performed either on a slide containing a freshly transferred pure specimen (Fig. 5-16) or on an agar slant containing viable young colonies (Fig. 5-17). When a drop of 3% hydrogen peroxide is placed on catalase-positive bacteria, oxygen gas bubbles form immediately. If the slide test is run, observation under low power on the microscope may be useful for observing weakly positive reactions. No bubbling is considered a negative result. Hydrogen peroxide is very unstable, so a positive control should be run to verify the chemical's reactivity.

Variations on the catalase test are used in identifying species of *Mycobacterium*. In a semiquantitative test, 30% hydrogen peroxide is added to a test tube containing growth. Since most mycobacteria are catalase-positive, the height of the bubbles produced is the differential characteristic. If the bubbles are higher than 45 mm, the test is considered positive. Another variation differentiates between heat-stable and heat-labile forms of catalase. Tubes containing growth are heated for 20 minutes, then the hydrogen peroxide is added. Any bubbling is considered a positive result for heat-stable catalase production.

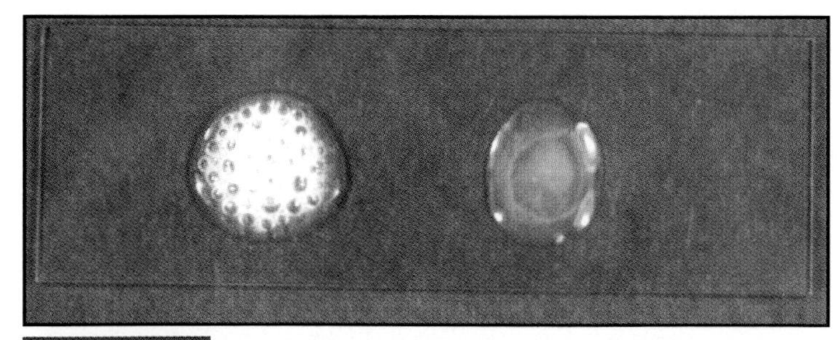

FIGURE 5-16 The catalase slide test in which visible bubble production indicates a positive result. *Staphylococcus aureus* (+) is on the left, *Enterococcus faecium* (−) is on the right.

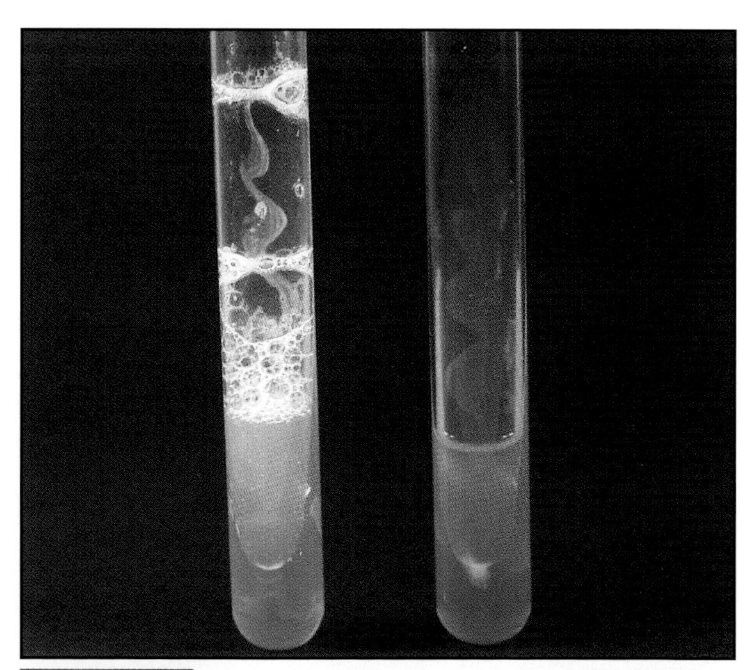

FIGURE 5-17 The catalase test may also be performed on an agar slant. *Staphylococcus aureus* (+) is on the left, *Enterococcus faecium* (−) is on the right.

CITRATE UTILIZATION TEST

PURPOSE The citrate utilization test is used to determine the ability of an organism to use citrate as its sole carbon source. This ability is brought about by the enzyme citrase. Citrate utilization is one part of a test series referred to as the *IMViC* (*I*ndole, *M*ethyl red, *V*oges-Proskauer and *C*itrate tests) which is used for differentiation among the *Enterobacteriaceae*.

PRINCIPLE Simmon's citrate agar is a defined medium in which sodium citrate is the sole carbon source and ammonium ion is the sole nitrogen source. Bromthymol blue is included as a pH indicator. The medium is formulated at an initial pH of 6.9 and is green. At a pH higher than 7.6, bromthymol blue turns the medium a deep blue color.

Citric acid typically is produced by the combination of acetyl CoA and oxaloacetic acid at the entry to the Krebs Cycle. However, some organisms are capable of using citrate as a carbon source if no fermentable carbohydrate is present. They catabolize citrate as shown in Figure 5-18.

Since citrate utilization is an aerobic process, agar slants are used to increase the surface area exposed to air. The slant is inoculated with a light streak across its surface. As CO_2 from citrate-positive organisms accumulates in the medium, it reacts with Na^+ and H_2O to produce alkaline compounds, such as sodium carbonate (Na_2CO_3). This makes the medium basic and turns the bromthymol blue pH indicator Prussian blue (Fig. 5-19). Thus, conversion of the medium to blue is a positive result.

A green color with heavy growth on the slant may also indicate a citrate positive organism since growth is visible only if the cells have entered log phase (possible only if the citrate has been utilized). It is customary, therefore, to use a light inoculum to avoid confusing a heavy inoculum with actual growth. Equivocal results such as this are typically incubated an additional 24 hours to see if blue color appears, or retested using a light inoculum to observe growth of the streak line.

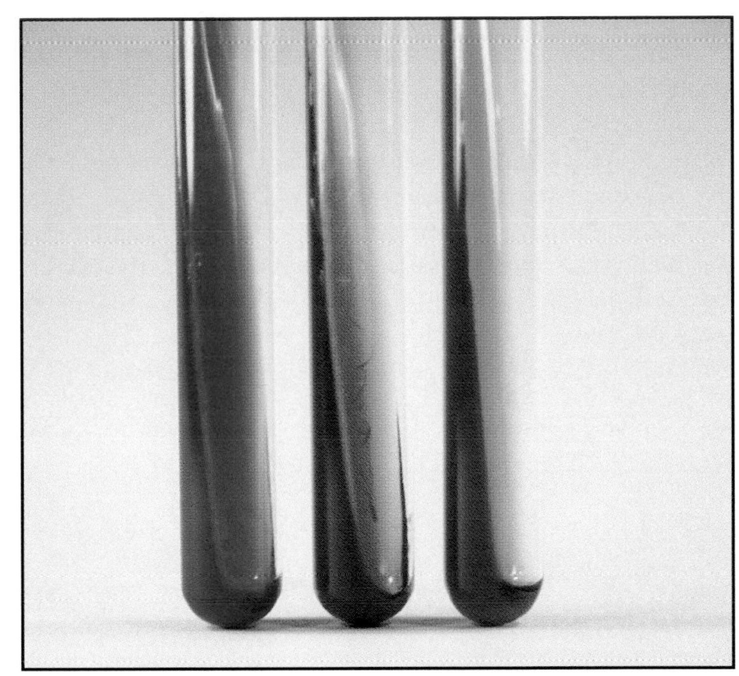

FIGURE 5-19 Simmon's citrate agar inoculated with *Citrobacter diversus* (+) on the left, *Bacillus cereus* (–) in the center, and an uninoculated control on the right.

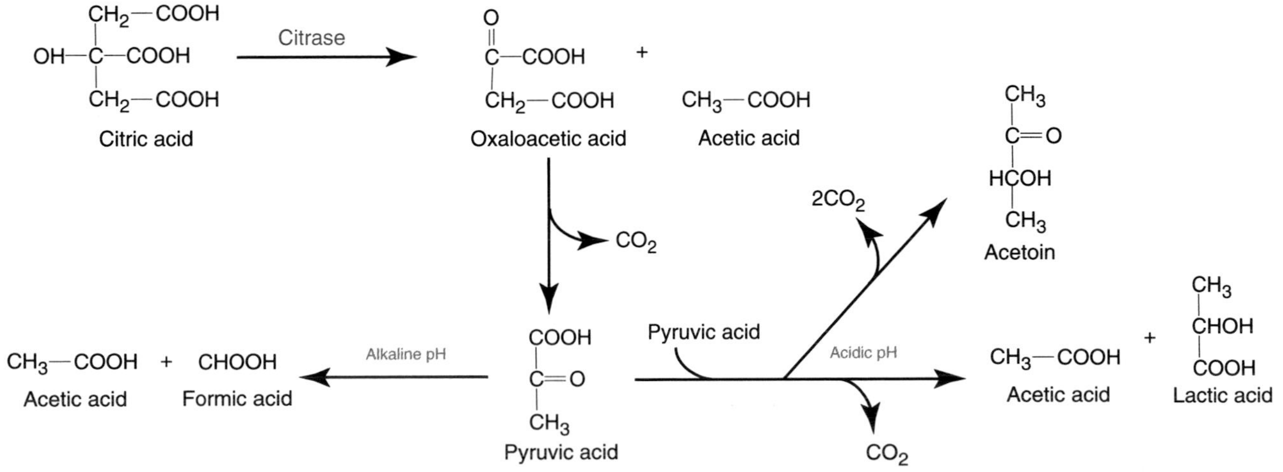

Citric acid (CH_2—COOH / OH—C—COOH / CH_2—COOH) —Citrase→ Oxaloacetic acid (O=C—COOH / CH_2—COOH) + Acetic acid (CH_3—COOH)

Oxaloacetic acid → CO_2 + Pyruvic acid (COOH / C=O / CH_3)

Pyruvic acid —Alkaline pH→ Acetic acid (CH_3—COOH) + Formic acid (CHOOH)

Pyruvic acid —Acidic pH→ CO_2 + Acetic acid (CH_3—COOH) + Lactic acid (CH_3 / CHOH / COOH)

Pyruvic acid → $2CO_2$ + Acetoin (CH_3 / C=O / HCOH / CH_3)

FIGURE 5-18 Metabolism in citrate-positive organisms. Once in the cell, citrate is hydrolyzed by citrase into oxaloacetic acid and acetic acid. The oxaloacetic acid is further hydrolyzed into pyruvic acid and CO_2. This CO_2 is responsible for the color change in the medium as described in the text. The fate of the pyruvic acid depends on the cellular pH as shown in the figure.

COAGULASE TEST

PURPOSE This test is used to detect the ability of a Gram-positive, catalase-positive coccus to clot plasma. More specifically, it is used to distinguish between the pathogenic *Staphylococcus aureus* and other *Micrococcaceae*, which includes other species of *Staphylococcus* and the genus *Micrococcus*. There are two forms of this test, each identifying different types of coagulase. The slide test identifies coagulase bound to the cells (also called *clumping factor*), whereas the tube test identifies free coagulase in the medium. The latter is usually run only when the isolate is negative for the slide test.

PRINCIPLE The coagulase tube test uses citrated plasma (*i.e.*, plasma treated with the anticoagulant sodium citrate to prevent normal clotting) and identifies free coagulase in the medium. A coagulase-positive organism activates the normal clotting mechanism with the enzyme coagulase (staphylocoagulase) in some unidentified way. This produces a clot in the medium in as little as 30 minutes (Fig. 5-20). Clotting may be complete or may be seen as fibrin threads. Any degree of clotting is considered positive. Failure to clot within 24 hours is considered a negative result.

To prevent a false negative result, a known coagulase-positive *S. aureus* should be run as a control to verify the quality of the coagulase medium. A false negative result may also occur when testing some *S. aureus* strains that produce fibrinolytic enzymes which break up the clot. To avoid a false negative, tubes should be observed for clotting every 30 minutes for the first couple of hours and the test should not be run more than 24 hours.

The slide test detects coagulase bound to the cells. Two emulsions of the test organism are made on a glass slide in drops of sterile saline. Coagulase plasma is added to one, sterile water is added to the other as a negative control. Each emulsion is mixed, then observed for agglutination (Fig. 5-21).

These tests are specific to the genus *Staphylococcus*. Other genera may clot the plasma, but this is probably due to the catabolism of the anticoagulant citrate. In the absence of citrate, the medium will coagulate on its own even without coagulase activity.

FIGURE 5-20 The coagulase tube test showing coagulase-negative *Staphylococcus epidermidis* above and the more pathogenic coagulase-positive *S. aureus* below. It is thought that coagulase increases virulence by surrounding infecting organisms with a clot which protects them from host defenses, such as phagocytosis and antibodies. This test was run for 24 hours.

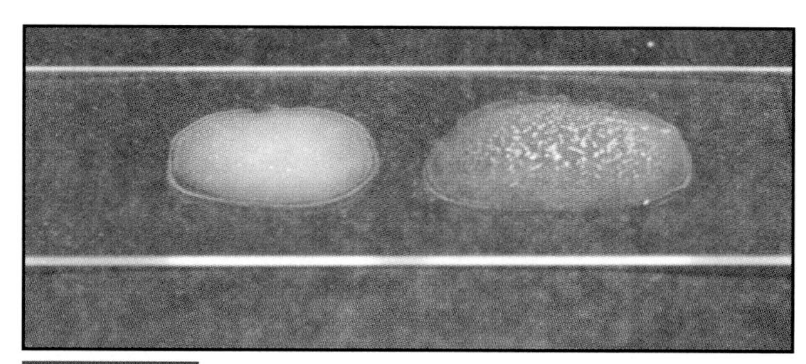

FIGURE 5-2 The coagulase slide test. Emulsions of *Staphylococcus aureus* (+) on the right and *S. epidermidis* (–) on the left were prepared in sterile saline. Agglutination of the coagulase plasma is indicative of a positive result.

DECARBOXYLATION TESTS

PURPOSE This test is used to detect the ability of an organism to decarboxylate an amino acid (typically lysine, ornithine or arginine). Decarboxylase tests are used to differentiate organisms in the family *Enterobacteriaceae*.

PRINCIPLE Decarboxylation is a general name given the process of removing the carboxyl group (COOH) of an amino acid, producing an amine and carbon dioxide. Decarboxylation of each amino acid substrate requires a specific enzyme. Presence of each decarboxylase enzyme can be identified with a base medium containing its amino acid substrate. Amino acids commonly used for clinical identification are lysine, arginine and ornithine. Decarboxylation in general is shown in Figure 5-22 and the specific reactions are shown in Figures 5-23 through 5-25.

Møller's decarboxylase medium contains peptone, glucose, bromcresol purple, and a coenzyme pyridoxal phosphate in addition to the specific amino acid substrate. Bromcresol purple, a pH indicator, is purple at pH greater than 6.8 and yellow at pH less than 5.2.

After inoculation, an overlay of mineral oil is used to seal the medium from external oxygen and promote fermentation. The accumulation of acid end products (from fermentation) is necessary because decarboxylase enzymes are inducible and are only produced in the presence of their substrate and an acidic environment. Glucose fermentation in the anaerobic medium initially turns it yellow due to the accumulation of acid end products. The low pH induces decarboxylase-positive organisms to produce the enzyme. Subsequent decarboxylation turns the medium purple due to the accumulation of alkaline end products (Fig. 5-26). If the organism is only capable of glucose fermentation, the medium will remain yellow and is considered negative.

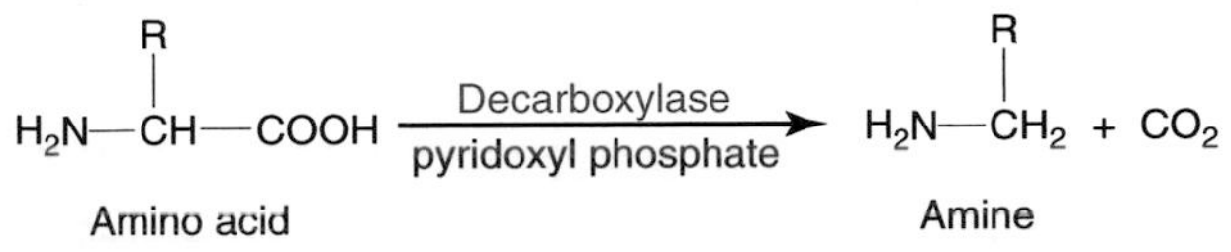

$$\underset{\text{Amino acid}}{H_2N-CH(R)-COOH} \xrightarrow[\text{pyridoxyl phosphate}]{\text{Decarboxylase}} \underset{\text{Amine}}{H_2N-CH_2(R)} + CO_2$$

FIGURE 5-22 Amino acid decarboxylation results in the formation of an amine and carbon dioxide.

$$\underset{\text{Lysine}}{H_2N-CH(COOH)-CH_2-CH_2-CH_2-CH_2-NH_2} \xrightarrow[\text{pyridoxyl phosphate}]{\text{Lysine decarboxylase}} \underset{\text{Cadaverine}}{H_2N-CH_2-CH_2-CH_2-CH_2-CH_2-NH_2} + CO_2$$

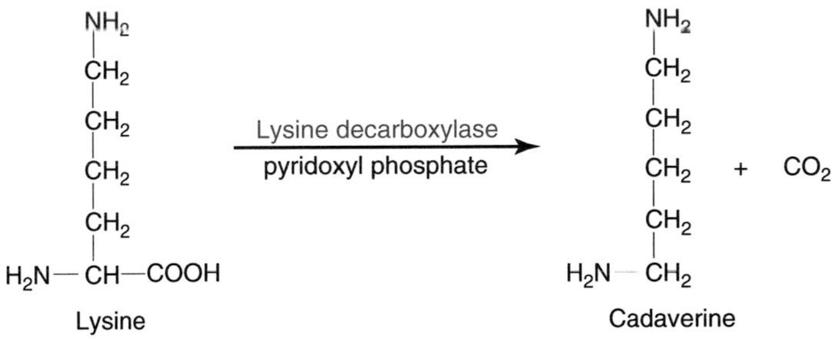

FIGURE 5-23 Decarboxylation of the amino acid lysine produces cadaverine and CO_2.

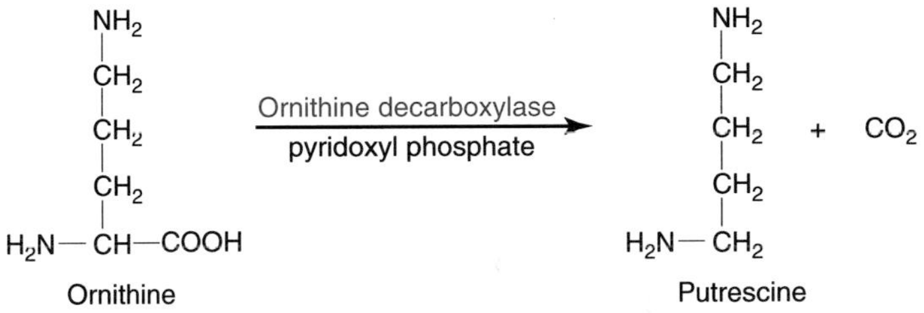

$$\underset{\text{Ornithine}}{H_2N-CH(COOH)-CH_2-CH_2-CH_2-NH_2} \xrightarrow[\text{pyridoxyl phosphate}]{\text{Ornithine decarboxylase}} \underset{\text{Putrescine}}{H_2N-CH_2-CH_2-CH_2-CH_2-NH_2} + CO_2$$

FIGURE 5-24 Decarboxylation of the amino acid ornithine produces putrescine and CO_2.

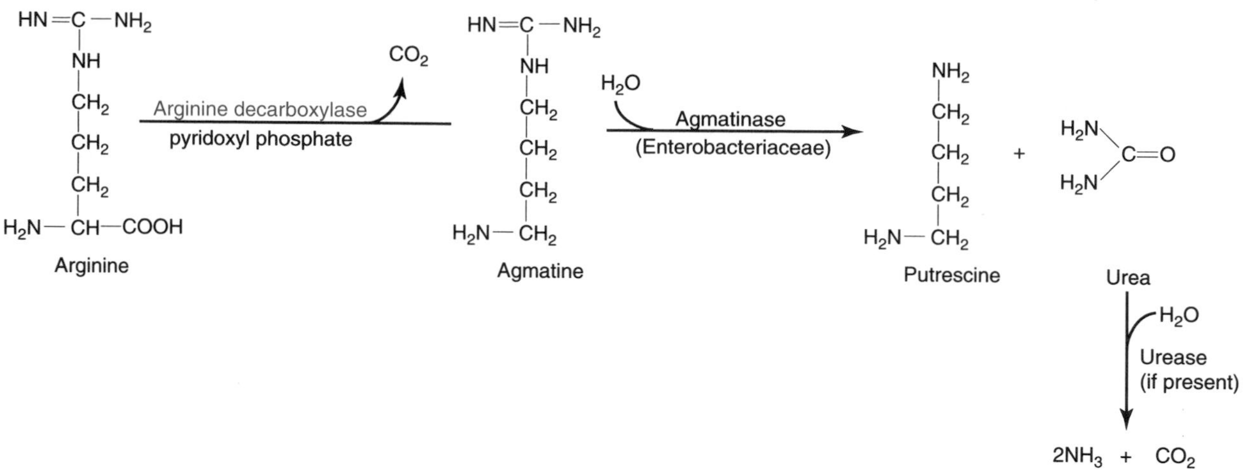

$HN{=}C{-}NH_2$ | NH | CH_2 | CH_2 | CH_2 | $H_2N{-}CH{-}COOH$

Arginine

$\xrightarrow[\text{pyridoxyl phosphate}]{\text{Arginine decarboxylase}}$ CO_2

$HN{=}C{-}NH_2$ | NH | CH_2 | CH_2 | CH_2 | $H_2N{-}CH_2$

Agmatine

H_2O $\xrightarrow[\text{(Enterobacteriaceae)}]{\text{Agmatinase}}$

NH_2 | CH_2 | CH_2 | CH_2 | $H_2N{-}CH_2$

Putrescine

\+

$(H_2N)_2C{=}O$

Urea

$\downarrow$ H_2O, Urease (if present)

$2NH_3 + CO_2$

FIGURE 5-25 Decarboxylation of the amino acid arginine produces the amine agmatine. Members of *Enterobacteriaceae* are capable of degrading agmatine into putrescine and urea. Those strains with urease can further break down the urea into ammonia and carbon dioxide. Thus, the end products of arginine catabolism are carbon dioxide, putrescine and urea, or carbon dioxide, putrescine and ammonia.

FIGURE 5-26 The lysine decarboxylase test results are shown here, but the colors are the same for all amino acids in Møller's medum. *Pseudomonas aeruginosa* (+) is on the left and *Proteus vulgaris* (−) is on the right. An uninoculated control is in the center.

DNASE TEST

PURPOSE DNase test agar is used to identify bacteria capable of producing the exoenzyme DNase. Important examples of DNase-positive organisms are *Staphylococcus aureus*, *Streptococcus pyogenes*, and *Serratia marcescens*.

PRINCIPLE An enzyme that catalyzes the depolymerization of DNA into small fragments is called a *deoxyribonuclease* or *DNase* (Fig. 5-27). Ability to produce this enzyme can be determined by culturing and observing a suspected organism on a DNase test agar plate.

One type of DNase test agar contains an emulsion of DNA, peptides as a nutrient source, and methyl green dye. The dye and polymerized DNA form a complex that gives the agar a blue-green color at pH 7.5. Bacterial colonies that secrete DNase will hydrolyze the DNA in the medium into smaller fragments unbound from the methyl green dye. This results in clearing around the growth (Fig. 5-28).

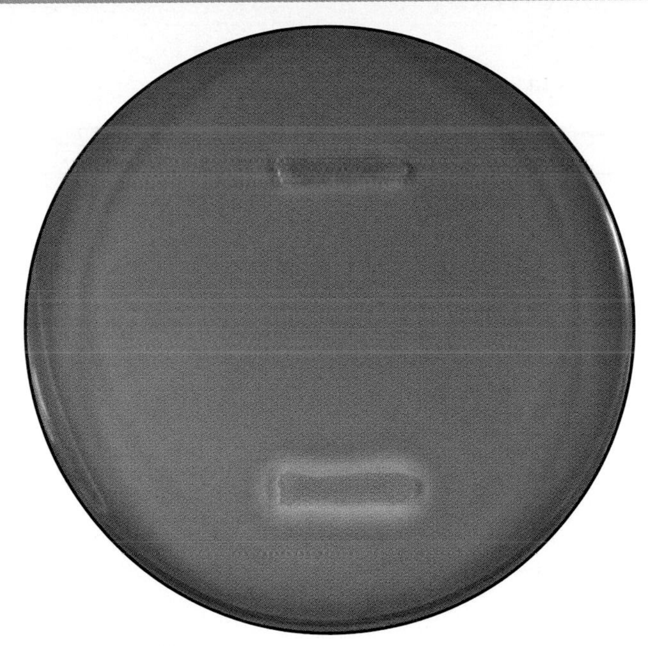

FIGURE 5-28 A DNase agar plate with methyl green. *Staphylococcus aureus* (+) is below and *S. epidermidis* (−) is above.

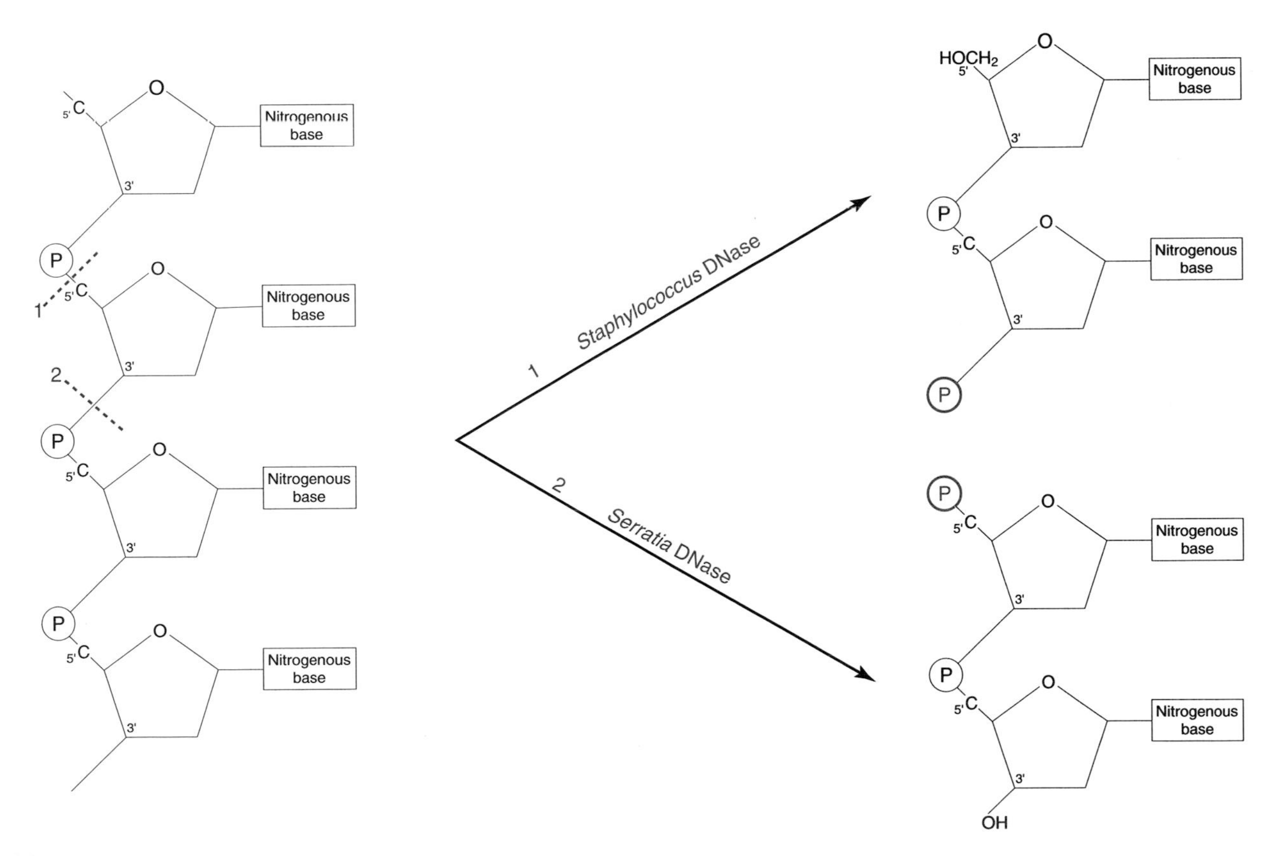

FIGURE 5-27 Two patterns of DNA hydrolysis. DNase from *Staphylococcus* hydrolyzes DNA at the bond between the 5'-carbon and the phosphate (illustrated by line 1), thereby producing fragments with a free 3'-phosphate (shown in red on the upper fragment). Most fragments are one or two nucleotides long. A dinucleotide is shown here. *Serratia* DNase cleaves the bond between the phosphate and the 3'-carbon (illustrated by line 2) and produces fragments with free 5'-phosphates (shown in red on the lower fragment). Most fragments are two to four nucleotides in length. A dinucleotide is shown here.

ENTEROTUBE® II

PURPOSE The Enterotube® II (Becton Dickinson Microbiology Systems, Sparks, MD) is used for rapid identification of enteric bacteria. With a single inoculation, 15 test results may be obtained within a day and identification of a suspected pathogen may be made.

PRINCIPLE The Enterotube® II consists of 12 compartments containing various media (Fig. 5-29) and allows determination of 15 different characteristics of the organism. (Each test performed with the Enterotube® II is explained in detail elsewhere in this section.)

The Enterotube® II contains a wire which is touched to a colony (isolated on appropriate plated medium) of the organism to be tested. The wire is then drawn through the tube inoculating the media in each chamber as it passes. After 18 to 24 hours incubation, the test results are read by comparing the inoculated tube to an uninoculated control. Two inoculated tubes and a control are shown in Figure 5-30.

Identification of the enteric bacterium requires scoring all tests but Voges-Proskauer (VP). The remaining tests are clustered into groups, and each positive result within a group is assigned a number — either 4, 2 or 1 (Fig. 5-31). The numbers within each group are added together, with their sums being used to make a five digit number. Thus, the combination of test results gives a unique (usually) ID value for each organism. The ID value is then located in the Computer Coding and Identification System (CCIS) to give the identity of the unknown organism. Should more than one organism have the same ID value, a test (often VP) to differentiate between them is suggested.

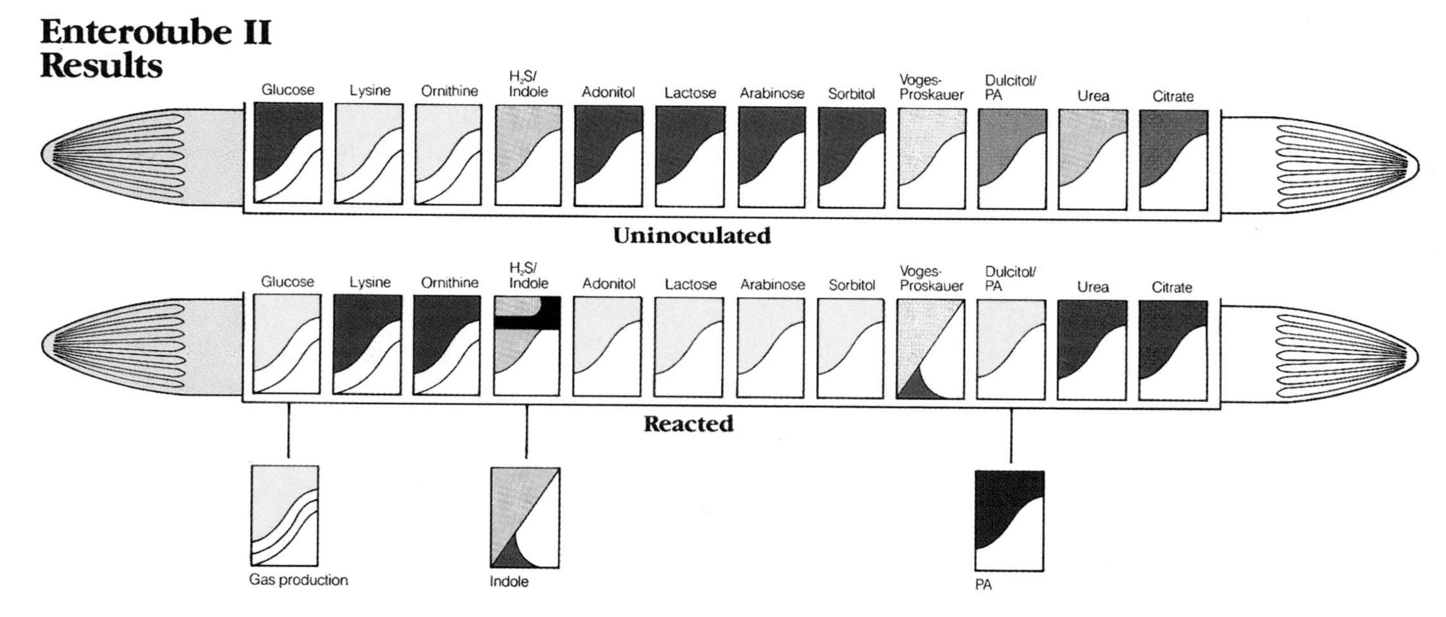

FIGURE 5-29 Enterotube® II tests and results. (Illustration courtesy Becton Dickinson and Company.)

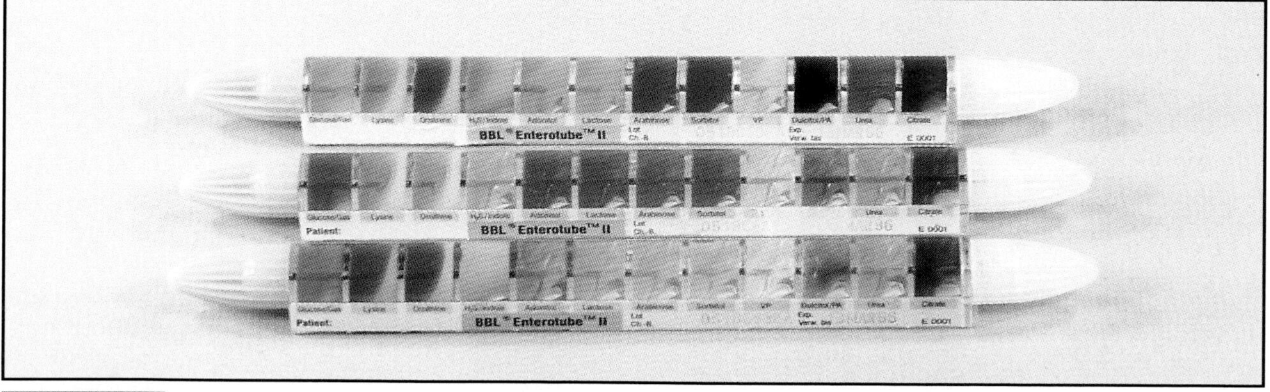

FIGURE 5-30 Two inoculated tubes, one in front of and one behind an uninoculated control tube.

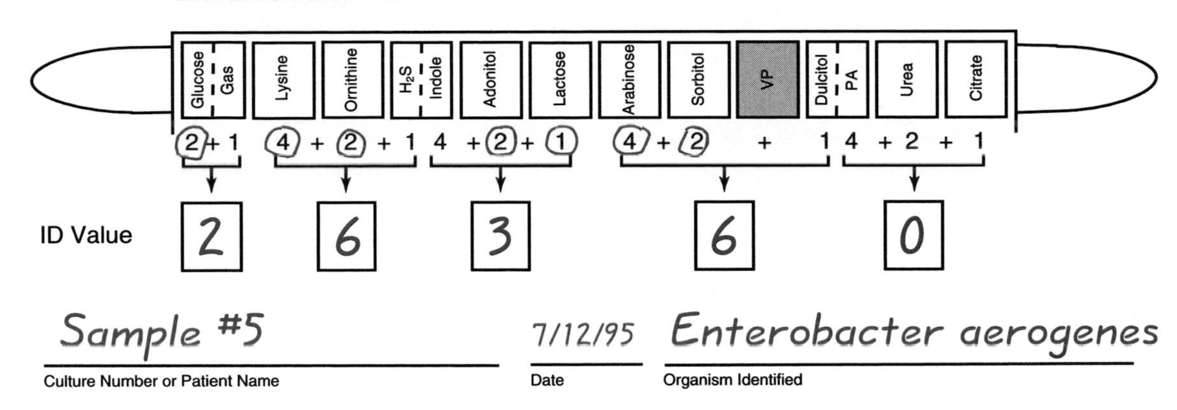

FIGURE 5-31 A sample score sheet for the front tube in Figure 5-30. In this case, the organism was positive for glucose fermentation, lysine decarboxylase, ornithine decarboxylase, adonitol fermentation, and lactose fermentation. The ID value obtained was 26360, which corresponds to *Enterobacter aerogenes*.

FERMENTATION TESTS (PURPLE BROTH AND PHENOL RED BROTH)

PURPOSE Each of these media is a family of differential tests used to detect the ability of an organism to ferment various carbohydrates. Fermentation characteristics are especially useful in identifying Gram-negative enteric bacteria.

PRINCIPLE Fermentation is a metabolic process in which the final electron acceptor is an organic molecule. Glucose fermentation typically begins with the production of pyruvic acid by glycolysis. (Some bacteria may alternatively use the pentose phosphate shunt or Entner-Doudoroff pathway, but they still produce pyruvic acid.) Various fermentation end products may be produced from pyruvic acid, including a variety of acids, H_2 or CO_2 gases, and alcohols. The specific end products depend on the organism and the substrate fermented (Fig. 5-32)

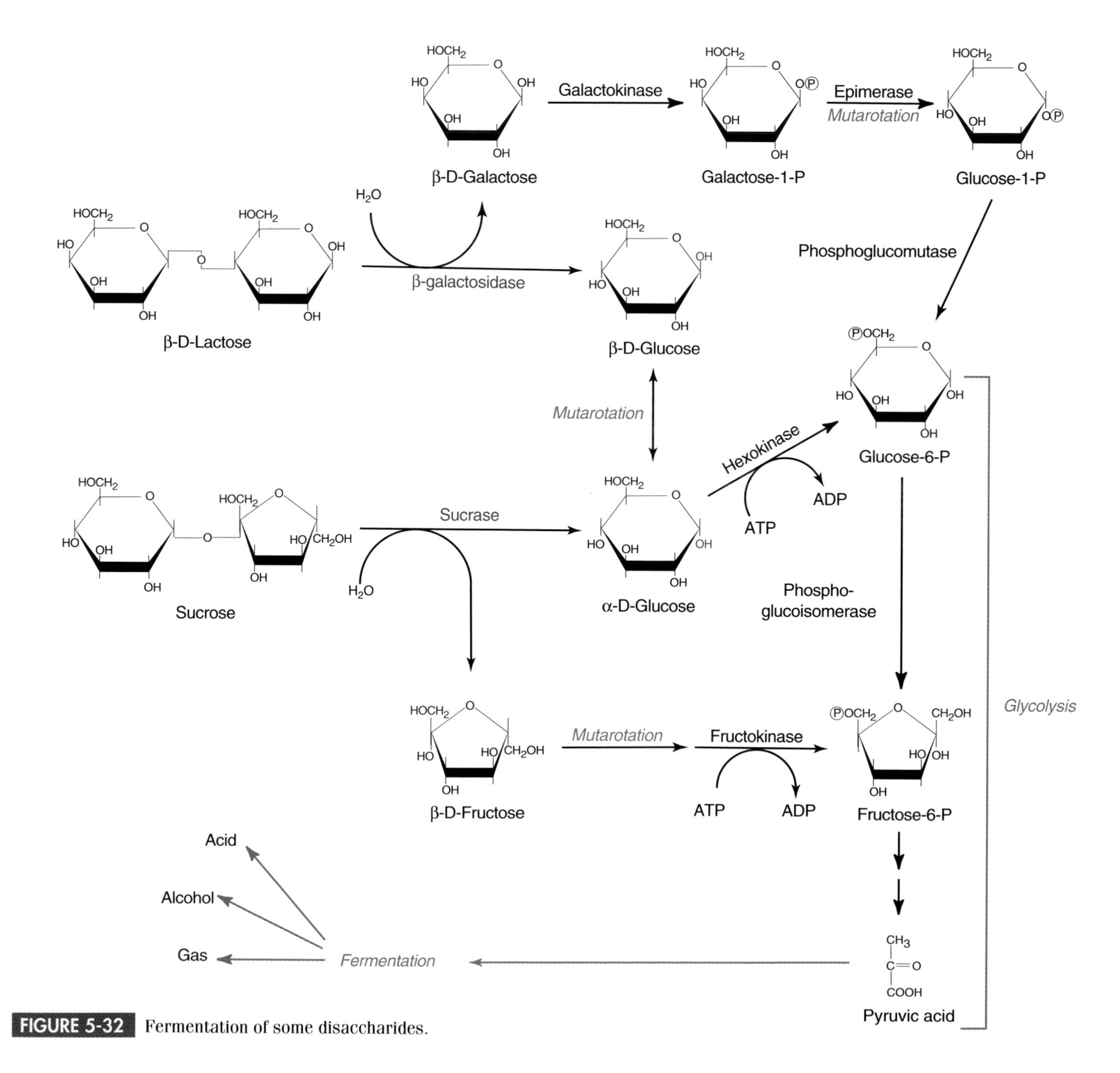

FIGURE 5-32 Fermentation of some disaccharides.

The principle behind purple broth and phenol red broth is the same. Each medium consists of a basal recipe to which a single fermentable carbohydrate is added. Both media include peptone, a pH indicator (either bromcresol purple, which is yellow below pH 6.8 and purple above, or phenol red which is yellow below pH 6.8 and red above pH 7.4) and a specific carbohydrate. Formulating these media with different carbohydrates allows determination of a species' fermentation end products and its ability to enzymatically convert other sugars (*e.g.*, disaccharides) into glucose. Acid production (A) will lower the pH and turn either pH indicator yellow (Figs. 5-33 and 34). An inverted tube (Durham tube) is included to trap any gas produced. A bubble is indicative of gas production (G).

The amino acids in peptone may be degraded by deamination which releases ammonia (NH_3) into the medium. Accumulation of NH_3 raises the pH and turns the pH indicator back to its original color or beyond. This is especially noticeable in phenol red which becomes a fuchsia color indicative of an alkaline reaction (K). This alkaline reaction may be produced by bacteria that do not ferment the carbohydrate or by those that exhaust the carbohydrate and change their metabolism to use another available resource. Because these latter organisms neutralize any acid they produce, it is important to read this test no later than 48 hours after inoculation to avoid missing their ability to produce acid.

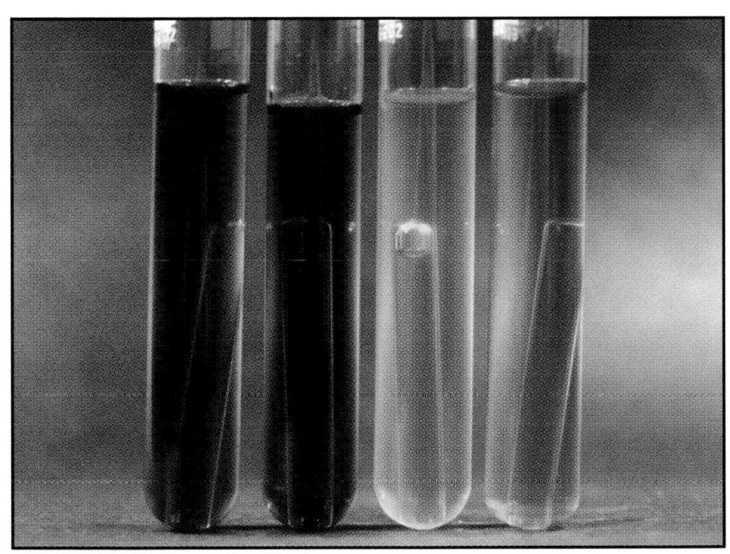

FIGURE 5-33 Purple glucose broth (bromcresol purple glucose broth) tubes. From left to right are: *Proteus vulgaris* (−/−), an uninoculated control, *Escherichia coli* (A/G), and *Staphylococcus aureus* (A/−).

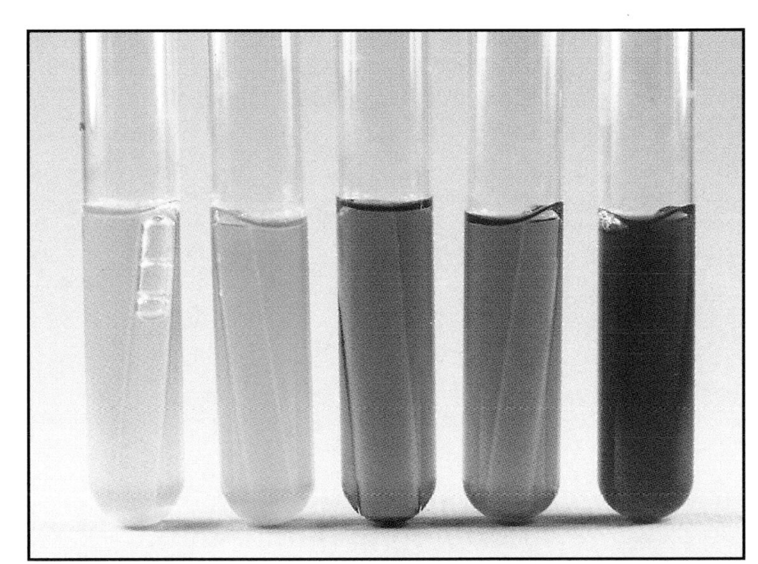

FIGURE 5-34 PR glucose tubes. From left to right are: *Escherichia coli* (A/G), *Staphylococcus aureus* (A/−), uninoculated control, *Micrococcus luteus* (−/−) and *Alcaligenes faecalis* (K).

GELATIN LIQUEFACTION TEST (NUTRIENT GELATIN)

PURPOSE This test is used to determine the ability of a microbe to produce hydrolytic exoenzymes called *gelatinases* that digest and liquefy gelatin. Gelatin liquefaction may be used to distinguish between the pathogenic *Staphylococcus aureus* which is positive and nonpathogenic *S. epidermidis*. Among the *Enterobacteriaceae*, *Serratia* and *Proteus* species are typically positive; the remainder of the group is generally negative. *Bacillus anthracis, B. cereus* and several other members of the genus are gelatinase-positive, as are *Clostridium tetani* and *C. perfringens*.

PRINCIPLE Many nutrient sources are too large to enter the cell. Some bacteria have the ability to produce and secrete enzymes that hydrolyze these compounds into smaller subunits that the cell can use.

Gelatin is a protein derived from collagen, a connective tissue found in vertebrates. Bacterial hydrolysis of gelatin occurs in two sequential reactions catalyzed by a family of exoenzymes referred to as *gelatinases*. These reactions are shown in Figure 5-35. Amino acids may then be used as an energy source for the cell or built back up into bacterial protein.

Nutrient gelatin tubes are stab inoculated, then incubated for up to one week. Gelatinase-positive organisms will liquefy the medium; it will remain solid when inoculated with gelatinase-negative organisms (Figs. 5-36 and 5-37). Care must be taken to distinguish between gelatin hydrolysis and gelatin melting. An uninoculated control should be incubated with the inoculated tubes because nutrient gelatin melts at 28°C. If the control is liquid after incubation, all tubes should be refrigerated until the control is solid.

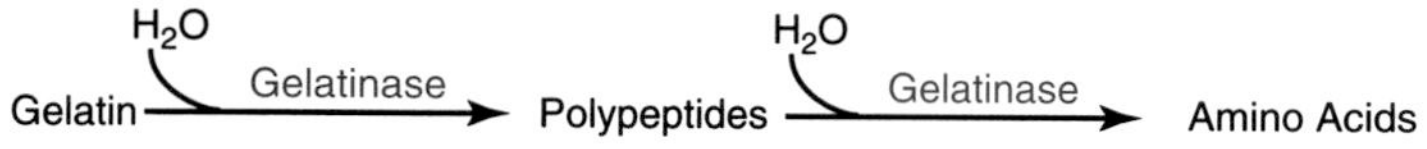

$$\text{Gelatin} \xrightarrow[\text{Gelatinase}]{H_2O} \text{Polypeptides} \xrightarrow[\text{Gelatinase}]{H_2O} \text{Amino Acids}$$

FIGURE 5-35 Hydrolysis of gelatin by the gelatinase group of enzymes.

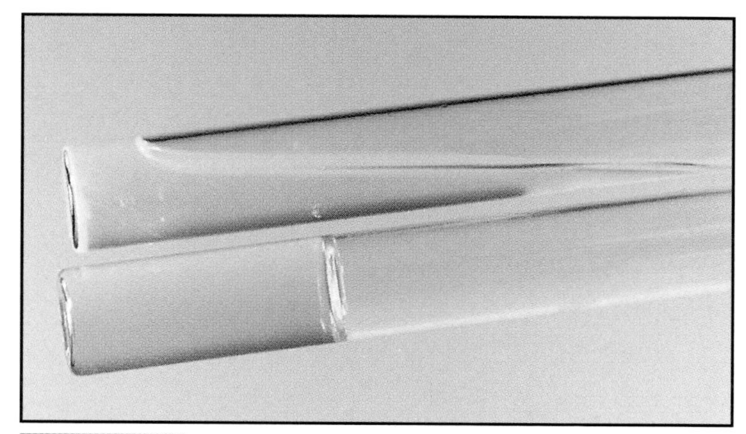

FIGURE 5-36 Nutrient gelatin stab tubes. *Aeromonas hydrophila* (+) above and *Micrococcus roseus* (−) below.

FIGURE 5-37 The form of liquefaction may also be of diagnostic use since not all gelatinase positive microbes completely liquefy the gelatin. Shown here is *Micrococcus luteus* liquefying the gelatin in the shape of a crater.

INDOLE PRODUCTION TEST (SIM MEDIUM)

PURPOSE The indole test is used to identify bacteria capable of producing indole using the enzyme *tryptophanase*. It is one component of the *IMViC* battery of tests (Indole, Methyl red, Voges-Proskauer, and Citrate) used to differentiate the *Enterobacteriaceae*.

PRINCIPLE The amino acid tryptophan can be converted by bacteria possessing the enzyme tryptophanase to indole, ammonia (by deamination), and pyruvic acid (Fig. 5-38). Pyruvic acid can be converted to energy in the Krebs cycle or it can enter glycolysis and be used to synthesize other compounds required by the cell. Ammonia can be used to synthesize amino acids. The byproduct indole is the metabolite identified by this test.

Indole medium, which may be solid or liquid, typically includes peptone, necessary growth factors, and the amino acid tryptophan. SIM medium, a semisolid combination medium also used to test for sulfur reduction and motility, was used for this example. After inoculation and appropriate incubation, a small amount of test reagent (either Kovac's or Erlich's) is added to the medium. Both test reagents contain HCl and a form of dimethylaminobenzaldehyde (DMABA) dissolved in amyl alcohol. Amyl alcohol is insoluble in water and will form a separate layer on top of the medium. In indole-positive organisms, the DMABA reacts with indole to produce a roseindole dye (Fig. 5-39) which makes the alcohol layer a cherry red color (Fig. 5-40). No color change is observed in the test reagent for indole-negative organisms.

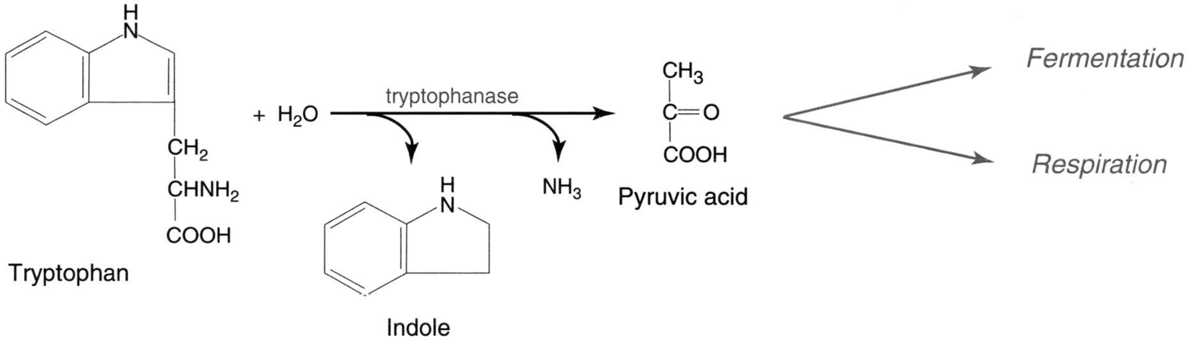

Tryptophan + H_2O —(tryptophanase)→ Indole + NH_3 + Pyruvic acid → Fermentation / Respiration

FIGURE 5-38 Tryptophan catabolism in indole-positive organisms.

2 Indole + p-Dimethylaminobenzaldehyde + HCl + Amyl Alcohol → Rosindole dye (cherry red)

(p-Dimethylaminobenzaldehyde + HCl + Amyl Alcohol: Kovac's Reagent)

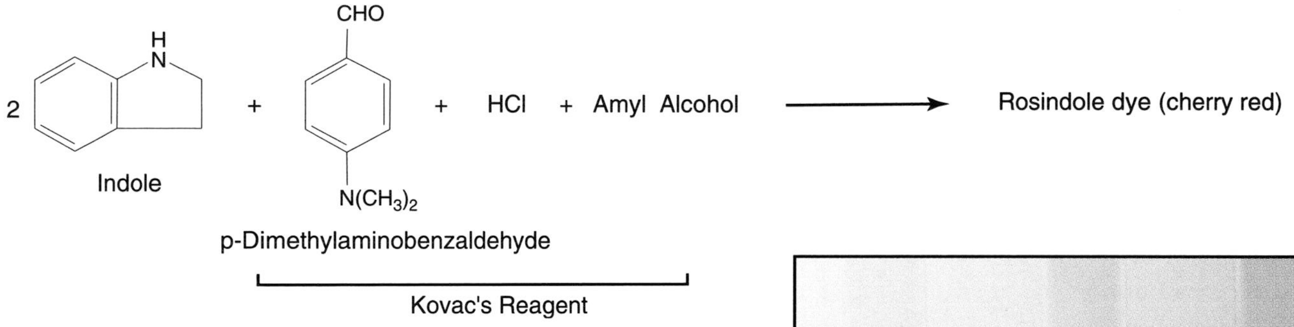

FIGURE 5-39 Indole reaction with Kovac's Reagent.

FIGURE 5-40 The indole test. This is SIM medium inoculated with *Morganella morganii* (indole-positive) on the left and *Enterobacter aerogenes* (indole-negative) on the right. An uninoculated control is in the center.

KLIGER'S IRON AGAR (KIA)

PURPOSE Kligler's iron agar (KIA) differentiates bacteria, especially the *Enterobacteriaceae*, based on their ability to ferment glucose and lactose to acid or acid and gas end products, and to produce hydrogen sulfide. Because all enterics ferment glucose to acid end products and are so similar morphologically, a single medium providing a variety of results is of great use in their identification. It may be used to separate the lactose fermenting coliforms, such as *Escherichia coli, Klebsiella pneumoniae and Enterobacter aerogenes,* from lactose nonfermenting enterics, such as *Salmonella, Shigella, Citrobacter and Proteus.*

PRINCIPLE Kligler's iron agar is formulated with 2% polypeptone, 1% lactose and 0.1% glucose. Thiosulfate is included as an electron acceptor for sulfur reducers and ferric ammonium citrate is added as the H_2S indicator. Phenol red (yellow at pH less than 6.8 and red above 6.8) is the pH indicator. It is prepared as an agar slant, with a deep butt to provide a region for anaerobic growth. It is inoculated by a stab in the agar butt followed by a fishtail streak of the slant.

Interpretations are made after an 18 to 24 hour incubation period. Many results are possible, including, glucose fermentation, lactose fermentation, gas production, no fermentation, alkaline reaction or hydrogen sulfide production. The various results are shown in Figure 5-41 and described in Table 5-1.

Typically, glucose fermentation takes place within the first few hours which lowers the pH and turns the medium yellow. The glucose, however, is in short supply and if the organism is to continue fermentation it must be able to ferment lactose. In the absence of this ability the organism will begin to break down the amino acids and proteins in the medium, producing NH_3 and raising the pH. The alkalinity produced will turn the slant red but will not overcome the acid in the butt which remains yellow.

If the test organism is able to ferment lactose, it will continue to do so after the supply of glucose is depleted. Due to the high concentration of lactose, however, this process will not be complete after 24 hours. Consequently, both slant and butt will appear yellow. Gas produced as a result of fermentation will appear as fissures in the medium or will lift the agar off the bottom of the tube. (It is important to interpret the test results at the appropriate time. An early reading could give a false lactose positive result due to the yellow color produced by glucose fermentation. A late reading could give a false lactose negative result if the slant has turned red due to secondary peptone metabolism.)

Hydrogen sulfide (H_2S) may be produced by the reduction of thiosulfate in the medium or by the breakdown of cysteine in the peptone. Ferric ammonium citrate reacts with the H_2S to form a black precipitate, usually seen in the butt. If the black precipitate obscures the color of the medium, *it is recorded as acid-positive.*

An organism that does not ferment glucose or lactose may utilize the peptone and turn the medium red. If the organism can catabolize the peptone aerobically and anaerobically, both the slant and butt will turn red. If the organism is an obligate aerobe, only the slant will turn red.

FIGURE 5-41 Representative results in KIA agar slants. Results are reported as (slant/butt, H_2S) From left to right: *Proteus mirabilis* (K/A,G, H_2S), *Escherichia coli* (A/A,G), uninoculated control, *Pseudomonas aeruginosa* (K/NC), and *Morganella morganii* (K/A,G).

TABLE 5-1 Results, Symbols and Interpretations of KIA Agar

RESULTS (Slant/Butt)	SYMBOL (Slant/Butt)	INTERPRETATION
red/yellow	K/A	glucose fermentation only peptone catabolized
yellow/yellow	A/A	glucose and lactose fermentation
red/red	K/K	no fermentation peptone catabolized
red/no color change	K/NC	no fermentation peptone used aerobically
yellow/yellow with bubbles	A/A,G	glucose and lactose fermentation gas produced
red/yellow with bubbles	K/A,G	glucose fermentation only gas produced
red/yellow with bubbles and black precipitate	K/A,G, H_2S	glucose fermentation only gas produced H_2S produced
red/yellow with black precipitate	K/A, H_2S	glucose fermentation only H_2S produced
yellow/yellow with black precipitate	A/A, H_2S	glucose and lactose fermentation H_2S produced
yellow/yellow with bubbles and black precipitate	A/A, G, H_2S	glucose and lactose fermentation gas produced H_2S produced
no change/no change	NC/NC	no fermentation (organism is growing very slowly or not at all)

A = acid production, G = gas production, K = alkaline reaction, H_2S = sulfur reduction

Standard symbols for reporting are: A = acid, K = alkaline, G = gas, H_2S = sulfur reduction positive, and NC = no change. These symbols are customarily written with the slant results first followed by a slash and the butt results.

LIPASE TEST (SPIRIT BLUE AGAR AND TRIBUTYRIN AGAR)

PURPOSE The lipase test is used to identify bacteria capable of producing the exoenzyme lipase.

PRINCIPLE Triglycerides in the environment may serve as a bacterial carbon and energy source, but are too large to enter the cell. Some bacteria produce and secrete exoenzymes called *lipases* that hydrolyze triglycerides to glycerol and three long chain fatty acids (Fig. 5-42). Glycerol may be converted into dihydroxyacetone phosphate, an intermediate of glycolysis. Fatty acids may be catabolized by a process called β-oxidation. Two carbon fragments from the fatty acid are combined with Coenzyme A to produce Acetyl CoA which may then be used in the Krebs cycle to produce energy. Each Acetyl CoA produced by this process also yields one NADH and one $FADH_2$. Glycerol and fatty acids may alternatively be used in anabolic pathways.

The spirit blue agar plate shown in Figure 5-43 contains an emulsion of olive oil and spirit blue dye in a complex which gives the agar plate an opaque blue appearance. Bacteria growing on the agar surface that secrete lipase will hydrolyze the olive oil in the medium resulting in a halo around the growth.

In another example, the tributyrin agar plate shown in Figure 5-44 contains the triglyceride tributyrin and is initially opaque. Lipase-positive organisms will exhibit a clear zone around their growth as the tributyrin is hydrolyzed.

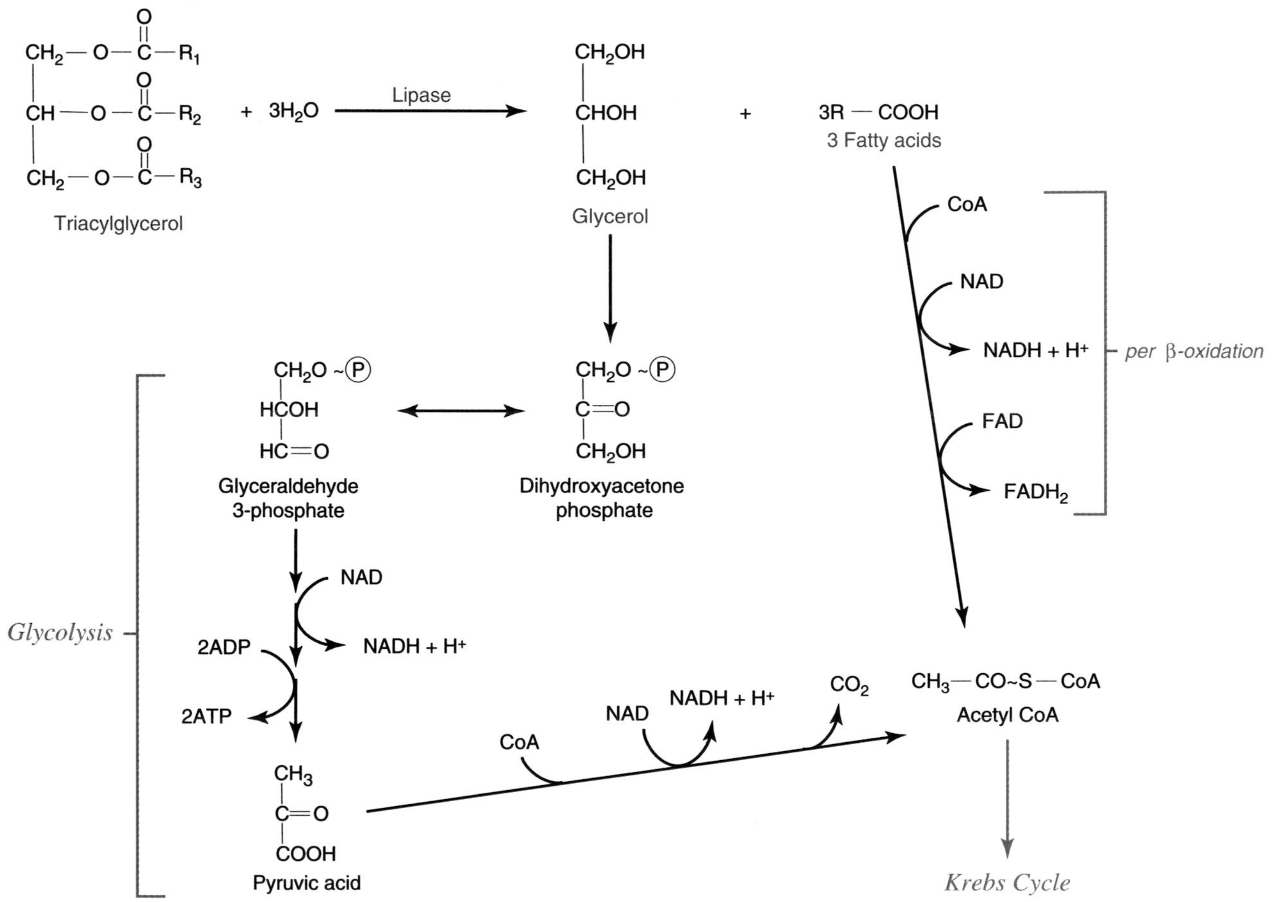

FIGURE 5-42 Lipid metabolism.

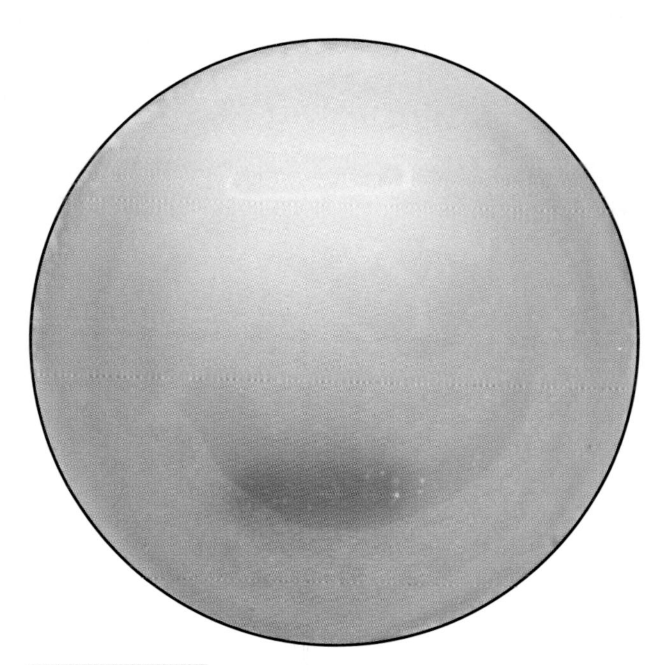

FIGURE 5-43 Spirit blue agar plate with *Serratia marcescens* (lipase-positive) above and *Lactobacillus plantarum* (lipase-negative) below.

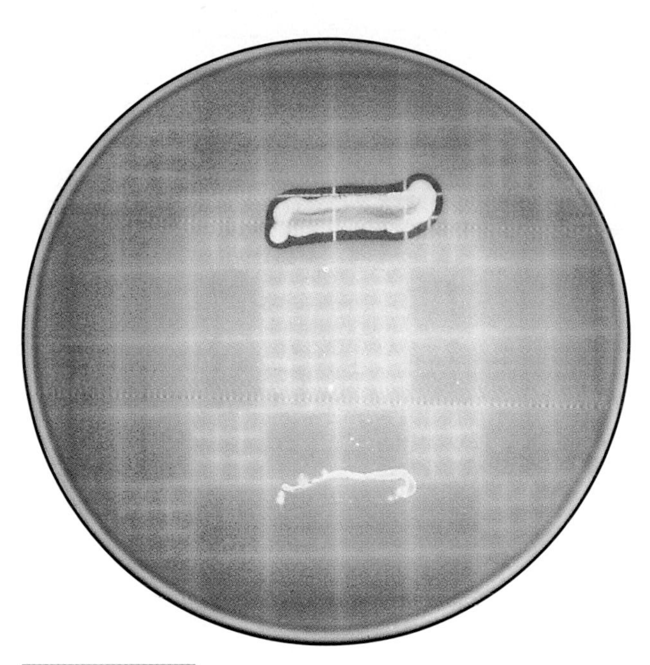

FIGURE 5-44 Tributyrin agar plate with *Citrobacter amalonaticus* (lipase-positive) above and *Erwinia amylovora* (lipase-negative) below.

LITMUS MILK MEDIUM

PURPOSE Bacteria may be differentiated based on the variety of reactions in litmus milk, especially species of the genus *Clostridium*. The medium is also used to maintain cultures of lactic acid bacteria.

PRINCIPLE Litmus milk is an undefined medium consisting of skim milk and azolitmin. Skim milk provides nutrients for growth, with lactose as the carbohydrate source and casein as the primary protein source. Azolitmin is a pH indicator that is pink at an acid pH (4.5) and blue at an alkaline pH (8.3). Between these extremes, it is purple. (The medium is formulated to an initial pH of 6.8, thus accounting for its purple color.)

Four basic categories of reaction may occur, either separately or in combination. These are: lactose fermentation, reduction of litmus, casein hydrolysis, and casein coagulation. Figures 5-45, 5-46 and 5-47 illustrate some reactions in litmus milk. Table 5-2 summarizes the reactions in litmus milk.

Lactose is a disaccharide which yields the monosaccharides glucose and galactose when hydrolyzed by the enzyme β-galactosidase (Fig. 5-48). Glucose may then be fermented to acid end products, lowering the pH and turning the litmus a pink color. This is an *acid reaction*. Accumulating acid may also cause precipitation of casein and form an *acid clot* (Fig. 5-49). If *gas* is produced, fissures may be visible in the clot (Fig. 5-47). Extreme gas production may break the clot and produce *stormy fermentation*, which is typical of *Clostridium perfringens*.

Some bacteria have proteolytic enzymes such as rennin, pepsin or chymotrypsin that digest casein and coagulate (curdle) the milk. The action of rennin (also known as *rennet*) is shown in Fig. 5-50. The insoluble rennet curd is soft and retracts from the sides of the tube (unlike an acid clot) leaving behind a grayish fluid called *whey*. This reaction is a *soft clot* or *soft curd*. *Proteolysis* of the soft curd occurs if bacteria have the appropriate proteolytic enzymes to break the protein down into its component amino acids. Proteolysis is evidenced by clearing of the fluid.

Some bacteria grow in litmus milk but do not ferment the lactose. Rather, they partially digest the casein. This releases NH_3 which raises the pH and turns the litmus blue. This is an *alkaline* reaction.

Litmus may act as the electron acceptor during lactose fermentation. Reduction of litmus is evidenced by its conversion to a white color in the lower portion of the tube (the anaerobic zone). Be careful not to confuse accumulation of sediment with reduction of litmus, which should turn all of the tube white except for the upper 2 cm.

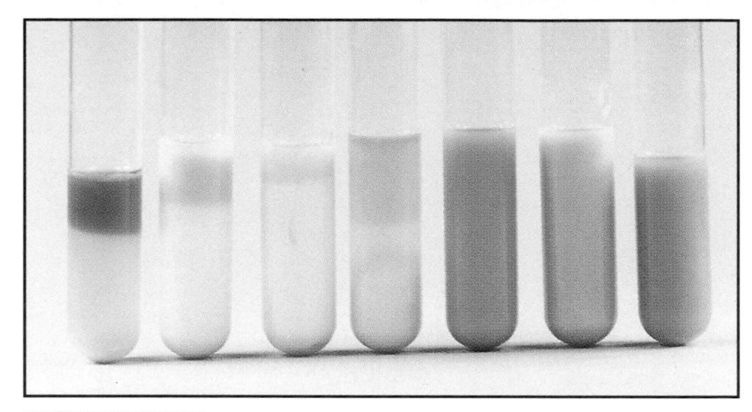

FIGURE 5-45 Reactions in litmus milk. From left to right: *Pseudomonas aeruginosa* (P), *Streptococcus lactis* (ACR), *Klebsiella pneumoniae* (AGCR-note small fissure in clot), *Aeromonas hydrophila* (CR), uninoculated control, *Staphylococcus aureus* (A), and *Alcaligenes faecalis* (K).

FIGURE 5-46 An acid clot in the top tube, an uninoculated control below.

FIGURE 5-47 *Klebsiella pneumoniae* in litmus milk with an obvious gas fissure in the clot. Compare the amount of gas production with the *K. pneumoniae* in Figure 5-45. Differences such as this may be due to using different strains of the organism, a different incubation time, or inoculating different amounts of the organism in the two tubes.

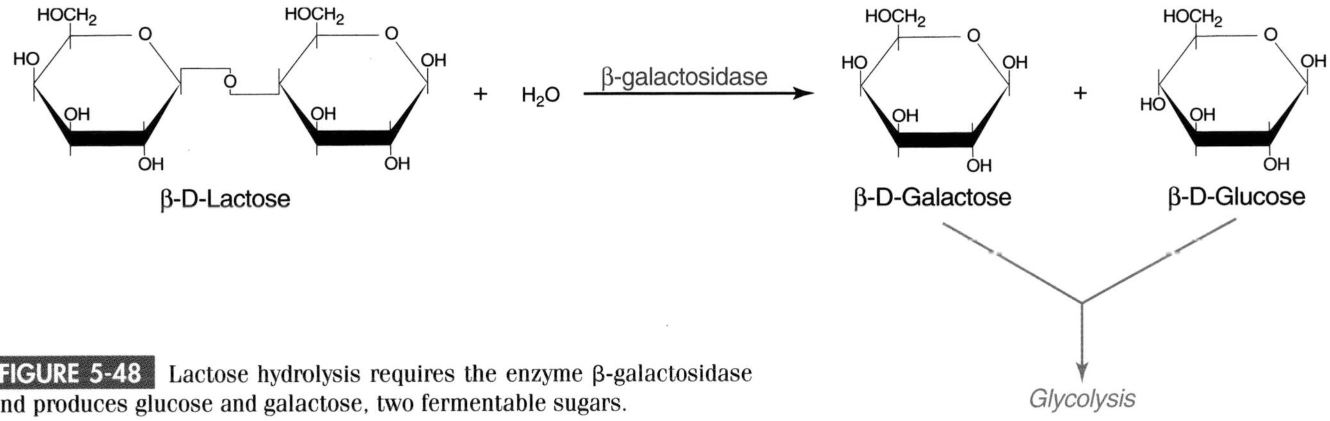

$$\beta\text{-D-Lactose} + H_2O \xrightarrow{\beta\text{-galactosidase}} \beta\text{-D-Galactose} + \beta\text{-D-Glucose} \rightarrow \text{Glycolysis}$$

FIGURE 5-48 Lactose hydrolysis requires the enzyme β-galactosidase and produces glucose and galactose, two fermentable sugars.

$$\text{calcium caseinate (soluble salt of casein)} \xrightarrow[\text{low pH}]{\text{casease}} \text{caseinogen (acid clot) (insoluble precipitate)}$$

FIGURE 5-49 An acid clot is due to casease catalyzing the formation of caseinogen, an insoluble precipitate, under acidic conditions.

$$\text{casein (soluble)} \xrightarrow[Ca^{++}]{\text{rennin}} \text{paracasein (soft curd) (insoluble precipitate)}$$

FIGURE 5-50 Rennin converts casein to paracasein to form a curd (soft clot).

TABLE 5-2 A Summary of Common Litmus Milk Reactions

RESULTS	SYMBOL	INTERPRETATION
purple (same as control)	NC	no reaction
pink throughout	A	lactose fermentation with acid end product
pink at surface, white below	AR	lactose fermentation with acid end product reduction of litmus
pink at surface, white below and solid	ACR	lactose fermentation with acid end product acid clot reduction of litmus
pink at surface, white below and solid with fissures	AGCR	lactose fermentation with acid *and* gas end products acid clot reduction of litmus
pink at surface, white below, clot broken apart	AGCS	lactose fermentation with acid *and* gas end products (stormy fermentation) acid clot reduction of litmus
blue	K	alkaline reaction
semisolid	C	rennet curd
clearing of the medium, often brownish in color (but color may be affected by pigmentation of the organism)	P	proteolysis (peptonization) of casein
white below, no change (lavender) at top	R	reduction of litmus

LYSINE IRON AGAR (LIA)

PURPOSE Lysine iron agar is a differential medium used to distinguish between enterics based on their ability to decarboxylate and/or deaminate lysine, and produce hydrogen sulfide (H_2S).

PRINCIPLE Lysine iron agar is an undefined, differential medium. It is undefined because peptone and yeast extract supply amino acids and vitamins, respectively. It also has a small amount of glucose (0.1%) as the fermentable carbohydrate. The amino acid lysine is the primary substrate for deamination and decarboxylation reactions. Bromcresol purple (yellow at a pH less than 6.8) detects changes in pH due to fermentation, lysine decarboxylase and lysine deaminase activity. Thiosulfate is an electron receptor which, if reduced, becomes hydrogen sulfide (H_2S). Ferric ion in the medium reacts with any H_2S to produce a black precipitate.

LIA is prepared as a slant with a deep butt. This results in an aerobic zone on the slant and an anaerobic zone in the butt. Two stabs of the butt and a fishtail streak of the slant are used to inoculate the medium. A variety of results may be observed (Fig. 5-51 and Table 5-3). Acid production due to fermentation lowers the pH, so bromcresol purple turns the butt yellow. Decarboxylase activity requires an anaerobic environment and results in a purple color in the butt. Deamination of lysine occurs in the slant and results in a reddish color. Hydrogen sulfide production is seen as a black color in the butt due to the reaction of H_2S and ferric ion. Gas formation, as evidenced by fissures in or lifting of the agar, is also a possible result.

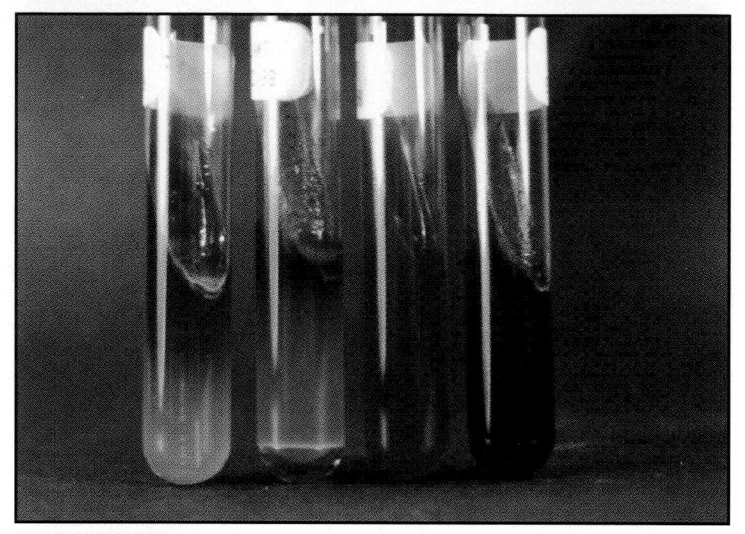

FIGURE 5-51 Lysine iron agar tubes that illustrate common results. From left to right with results in parentheses (slant/butt H_2S): *Proteus mirabilis* (red/yellow/H_2S–) is interpreted as deaminase-positive, decarboxylase-negative, and H_2S-negative; *Citrobacter freundii* (purple/yellow/H_2S+ [just a little black is visible near the middle]) is interpreted as deaminase-negative, decarboxylase-negative, and H_2S-positive; uninoculated control; and *Salmonella typhimurium* (purple/purple [obscured by the black]/H_2S+) is interpreted as deaminase-negative, decarboxylase-positive, and H_2S-positive). Notice that the agar has been lifted due to gas production in the *Citrobacter freundii* tube (second from the left).

TABLE 5-3 A Table of Lysine Iron Agar Results

TEST	SLANT	BUTT
Lysine deamination positive	red	not read
Lysine deamination negative	purple	not read
Lysine decarboxylation positive	purple	purple
Lysine decarboxylation negative	purple	yellow

METHYL RED (MR) TEST

PURPOSE The methyl red test is used to identify bacteria that produce stable acid end products by means of a mixed-acid fermentation of glucose. It is one component of the *IMViC* battery of tests (*I*ndole, *M*ethyl red, *V*oges-Proskauer, and *C*itrate) used to differentiate the *Enterobacteriaceae.*

PRINCIPLE The combination medium used for this test — MR/VP (methyl red/Voges-Proskauer) broth — includes peptone, glucose and a phosphate buffer. Some bacteria (especially enterics) are able to perform a *mixed-acid fermentation* of glucose and produce large amounts of stable acids (Fig. 5-52). Organisms capable of producing enough acid end products to overcome the buffering system and sufficiently lower the pH are methyl red-positive. Because many microbes produce acids within 18–24 hours but continue to catabolize them to more neutral compounds, test organisms are typically incubated for 2 to 5 days to assure the presence of *stable* acids.

After the appropriate incubation time, a small aliquot of broth is removed and methyl red indicator is added. Methyl red is red at a pH less than 4.4 and yellow at a pH greater than 6.0. Between these two pH values, methyl red is various shades of orange. A red color indicates a positive methyl red result and yellow is negative (Fig. 5-53). An orange color is not considered positive and indicates further incubation is required.

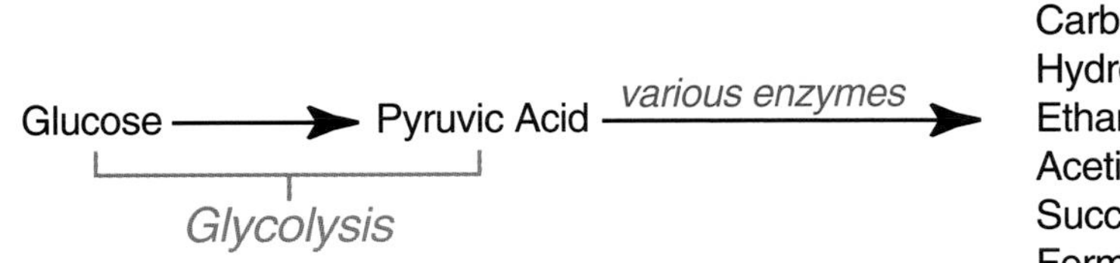

FIGURE 5-52 The mixed-acid fermentation of *Escherichia coli*, a representative methyl red-positive enteric bacterium. Products are listed in order of abundance. Most of the formic acid is converted to H_2 and CO_2 gases.

FIGURE 5-53 The methyl red test. *Enterobacter aerogenes* (MR-negative) on the left and *Escherichia coli* (MR-positive) on the right.

MOTILITY TEST

PURPOSE This test is used to detect bacterial motility. Notable among the motile pathogens are *Borrelia burgdorferi*, *Campylobacter jejuni*, *Legionella pneumophila*, *Listeria monocytogenes*, *Pseudomonas aeruginosa*, *Vibrio cholerae*, *Salmonella typhi*, *Proteus spp.*, *Escherichia coli*, and *Aeromonas hydrophila*.

PRINCIPLE Motility test medium is formulated with a low agar concentration to allow limited movement of motile bacteria. The agar tube is inoculated by stabbing with a straight transfer needle. Any motility will be detectable as diffuse growth radiating from the central stab line.

A tetrazolium salt (TTC) may be included in the medium to make interpretation easier. TTC is used by the bacteria as an electron acceptor. In its oxidized form, TTC is colorless and soluble; when reduced it is red and insoluble (Fig. 5-54). A positive result for motility is indicated when the red (reduced) TTC is seen radiating in all directions from the central stab. A negative result shows red only along the stab line. Figures 5-55 and 5-56 illustrate motility test medium with and without TTC.

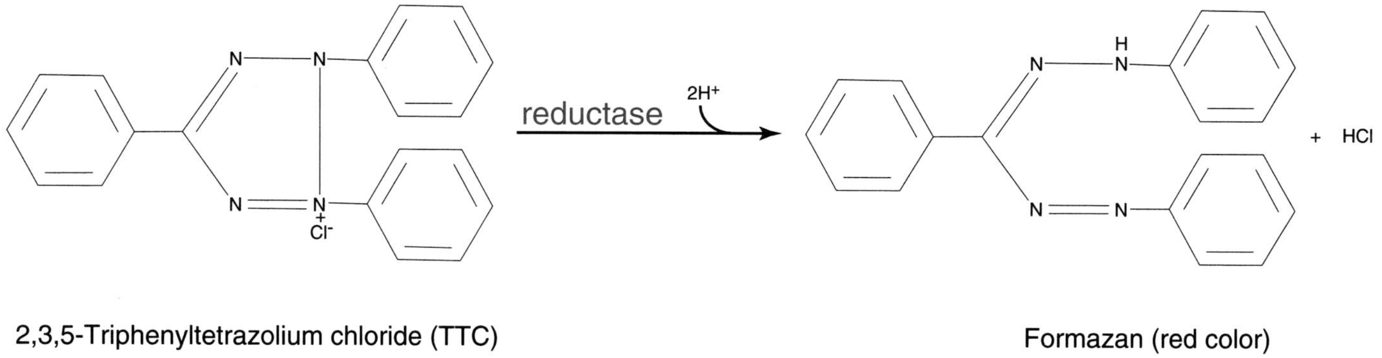

2,3,5-Triphenyltetrazolium chloride (TTC) $\xrightarrow[\text{reductase}]{2H^+}$ Formazan (red color) + HCl

FIGURE 5-54 Reduction of the colorless and soluble 2,3,5-Triphenyltetrazolium chloride by metabolizing bacteria results in its conversion to a red and insoluble formazan. The location of the growing bacteria can be easily determined by the location of the red color in the medium.

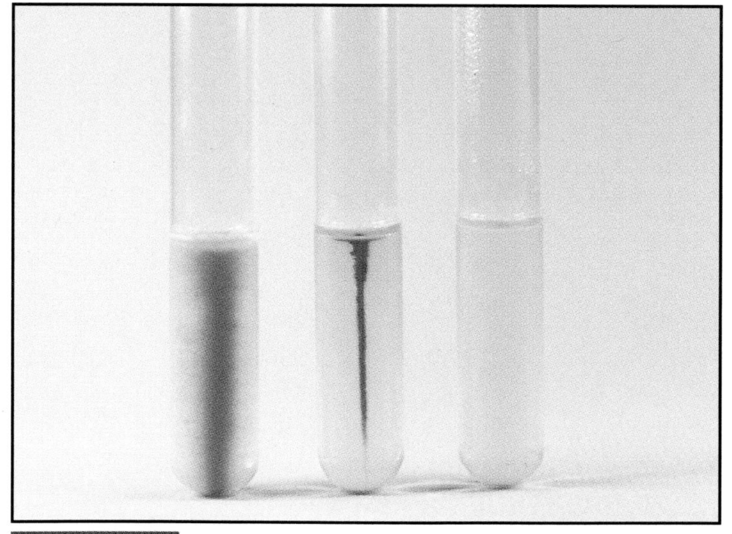

FIGURE 5-55 Motility test medium tubes containing TTC inoculated with *Aeromonas hydrophila* (motile) on the left, *Micrococcus luteus* (nonmotile) in the center, and an uninoculated control on the right.

FIGURE 5-56 Motility test medium tubes without TTC inoculated with *Aeromonas hydrophila* (motile) on the left, an uninoculated control in the center, and *Micrococcus luteus* (nonmotile) on the right. Compare with Figure 5-55.

NITRATE REDUCTION TEST

PURPOSE This test is used to detect the ability of an organism to reduce nitrate (NO_3) to nitrite (NO_2) or some other nitrogenous compound, such as molecular nitrogen (N_2), using the enzyme *nitrate reductase*.

PRINCIPLE Nitrate (NO_3) may be reduced to several different compounds (Fig. 5-57) via two bacterial metabolic processes — *anaerobic respiration* and *denitrification*. In anaerobic respiration, the organism uses nitrate as the final electron acceptor of the cytochrome system, producing nitrite, ammonia, molecular nitrogen, nitric oxide, or some other reduced nitrogenous compound, depending on the species. Denitrification, an essential component of the nitrogen cycle, reduces nitrate to molecular nitrogen.

Nitrate broth is an undefined medium of beef extract, peptone, and potassium nitrate (KNO_3). Some formulations include a small amount of agar to slow oxygen diffusion and encourage anaerobic growth (since nitrate reductase is inhibited by oxygen). An inverted Durham tube is included in each broth to trap liberated nitrogen gas.

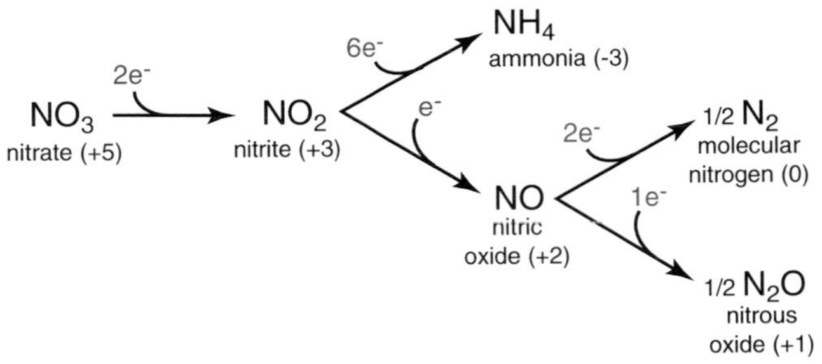

FIGURE 5-57 Possible end products of nitrate reduction. The oxidation state of nitrogen in each compound is shown in parentheses.

After appropriate incubation, the tubes are inspected for trapped gas (Fig. 5-59). If no gas is present, or if there is visible gas, but the organism is known to produce gas by fermentation (as evidenced by a fermentation test), the test is continued with the addition of reagents (see below). If the organism is a known nonfermenter, any gas produced is an indication of denitrification (*i.e.,* nitrate was reduced to nitrogen gas). No further testing is necessary.

In the absence of gas production in the Durham tube, sulfanilic acid and α-naphthylamine are added to the broth to detect the presence of nitrite (phase 1 reaction). In this aqueous environment, nitrite will form nitrous acid (HNO_2) which reacts with the sulfanilic acid and α-naphthylamine to produce a red, water soluble compound (Fig. 5-58). A red color, therefore, indicates the reduction of nitrate to nitrite and is a positive result (Fig. 5-60).

If no color change occurs after the addition of reagents, it is because the organism either does not reduce nitrate or reduces it to a compound other than nitrite. To discriminate between these two possibilities, zinc dust is added to the medium (phase 2 reaction). If nitrate remains in the medium, zinc will reduce it to nitrite, and a pink color will be observed. *A pink color upon addition of zinc is interpreted as a negative for nitrate reduction.* No color change means nitrate has been reduced and is interpreted as positive (Fig. 5-61).

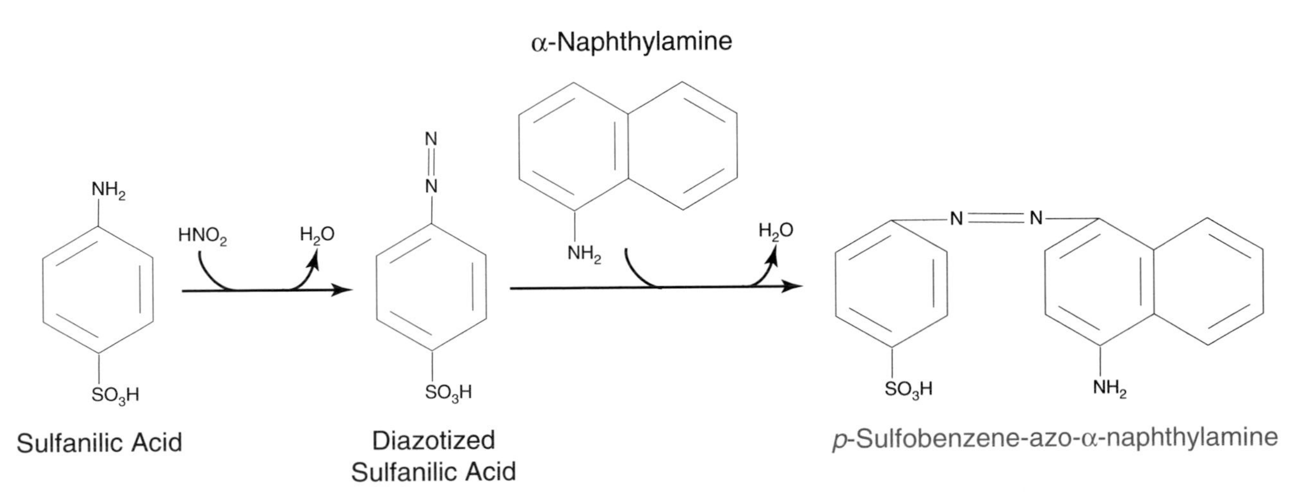

FIGURE 5-58 Phase 1 indicator reaction. If nitrate is reduced to nitrite, nitrous acid will form in the medium. Nitrous acid then reacts with sulfanilic acid to form diazotized sulfanilic acid, which reacts with the α-naphthylamine to form *p*-sulfobenzene-azo-α-naphthylamine, which is red. Thus, a red color indicates the presence of nitrite and is considered a positive result for nitrate reduction.

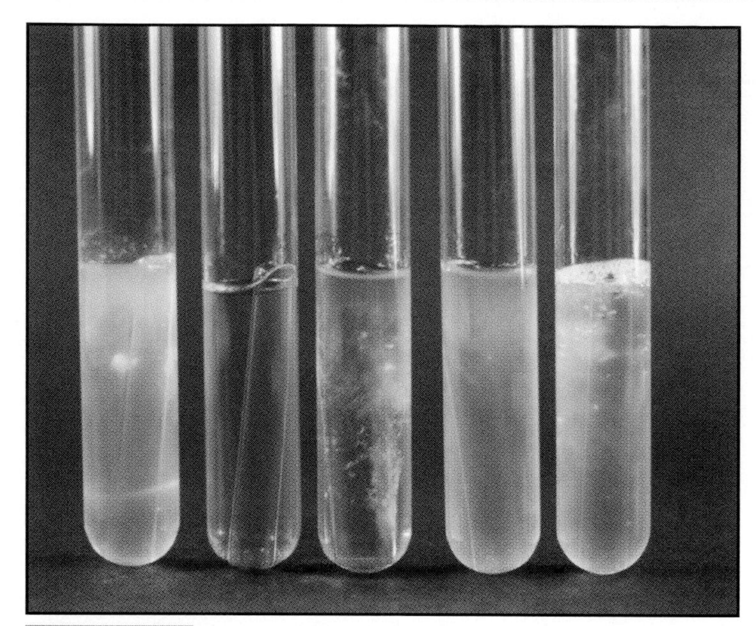

FIGURE 5-59 Nitrate broth tubes after incubation before the addition of reagents. From left to right: *Enterobacter aerogenes*, an uninoculated control, *Enterococcus faecalis*, and two different strains of *Pseudomonas aeruginosa*. Note the gas produced by the *P. aeruginosa* strain on the far right indicating a positive result (*P. aeruginosa* is a nonfermenter, therefore, the gas produced is an indication of denitrification.) The four tubes on the left must now proceed to the phase 1 reaction.

FIGURE 5-60 Nitrate broth tubes after addition of sulfanilic acid and α-naphthylamine (phase 1 reaction). From left to right: *Enterobacter aerogenes* (+1), an uninoculated control, *Enterococcus faecalis* (−), and the two *P. aeruginosa* strains (no reagents were added to the one on the right since it was positive for denitrification). A positive result indicates reduction of nitrate to nitrite; a negative result at this point must be checked for the presence of nitrate in the medium (phase 2 reaction).

FIGURE 5-61 Nitrate broth tubes after addition of zinc (phase 2 reaction) to tubes that have been negative up to this point. A red color indicates the presence of unreduced nitrate — a *negative* result. No color change is considered a positive result for nitrate reduction. From left to right: *Enterobacter aerogenes*, an uninoculated control, *Enterococcus faecalis* (−), *Pseudomonas aeruginosa* (+2), and the second strain of *P. aeruginosa*.

NOVOBIOCIN SUSCEPTIBILITY TEST

PURPOSE The novobiocin test is used to differentiate coagulase-negative staphylococci. Most frequently it is used to presumptively identify the novobiocin-resistant *Staphylococcus saprophyticus*, a common urinary tract pathogen in young, sexually active females.

PRINCIPLE With the exception of *Staphylococcus saprophyticus*, most clinically important staphylococci are susceptible to the antibiotic novobiocin. When agar plates are cultured with a novobiocin-susceptible organism and a novobiocin impregnated disk is placed on it, a large clearing around the disk will appear (Fig. 5-62). Conversely, organisms resistant to novobiocin will produce a small zone or no zone at all, depending on several factors.

The factors affecting zone size are the susceptibility of the organism to novobiocin, the concentration of the inoculum, the concentration of diffuse antibiotic in the agar, and the temperature and duration of incubation. The rate and amount of diffusion of the antibiotic are standardized by using 5 μg novobiocin disks on 5% sheep blood agar. The test organism concentration is controlled by diluting to 0.5 McFarland turbidity standard (Fig. 7-10) immediately before inoculation. Incubation takes place at 35°C for 24 hours. An isolate producing a zone of 12 mm or less is considered novobiocin-resistant (R). A zone greater than 12 mm indicates susceptibility (S).

FIGURE 5-62 Novobiocin disk test on a sheep blood agar plate. *Staphylococcus saprophyticus* (R) is on the left; *Staphylococcus epidermidis* (S) is on the right.

ONPG (o-NITROPHENYL-β-D-GALACTOPYRANOSIDE) TEST

PURPOSE The ONPG test is used to identify bacteria capable of producing the enzyme *β-galactosidase*.

PRINCIPLE In order for bacteria to ferment lactose, they must possess two enzymes: β-galactoside permease, a membrane-bound transport protein, and β-galactosidase, an intracellular enzyme that splits the disaccharide into β-glucose and β-galactose (Fig. 5-63).

Bacteria possessing both enzymes are active β-lactose fermenters. Bacteria possessing neither enzyme never ferment β-lactose. Bacteria that possess β-galactosidase but no β-galactoside permease may mutate and, over a period of days or weeks, begin to produce the permease. This third group, called *late lactose fermenters*, can be differentiated from the nonfermenters using a compound called *o*-nitrophenyl-β-D-galactopyranoside (ONPG). This compound can enter the cell without the assistance of β-galactoside permease and will react with β-galactosidase, if present, to produce a yellow color (Fig. 5-64).

Because β-galactosidase is an inducible enzyme, it will be produced only in the presence of its inducer (the substrate β-lactose). To avoid false negative results, organisms being tested for the presence of β-galactosidase are cultured in a lactose rich medium (*e.g.*, Kligler's iron agar or triple sugar iron agar) prior to the test.

Many variations of this test are available. In the version illustrated here, water containing a dissolved ONPG tablet is inoculated with the test organism. A color change to yellow within a few hours is a positive result (Fig. 5-65).

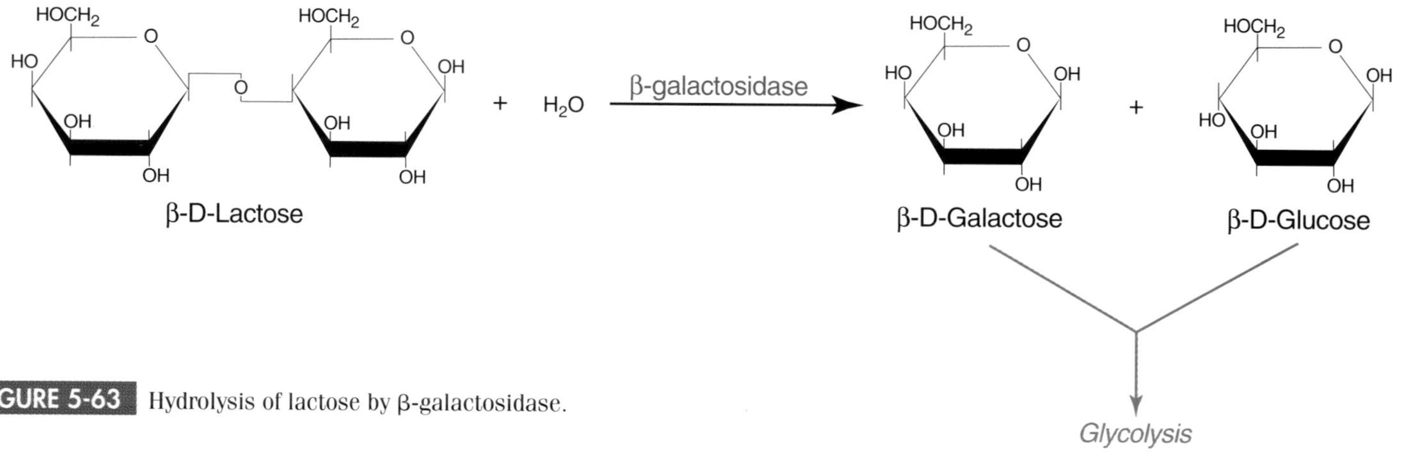

β-D-Lactose + H_2O —β-galactosidase→ β-D-Galactose + β-D-Glucose → Glycolysis

FIGURE 5-63 Hydrolysis of lactose by β-galactosidase.

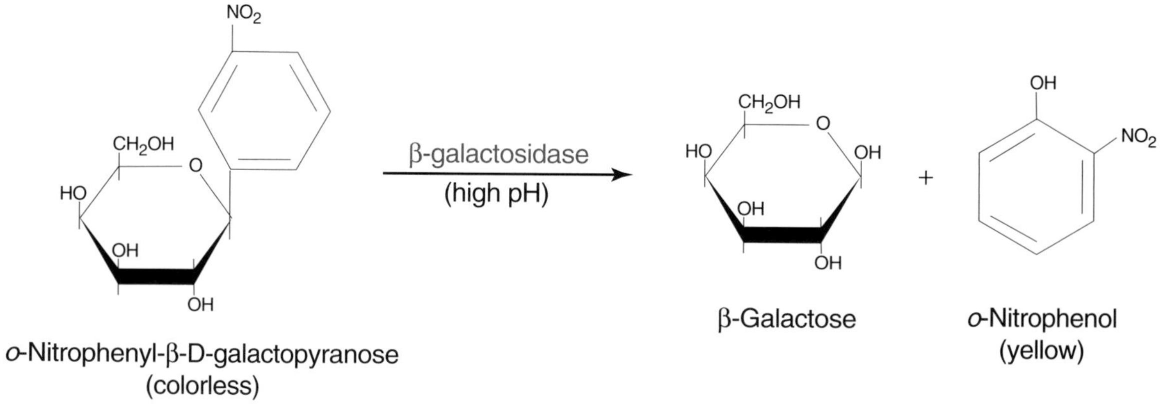

o-Nitrophenyl-β-D-galactopyranose (colorless) —β-galactosidase (high pH)→ β-Galactose + *o*-Nitrophenol (yellow)

FIGURE 5-64 Conversion of ONPG to β-galactose and *o*-nitrophenol by β-galactosidase.

FIGURE 5-65 The ONPG test. *Escherichia coli* (ONPG-positive) is on the left and *Proteus vulgaris* (ONPG-negative) is on the right. An uninoculated control is in the middle.

OPTOCHIN SUSCEPTIBILITY TEST

PURPOSE The optochin test is used to presumptively differentiate *Streptococcus pneumoniae* from other α-hemolytic streptococci.

PRINCIPLE *Streptococcus pneumoniae* is the only streptococcus susceptible to small concentrations of the antibiotic optochin. Therefore, to eliminate the few streptococci which show susceptibility to large concentrations of the antibiotic, the optochin impregnated disks used in this procedure contain a scant 5 μg.

Three or four colonies of the organism to be tested are transferred and streaked on a sheep blood agar plate in such a way as to produce confluent growth over approximately one half of the surface. The optochin impregnated disk is then placed in the center of the inoculum and allowed to incubate at 35°C for 24 hours in a candle jar or 5% to 7% CO_2.

The antibiotic will diffuse through the agar and inhibit growth of susceptible organisms in the area immediately surrounding the disk. This creates a clearing in the growth or *zone of inhibition* (Fig. 5-66). A zone ≥14 mm in diameter surrounding a 6 mm disk or a zone ≥16 mm surrounding a 10 mm disk is considered presumptively positive identification of *Streptococcus pneumoniae*. Smaller zones indicate further testing is required.

FIGURE 5-66 Optochin susceptibility test on a sheep blood agar plate. The zone of inhibition surrounding the disk indicates susceptibility to optochin and presumptive identification of *S. pneumoniae*.

OXIDASE TEST

PURPOSE This test is used to identify bacteria containing the respiratory enzyme *cytochrome oxidase*. Among its many uses is the presumptive identification of the oxidase-positive *Neisseria*. It can also be useful in differentiating the oxidase-negative *Enterobacteriaceae* from the oxidase-positive *Pseudomonadaceae*.

PRINCIPLE Bacterial electron transport chains (ETCs), although diverse in specific composition, have the same basic purpose — to produce energy (in the form of high energy bonds in ATP) via the oxidation of the NADH and $FADH_2$ produced in the Krebs cycle and glycolysis. Aerobes, facultative anaerobes, and microaerophiles have a cytochrome oxidase which uses electrons from the ETC to reduce oxygen (the final electron acceptor) to water.

In the oxidase test, an artificial electron donor (either dimethyl-*p*-phenylenediamine or tetramethyl-*p*-phenylenediamine) is used to reduce cytochrome oxidase. If present, cytochrome oxidase will oxidize the colorless reagent and turn it dark purple or blue (Fig. 5-67). If the test organism possesses no cytochrome oxidase the reagent will remain reduced and colorless. Thus, formation of a purple to blue color is positive; color change is negative.

The oxidase test may be run by placing a few drops of the phenylenediamine solution on bacterial growth (Fig. 5-68) or by transferring a large inoculum to a filter paper saturated with the reagent (Fig. 5-69). In each case, a deep purple/blue color is a positive result. The oxidase reagent is unstable and will eventually oxidize even in the absence of cytochrome oxidase. A false positive can be avoided by reading the test within 10 seconds of reagent application.

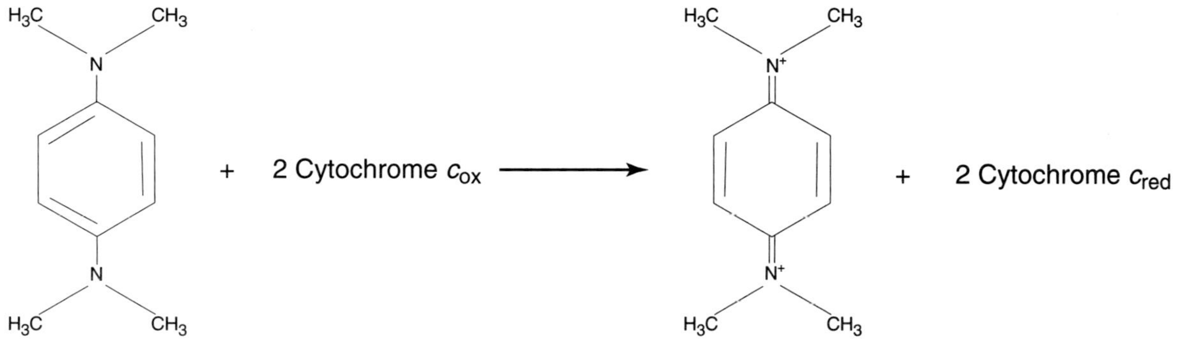

Tetramethyl-p-phenylenediamine$_{red}$ (colorless) + 2 Cytochrome c_{ox} → Tetramethyl-p-phenylenediamine$_{ox}$ (deep purple/blue) + 2 Cytochrome c_{red}

FIGURE 5-67 Chemistry of the oxidase reaction.

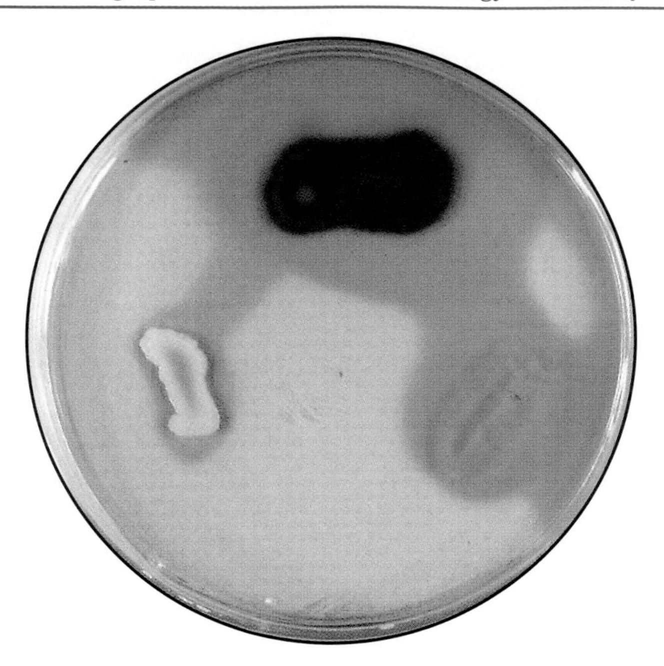

FIGURE 5-68 The oxidase test done on an agar plate. Clockwise from the top are *Pseudomonas aeruginosa* (+), *Clostridium sporogenes* (−), and *Staphylococcus aureus* (−). The reagent is very unstable and turns purple when exposed to air (as evidenced by the discoloration of the agar surface). The test should be read within 10 seconds to avoid false positive results.

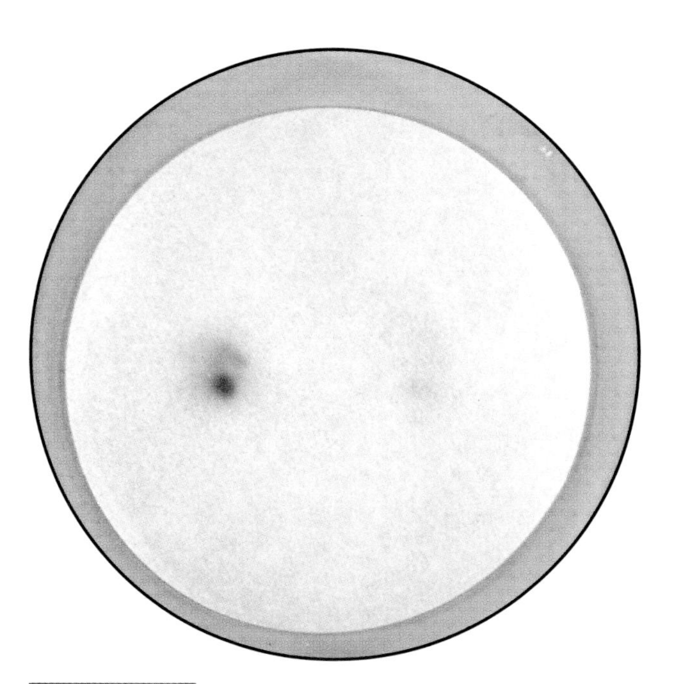

FIGURE 5-69 The oxidase test done on paper saturated in oxidase reagent. *Pseudomonas aeruginosa* (+) is on the left and *Staphylococcus aureus* (−) is on the right.

OXIDATION-FERMENTATION (O-F) TEST

PURPOSE This test is used to differentiate bacteria based on their ability to oxidize or ferment specific sugars. It allows presumptive separation of the fermentative *Enterobacteriaceae* from the oxidative *Pseudomonas* and *Bordetella*, and the inert *Alcaligenes* and *Moraxella*. *Staphylococcus* (fermentative) and *Micrococcus* (usually oxidative) can also be differentiated.

PRINCIPLE O-F test medium is a semisolid medium (due to a low agar content) with a high sugar to peptone ratio. The sugar/peptone imbalance favors weak acid production and minimizes the production of ammonia (via deamination) and amines which neutralize the acids. The pH indicator is bromthymol blue. Greater reactivity between the indicator and the acids is promoted by the increased permeability of the medium created by the reduced agar concentration.

O-F test medium is prepared to include any one of a variety of sugars. Typically, the medium used for testing enterics includes glucose, lactose, sucrose, maltose, mannitol or xylose. Staphylococcal O-F medium uses only glucose or mannitol.

Two tubes of the specific sugar medium are stabbed with the organism being tested. After inoculation, one tube is sealed with a layer of sterile mineral oil and the other is left unsealed. Mineral oil retards oxygen diffusion into the medium and thus promotes anaerobic growth. The unsealed tube allows aerobic growth.

Bromthymol blue is yellow at pH 6.0, green at pH 7.1 and blue at pH 7.6. Organisms able to ferment or ferment *and* oxidize the sugar will turn the sealed and unsealed media yellow. Organisms able only to oxidize the sugar will turn the unsealed medium yellow and leave the sealed medium green or blue. Organisms not able to metabolize the sugar will either produce no color change or will turn the medium blue. The results are summarized in Table 5-4 and shown in Figure 5-70.

TABLE 5-4 **Summary of O-F Medium Results**

SEALED MEDIA	UNSEALED MEDIA	INTERPRETATION
green or blue	green or blue	The organism performs no sugar metabolism; it is *nonsaccharolytic*
green or blue	yellow	The organism performs oxidative metabolism (O)
yellow	yellow	The organism performs fermentation (F) or fermentation *and* oxidation (O/F)

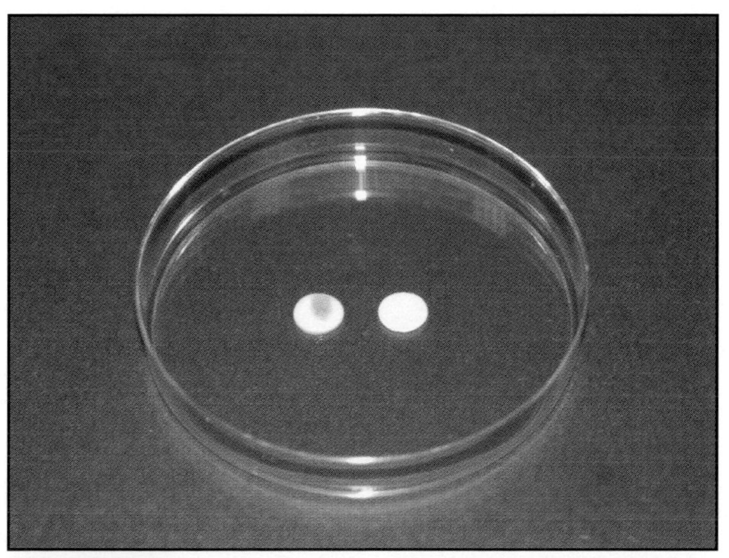

FIGURE 5-70 *Pseudomonas aeruginosa* (O) is in the two tubes on the left and *Shigella flexneri* (O/F) is in the two tubes on the right. Mineral oil in the first and third tubes prevents aerobic growth.

PHENYLALANINE DEAMINASE TEST

PURPOSE This medium is used to identify bacteria possessing the enzyme *phenylalanine deaminase*. It is used to differentiate the genera *Morganella*, *Proteus* and *Providencia* (positive) from other members of the *Enterobacteriaceae* (negative).

PRINCIPLE Among the ingredients of phenylalanine agar are yeast extract (used as the carbon and nitrogen source) and DL-phenylalanine. Organisms that produce phenylalanine deaminase are able to deaminate DL-phenylalanine to the keto acid, phenylpyruvic acid, and release the amine group as free ammonia (Fig. 5-71). Following incubation, the normally colorless phenylpyruvic acid may be detected by adding an oxidizing reagent to the medium. This example uses 10% ferric chloride ($FeCl_3$). Ferric chloride reacts with phenylpyruvic acid (if present) and changes from yellow to green almost immediately (Fig. 5-72). Formation of green color is positive; yellow is negative (Fig. 5-73).

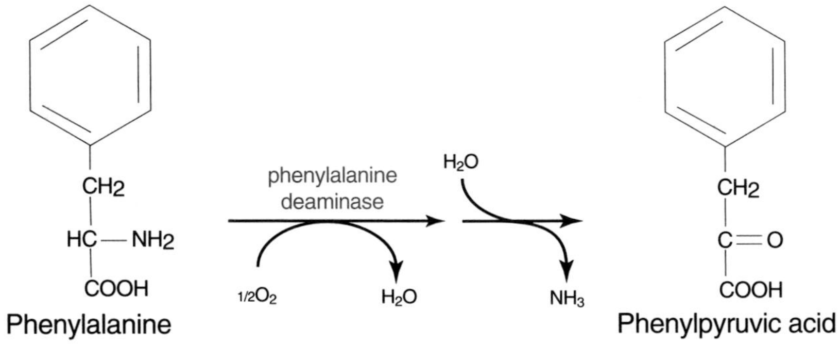

CH2
HC — NH2
COOH
Phenylalanine

phenylalanine deaminase
$1/2\,O_2$ → H_2O
H_2O → NH_3

CH2
C=O
COOH
Phenylpyruvic acid

FIGURE 5-71 Deamination of the amino acid phenylalanine by phenylalanine deaminase.

Phenylpyruvic Acid + $FeCl_3$ ⟶ Green Color

FIGURE 5-72 The reaction of phenylpyruvic acid with $FeCl_3$ to indicate a positive phenylalanine test.

FIGURE 5-73 Phenylalanine agar tubes. *Proteus mirabilis* (+) is on the left and *Escherichia coli* (–) is on the right. An uninoculated control is in the center.

PYR (PYRROLIDONYL ARYLAMIDASE) TEST

PURPOSE The PYR (pyrrolidonyl arylamidase) test is designed for presumptive identification of group A streptococci and enterococci by determining the presence of the enzyme *L-pyrrolidonyl arylamidase*.

PRINCIPLE Group A streptococci and enterococci produce the enzyme L-pyrrolidonyl arylamidase. L-pyrrolidonyl arylamidase hydrolyzes the amide pyroglutamyl-β-naphthylamide to produce L-pyrrolidone and β-naphthylamine. The presence of β-naphthylamine can be determined by the formation of a deep red color when the indicator reagent p-dimethylaminocinnamaldehyde is added.

PYR may be performed as an 18 hour agar test, a four hour broth test or, as used in this example, a rapid disk test. In each case the medium (or disk) contains pyroglutamyl-β-naphthylamide to which is added a heavy inoculum of the test organism. After the appropriate incubation or waiting period, a 0.01% p-dimethylaminocinnamaldehyde solution is added. Formation of a deep red color within a few minutes is interpreted as PYR-positive. Yellow or orange is PYR-negative (Fig. 5-74).

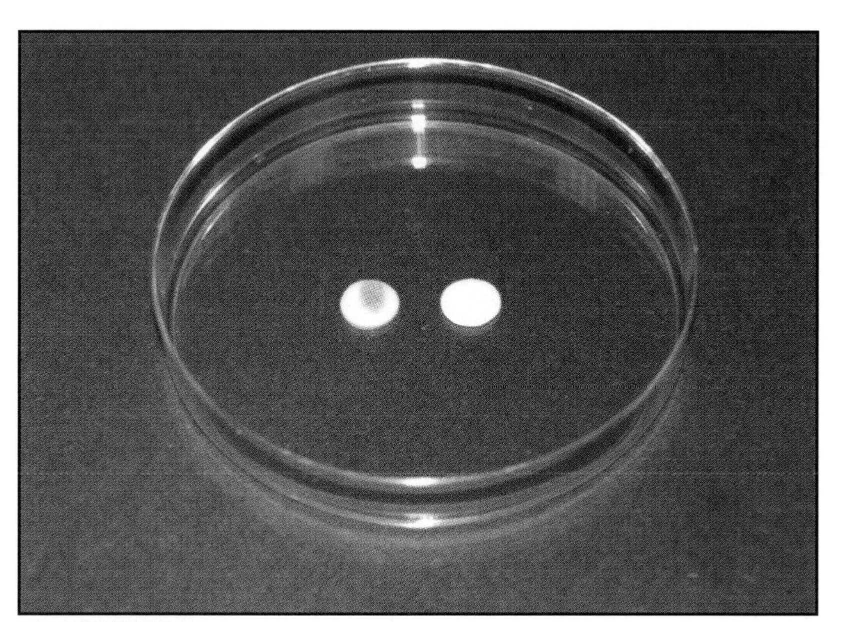

FIGURE 5-74 PYR disk test. The disk on the left was inoculated with *Streptococcus pyogenes* (PYR-positive); the disk on the right contains *Streptococcus agalactiae* (PYR–negative).

STARCH HYDROLYSIS TEST

PURPOSE This test is used to differentiate bacteria based on their ability to hydrolyze starch with the enzyme *amylase*. It aids in the differentiation of species from the genera *Corynebacterium*, *Clostridium*, *Bacillus*, *Bacteroides*, *Fusobacterium*, and members of group D streptococci.

PRINCIPLE Starch is a polysaccharide made up of α-D-glucose subunits. It exists as a mixture of two forms, linear (amylose) and branched (amylopectin), with the branched configuration being the predominant form. The α-D-glucose molecules in both amylose and amylopectin are bonded by 1,4-α-glycosidic (acetal) linkages (Fig. 5-75). The two forms differ in that the amylopectin contains polysaccharide side chains connected to approximately every 30th glucose in the main chain. These side chains are identical to the main chain except that the number 1 carbon of the first glucose in the side chain is bonded to carbon number 6 of the main chain glucose. The bond is, therefore, a 1,6-α-glycosidic linkage.

Starch is too large to pass through the bacterial cell membrane. Therefore, to be of metabolic value to the bacteria it must first be split into smaller fragments or individual glucose molecules. Organisms that produce and secrete the extracellular enzymes α-amylase and oligo-1,6-glucosidase are able to hydrolyze starch by breaking the glucosidic linkages between the sugar subunits. Although there usually are intermediate steps and additional enzymes utilized, the overall reaction is the complete hydrolysis of the polysaccharide to its individual α-glucose subunits (Fig. 5-75).

Starch agar is a simple plated medium of beef extract, soluble starch and agar. When organisms that produce α-amylase and oligo-1,6-glucosidase are grown on starch agar they hydrolyze the starch in the medium surrounding the bacterial growth. Because both the starch and its sugar subunits are soluble (clear) in the medium, the reagent iodine is used to detect the presence or absence of starch in the vicinity around the bacteria. Since iodine reacts with starch and produces a blue color, any microbial starch hydrolysis will be revealed as a clear zone surrounding the growth (Fig. 5-76).

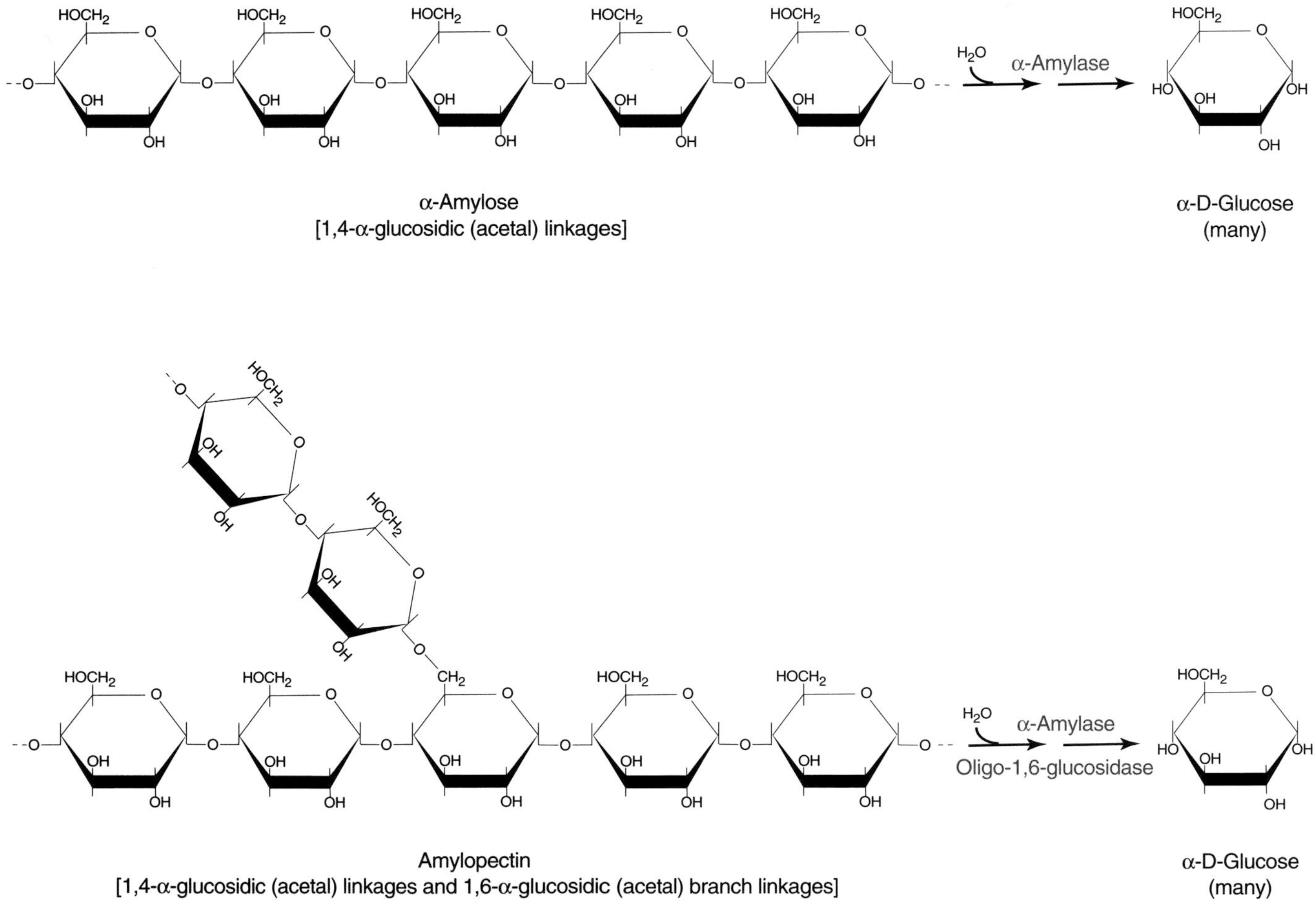

H_2O α-Amylase

α-Amylose
[1,4-α-glucosidic (acetal) linkages]

α-D-Glucose
(many)

H_2O α-Amylase Oligo-1,6-glucosidase

Amylopectin
[1,4-α-glucosidic (acetal) linkages and 1,6-α-glucosidic (acetal) branch linkages]

α-D-Glucose
(many)

FIGURE 5-75 Starch hydrolysis by α-amylase and oligo-1,6-glucosidase.

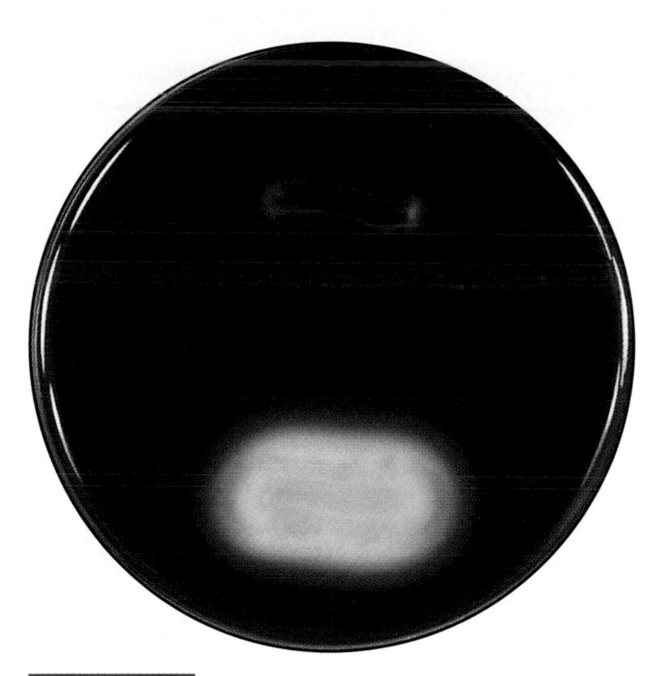

FIGURE 5-76 A starch agar plate with iodine added to detect amylase activity. *Escherichia coli* (negative) is above and *Bacillus cereus* (positive) is below.

SULFUR REDUCTION TEST (SIM MEDIUM)

PURPOSE The sulfur reduction test is used to identify bacteria capable of reducing sulfur. It is used to differentiate positive members of *Enterobacteriaceae*, especially members of the genera *Salmonella, Francisella,* and *Proteus* from the negative *Morganella morganii* and *Providencia rettgeri.*

PRINCIPLE Sulfur reduction produces H_2S either from the catabolism of the amino acid cysteine by the enzyme *cysteine desulfurase* during putrefaction, or by the reduction of thiosulfate in anaerobic respiration. In putrefaction sulfur containing amino acids (*e.g.*, cysteine) are broken down by the enzyme cysteine desulfurase producing pyruvic acid, ammonia and H_2S (Fig. 5-77). The ammonia and hydrogen sulfide are excreted from the cell and the pyruvic acid is retained for energy production via the Krebs cycle. In a type of anaerobic respiration, inorganic sulfur (in this case thiosulfate) becomes the final electron acceptor of the electron transport chain (Fig. 5-78).

Several brands of media are used for the detection of sulfur reduction. This example uses SIM medium. The ingredients in this combination medium relevant to sulfur reduction are peptone, beef extract, sodium thiosulfate, and peptonized iron or ferrous sulfate. Peptone provides the cysteine, sodium thiosulfate provides the reducible sulfur for anaerobic respiration, and peptonized iron or ferrous sulfate serves as an indicator by reacting with any H_2S produced to form a black precipitate (Fig. 5-79). The formation of a black color in the medium is characteristic of H_2S-positive organisms. No color change to black is indicative of H_2S-negative organisms. Results are shown in Figure 5-80.

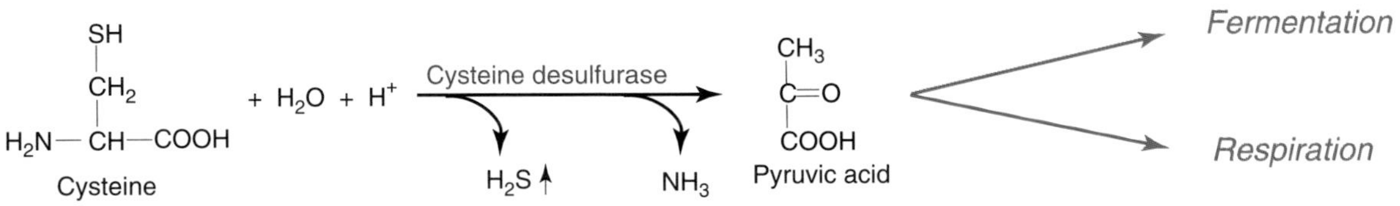

$$\underset{\text{Cysteine}}{H_2N{-}CH(CH_2SH){-}COOH} + H_2O + H^+ \xrightarrow{\text{Cysteine desulfurase}} \underset{\text{Pyruvic acid}}{CH_3{-}C({=}O){-}COOH} + H_2S\uparrow + NH_3$$

Pyruvic acid → Fermentation / Respiration

FIGURE 5-77 Putrefaction involving cysteine desulfurase produces H_2S.

$$3S_2O_3^{=} + 4H^+ + 4e^- \xrightarrow{\text{Thiosulfate reductase}} 2SO_3^{=} + 2H_2S\uparrow$$

FIGURE 5-78 Anaerobic respiration with thiosulfate as the final electron acceptor also produces H_2S.

$$H_2S + FeSO_4 \longrightarrow H_2SO_4 + FeS\downarrow$$

FIGURE 5-79 Indicator reaction for H_2S, a colorless gas. The FeS produced is a black precipitate and indicates the presence of H_2S.

FIGURE 5-80 Sulfur reduction in SIM medium. On the left is *Proteus mirabilis* (H_2S-positive); *Pseudomonas fluorescens* (H_2S-negative) is on the right.

SXT (TRIMETHOPRIM-SULFAMETHOXAZOLE) SUSCEPTIBILITY TEST

PURPOSE The SXT susceptibility test is used to differentiate groups A and B streptococci (SXT resistant) from other β-hemolytic streptococci (SXT susceptible). The bacitracin susceptibility test (see "Bacitracin Susceptibility Test" in this section) alone is sufficient to separate group A from group B streptococci, but the SXT test is necessary to distinguish between group A and some members of groups C, F, and G (which may also be bacitracin sensitive). Therefore, the SXT and bacitracin tests are often run on the same plate to help screen out non-group A β-hemolytic streptococci that are susceptible to bacitracin. Table 5-5 summarizes SXT-Bacitracin susceptibilities.

PRINCIPLE Sulfamethoxazole and trimethoprim, when used together, act synergistically to disrupt bacterial folic acid metabolism. The antibiotic disks typically used in this test contain 23.75 μg of sulfamethoxazole and 1.25 μg of trimethoprim.

The test is performed by placing a disk saturated with SXT on the surface of a sheep blood agar plate inoculated to form a bacterial lawn. Any clearing around the disk after incubation indicates susceptibility (S). Growth up to the edge of the disk indicates resistance (R).

The combination SXT and bacitracin susceptibility test is performed by placing one of each disk on the plate at least four centimeters apart (Fig. 5-81). Any clearing around either disk is interpreted as susceptibility.

TABLE 5-5 **Reactions of β-Hemolytic Streptococci to the Antibiotics Bacitracin and SXT (Sulfamethoxazole and Trimethoprim)**

ORGANISM	BACITRACIN	SXT
Group A	S	R
Group B	R	R
Groups C, F, and G	S or R	S

FIGURE 5-81 A bacitracin–SXT susceptibility test on a sheep blood agar plate containing *Streptococcus pyogenes* (group A). Bacitracin is on the left and SXT is on the right.

TRIPLE SUGAR IRON (TSI) AGAR

PURPOSE Triple sugar iron (TSI) agar differentiates bacteria based on their ability to ferment glucose, lactose and/or sucrose, and to reduce sulfur to hydrogen sulfide. It is used to distinguish between the morphologically similar members of *Enterobacteriaceae*, all of which ferment glucose to an acid end product.

PRINCIPLE Triple sugar iron agar is the same as Kligler's iron agar (KIA) except that TSI includes sucrose as a third sugar. For a description of the biochemistry of TSI, refer to "Kligler's Iron Agar" in this section. The various reactions in TSI are shown in Figure 5-82 and summarized in Table 5-6.

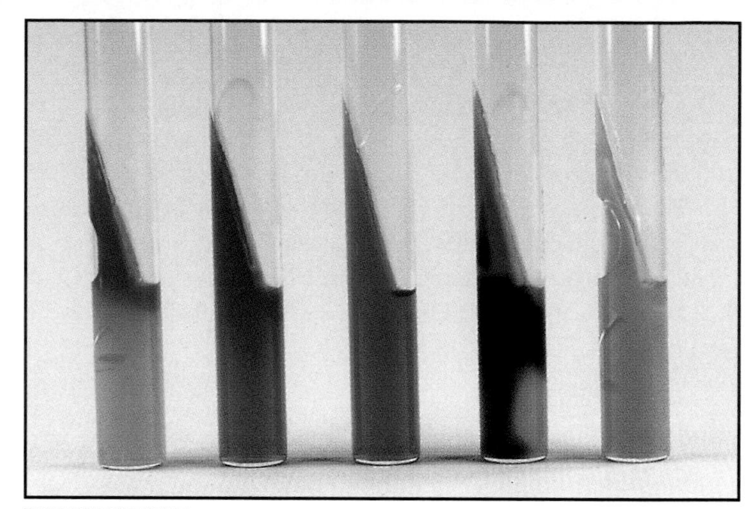

FIGURE 5-82 TSI agar slants. From left to right: *Morganella morganii* (K/A,G), *Pseudomonas aeruginosa* (K/NC), uninoculated control, *Proteus mirabilis* (K/A,H_2S), and *Escherichia coli* (A/A,G).

TABLE 5-6 **Results, Symbols and Interpretations of TSI Agar**

RESULTS	SYMBOL	INTERPRETATION
red/yellow	K/A	glucose fermentation only peptone catabolized
yellow/yellow	A/A	glucose and lactose and/or sucrose fermentation
red/red	K/K	no fermentation peptone catabolized
red/no color change	K/NC	no fermentation peptone used aerobically
yellow/yellow with bubbles	A/A,G	glucose and lactose and/or sucrose fermentation gas produced
red/yellow with bubbles	K/A,G	glucose fermentation only gas produced
red/yellow with bubbles and black precipitate	K/A,G, H_2S	glucose fermentation only gas produced H_2S produced
red/yellow with black precipitate	K/A, H_2S	glucose fermentation only H_2S produced
yellow/yellow with black precipitate	A/A, H_2S	glucose and lactose and/or sucrose fermentation H_2S produced
no change/no change	NC/NC	no fermentation (organism is growing very slowly or not at all)

A = acid production, G = gas production, K = alkaline reaction, H_2S = sulfur reduction

UREASE TEST

PURPOSE This test is used to differentiate organisms based on their ability to hydrolyze urea with the enzyme *urease*. Urinary tract pathogens from the genus *Proteus* may be distinguished from other enteric bacteria by their rapid urease activity.

PRINCIPLE Urease test media (both solid and liquid) contain urea, essential growth factors, and the pH indicator phenol red. Phenol red, is yellow or orange/yellow below pH 8.4 and red above pH 8.4. The formulation is designed to determine microbial ability to convert urea to ammonia (Fig. 5-83) by detecting changes in pH. Production of ammonia (used by bacteria as a nitrogen source to produce amino acids and nucleotides) raises the pH. When the pH reaches 8.4, the phenol red turns from yellow to a pink. A pink color is a positive result.

Interpretations for both media begin after a one day incubation at 37°C. Pink color throughout in liquid media after 24 hours indicates a rapid urease-positive organism (Fig 5-84). Delayed or slow urease activity may take several days to detect.

Solid media allows a slightly more precise measurement of urease activity. Pink color throughout the slant *and* butt after 24 hours (*Proteus*, *Morganella*, and *Providencia*) indicates rapid activity (Fig. 5-85). Pink color in the slant *only* indicates a delayed urease-positive organism (some *Citrobacter*, *Enterobacter*, and *Klebsiella*). Media that remains yellow after several days indicates the test organism is urease-negative.

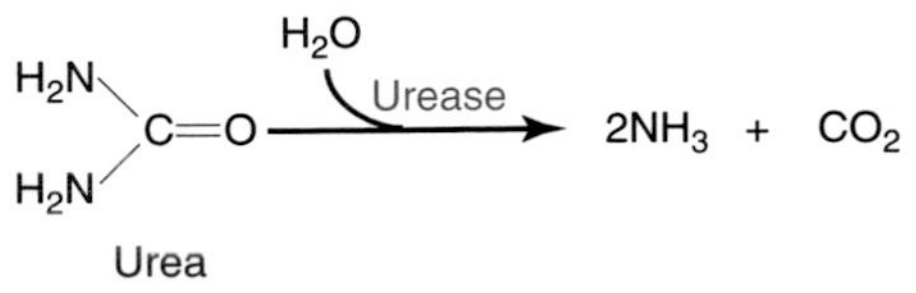

$$(H_2N)_2C{=}O \xrightarrow[\text{Urease}]{H_2O} 2NH_3 + CO_2$$

Urea

FIGURE 5-83 Urea hydrolysis.

FIGURE 5-84 Urea broth tubes with *Morganella morganii* (urease-positive) on the left and *Hafnia alvei* (urease-negative) on the right. An uninoculated control is in the center.

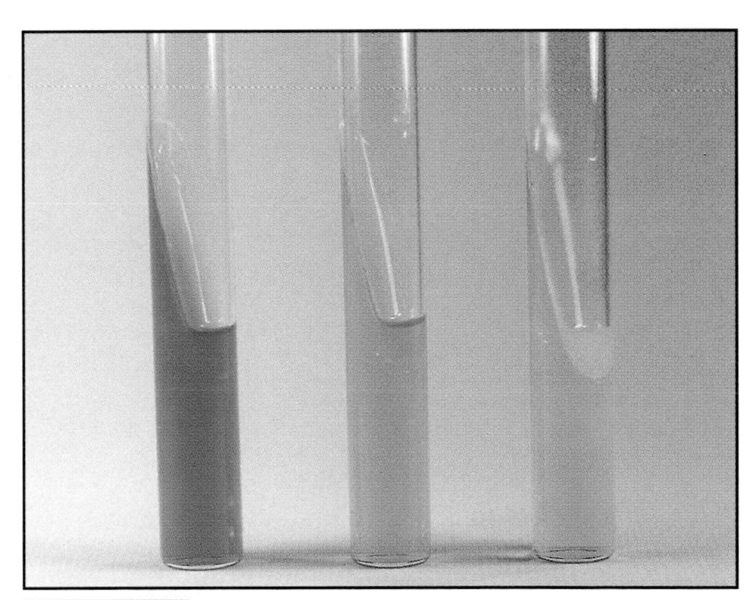

FIGURE 5-85 Urea agar tubes after a 24 hour incubation. *Morganella morganii* (urease-positive), a rapid urea splitter, is on the left and *Hafnia alvei* (urease-negative) is on the right. An uninoculated control is in the center.

VOGES-PROSKAUER (VP) TEST

PURPOSE The Voges-Proskauer test is used to identify organisms able to produce acetoin from the degradation of glucose during a 2,3-butanediol fermentation. It is one component of the *IMViC* battery of tests (*I*ndole, *M*ethyl red, *V*oges-Proskauer, and *C*itrate) used to differentiate the *Enterobacteriaceae*. It is useful in differentiating the VP-positive *Klebsiella pneumoniae* and species of *Enterobacter* from *Escherichia coli* which is VP-negative. Micrococci, which are VP-negative, can also be separated from the staphylococci which are VP-positive.

PRINCIPAL The Voges-Proskauer test is performed using MR/VP broth, a simple mixture of peptone, glucose and a phosphate buffer. In this medium many enteric bacteria perform mixed-acid fermentation. (See "Methyl Red Test" in this section) However, not all enterics produce enough stable acids to overcome the phosphate buffer sufficiently to lower the pH of the medium. For this latter group of bacteria, the chief end products of glucose metabolism are acetoin and 2,3-butanediol (Fig. 5-86).

After appropriate incubation, Barritt's Reagents A (α-naphthol) and B (KOH) are added to the sample. The tube is gently mixed to aerate the medium and oxidize any acetoin to diacetyl (Fig. 5-87). Diacetyl further reacts with components of the peptone (nitrogenous compounds called *guanidine nuclei*) to form a red color. Red is, therefore, a positive result for the VP test (Fig. 5-88). No color change or a copper color (due to the reaction of KOH and α-naphthol) are negative results.

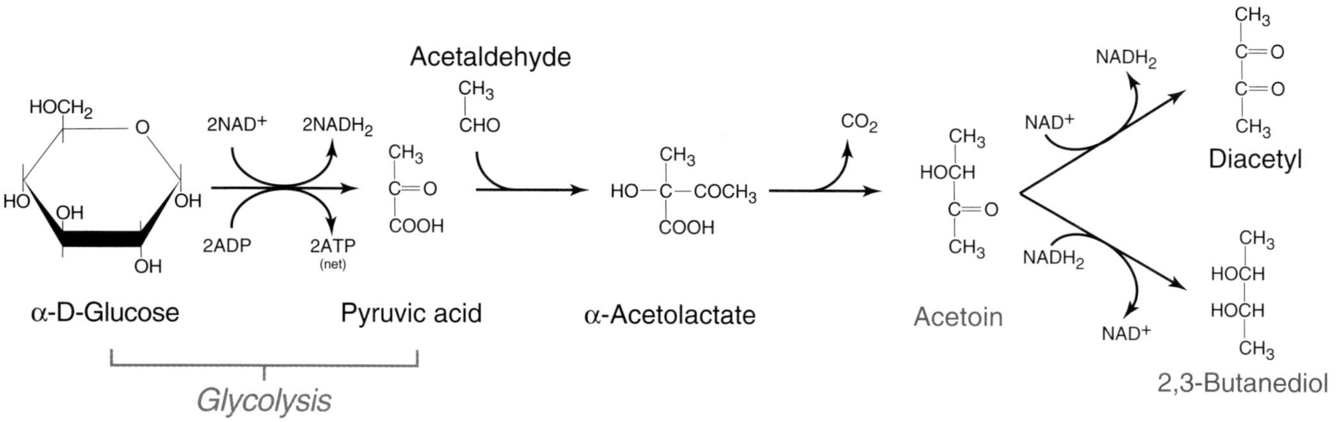

α-D-Glucose → ($2NAD^+$ → $2NADH_2$; 2ADP → 2ATP (net)) → Pyruvic acid → Acetaldehyde → α-Acetolactate → (CO_2) → Acetoin → (NAD^+ → $NADH_2$) Diacetyl; ($NADH_2$ → NAD^+) 2,3-Butanediol

Glycolysis

FIGURE 5-86 Production of acetoin during butanediol fermentation.

α-Naphthol (Barritt's Reagent A) + Acetoin → KOH (Barritt's Reagent B), O_2 ($2H_2O$) → Diacetyl + Guanidine (in peptone) → Red color

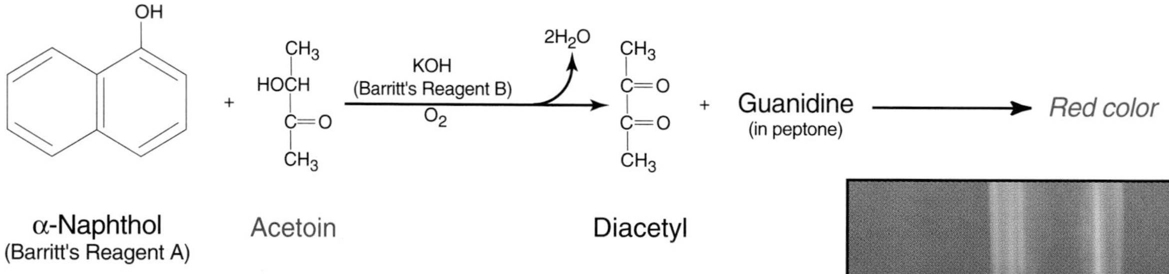

FIGURE 5-87 Chemistry of the Voges-Proskauer test.

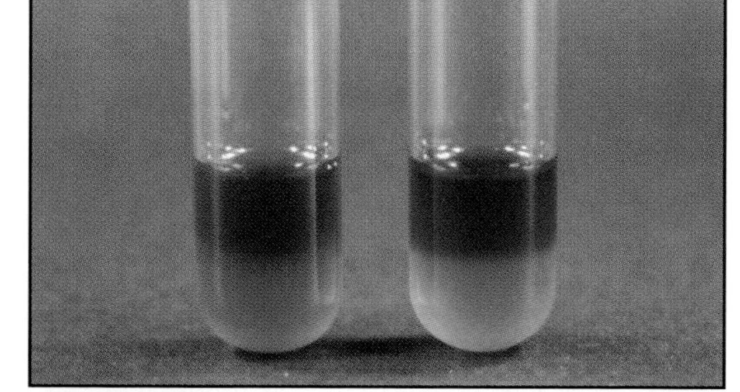

FIGURE 5-88 The Voges-Proskauer test with *Escherichia coli* (VP-negative) on the left and *Enterobacter aerogenes* (VP-positive) on the right. The copper color at the top of the VP-negative tube is due to the reaction of KOH and α-naphthol and should not be confused with a positive result.

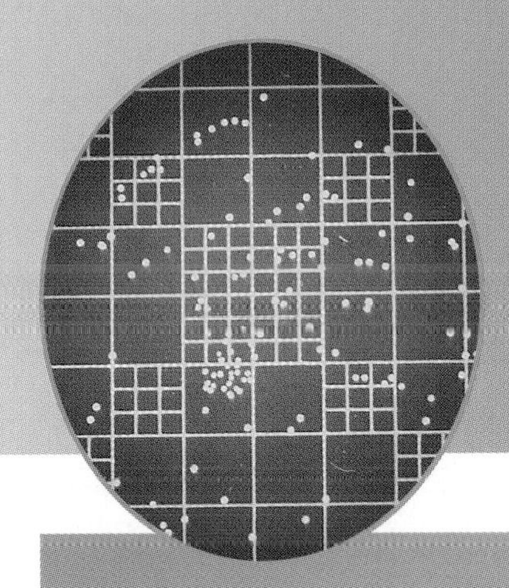

Quantitative Techniques

6

VIABLE (PLATE) COUNT

PURPOSE In the microbiology laboratory it is sometimes necessary to determine the density of living (viable) cells in a culture. The viable count is one method used to make this determination. It may also be used to calculate generation time and growth rate by measuring cell density at regular intervals.

PRINCIPLE The viable count involves plating a bacterial sample and counting the resulting colonies after incubation. As is implied by the name, the viable count measures only living cells and is an indirect method in that it employs a colony count rather than a cell count. It must be performed on young, actively growing cultures.

A *serial dilution* of the original broth is typically performed prior to plating (Fig. 6-1). The serial dilution is necessary to reduce the density of organisms in order to produce *countable* plates. Only plates containing between 30 and 300 colonies are considered countable. Plates that have fewer than 30 colonies have insufficient numbers to be reliable and plates with more than 300 are too crowded to count accurately (Figs. 6-2 through 6-4). The density of the original sample can be calculated using values obtained from the dilution series and the number of colonies on the countable plate (see below).

In order to maintain organism viability, the preferred method of plating dilutions is the *spread plate technique*. Platings from each of several dilutions are performed. If growth dynamics of the population are being measured, dilutions and multiple platings are performed at predetermined intervals.

When a diluted cell sample has been plated, the number of colonies produced can be used to calculate the original cell density of the sample using the following formula:

$$\text{Original cell density} = \frac{\text{\# Colonies counted}}{(\text{Volume plated})\ (\text{Dilution factor})}$$

Because it is not known if a colony formed on a plate developed from a single cell or multiple cells (as would occur with *Staphylococcus*, for example), the term *CFU (colony forming unit)* is conventionally used instead of *cell*. Therefore, cell densities are traditionally recorded as CFU/mL. Furthermore, since each colony is believed to have developed from a single colony forming unit, "CFU" replaces "colonies counted," so the formula becomes:

$$\text{Original cell density} = \frac{\text{\# CFU}}{(\text{Volume plated})\ (\text{Dilution factor})}$$

If 120 colonies are counted on a plate inoculated with 0.1 mL of a solution having a dilution factor (DF) of 10^{-5}, the original cell density would be calculated as follows:

$$\text{Original cell density} = \frac{\text{\# CFU}}{(\text{Volume plated})\ (\text{Dilution factor})}$$

$$\text{Original cell density} = \frac{120\ \text{CFU}}{(0.1\ \text{mL})\ (10^{-5})}$$

$$\text{Original cell density} = 1.2 \times 10^{8}\ \text{CFU/mL}$$

The advantage of the viable count over other counting methods is its measurement of living cells exclusively. Avoiding the counting of nonliving cells gives a better representation of population growth rate or degree of contamination. The major disadvantage is that it is dependent upon very careful technique. Although this procedure can be quite accurate, small errors made through careless technique can yield grossly inaccurate results.

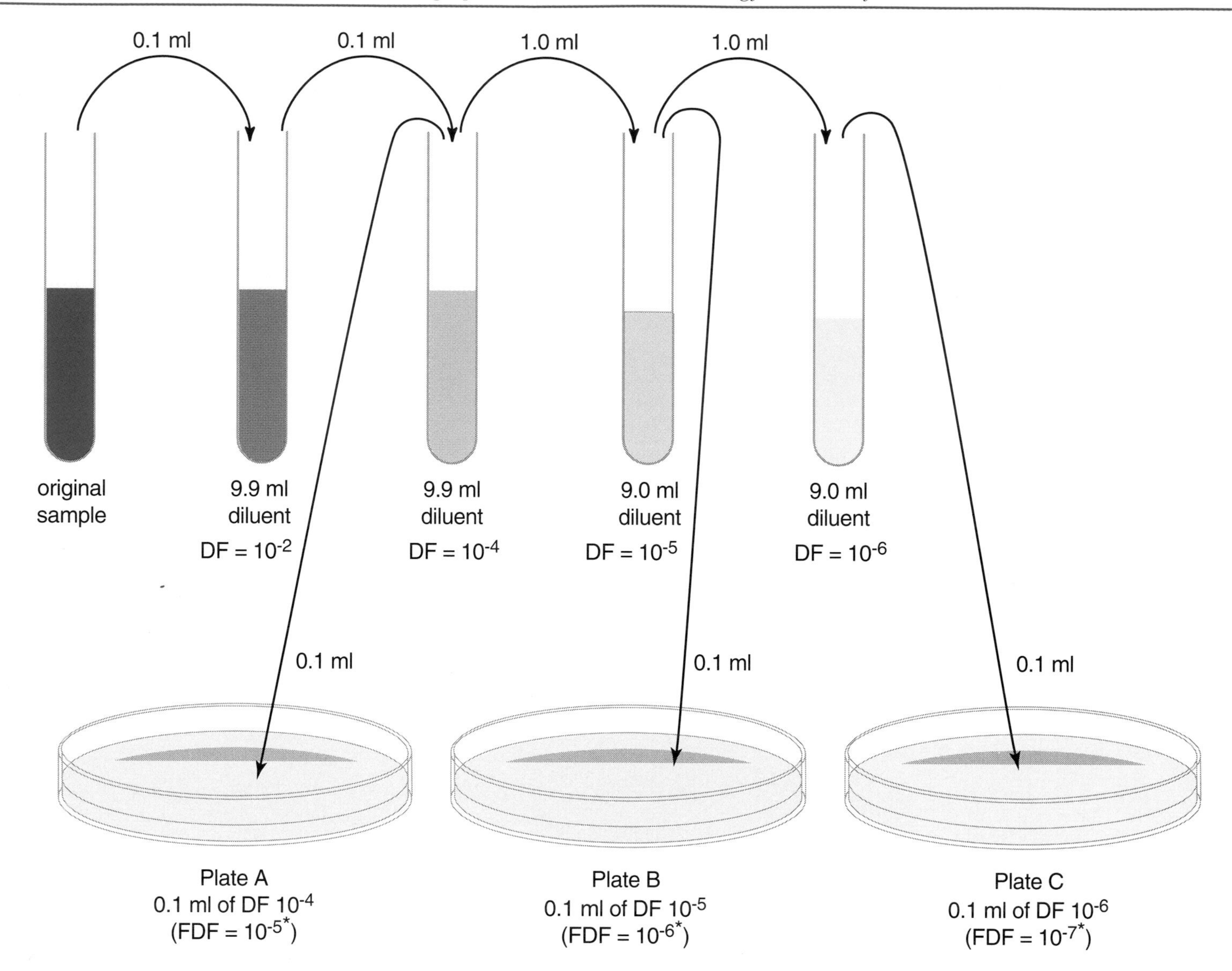

FIGURE 6-1 A sample dilution series. A small portion (0.1 mL) of the original sample is diluted in a larger volume (9.9 mL of diluent) of the first tube. This is followed by transfers from the first tube into the second, the second into the third, *etc.* Each specific dilution is assigned a dilution factor. Dilution factors are calculated using the formula, $D_2 = (V_1)(D_1)/V_2$ where D_2 is the new dilution factor being calculated, D_1 is the dilution factor of the sample being diluted (undiluted samples have a dilution factor of 1), V_1 is the volume of sample to be diluted, and V_2 is the combined volume of sample and diluent. For example, to calculate the new dilution factor when adding 0.1 mL of DF 10^{-2} to 9.9 mL of diluent (producing a combined volume of 10.0 mL), use the formula as illustrated below:

$$D_2 = \frac{(V_1)\,(D_1)}{V^2}$$

$$D_2 = \frac{(0.1\text{ mL})\,(10^{-2})}{10.0\text{ mL}} = 10^{-4}$$

*By convention, when 0.1 mL of a sample is inoculated onto a plate, the dilution factor is recorded as tenfold greater and the "volume plated" term in the equation is omitted. This is because 0.1 mL has one-tenth the cells of 1.0 mL, which is comparable to another tenfold dilution at the time of plating. For example, if 0.1 mL of DF 10^{-2} is plated, then the DF on the plate (*the final dilution factor*) is recorded as FDF 10^{-3} and the volume plated is ignored since it has already been taken into account. The original density can then be calculated using the following formula.

$$\text{Original cell density} = \frac{\#\text{ CFU}}{\text{Dilution factor}}$$

(It should be noted that this formula is a shortcut and will not provide the necessary units. If you choose to use it instead of the full version shown on the previous page, do not forget to record your results in CFU/mL.)

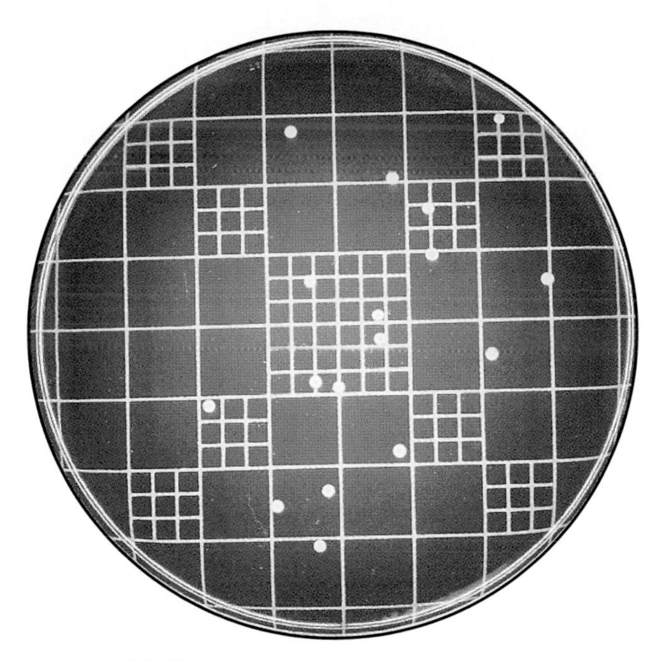

FIGURE 6-2 This is a viable count plate with too few colonies to provide a statistically reliable estimate of cell density. The plate, which contains 17 colonies, is recorded as TFTC ("too few to count").

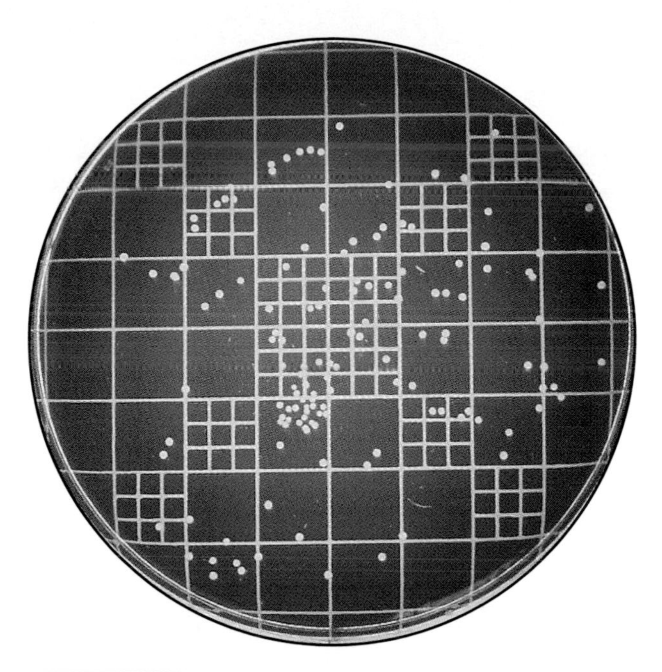

FIGURE 6-3 A countable plate has between 30 and 300 colonies. Therefore, this plate with approximately 130 colonies is countable and can be used to calculate cell density in the original sample.

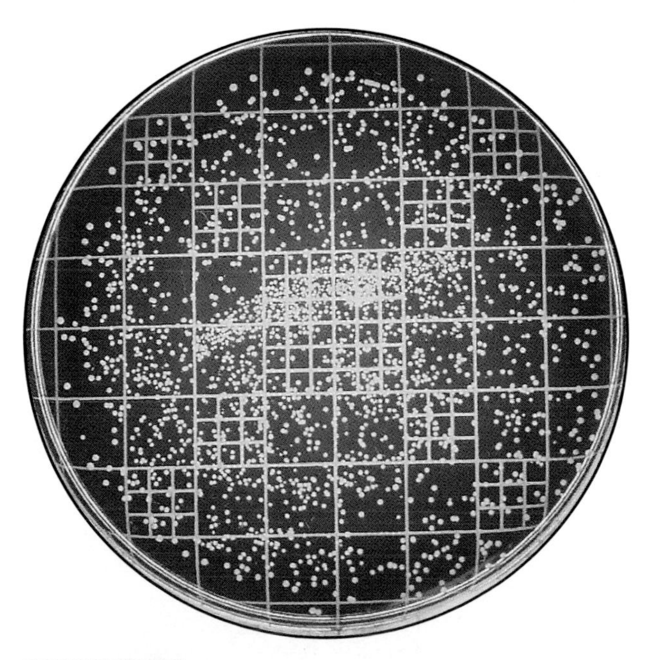

FIGURE 6-4 This viable count plate is recorded as TMTC ("too many to count") or TNTC ("too numerous to count") because there are more than 300 colonies on it.

DIRECT COUNT

PURPOSE The direct count method is used to determine bacterial cell density in a sample. A series of measurements of a culture taken over time can be used to calculate the organism's generation time and growth rate.

PRINCIPLE Counting all the bacteria, even in a small sample, would be virtually impossible. However, a direct count of cells in an extremely small sample volume may be done easily with a Petroff-Hausser counting chamber viewed under the microscope. This modified microscope slide contains a chamber or "well" of known depth (usually 0.02 mm) in the center with an etched grid on the bottom (Fig. 6-5). The grid has an area of 1 mm^2 and consists of 25 large squares marked by double lines. Inside each of these larger squares is a grid of 16 small squares marked by single lines. Therefore, the total number of squares in 1 mm^2 is 16×25 or 400 squares. This grid of small squares is the area used for counting (Fig. 6-6).

When the well is covered with a cover glass and filled with a suspension of cells the volume above each small square is 5×10^{-8} mL (The caption of Fig. 6-5 explains how this volume is determined). Since cell density is usually reported in cells/mL, calculation of original cell density must extrapolate from the volume above one small square up to an entire milliliter. This is done by dividing the average number of cells counted per square by the volume of sample above one square.

$$\text{Original cell density} = \frac{\text{Cells/small square}}{\text{Volume above a small square}}$$

Any dilution of the original sample must also be taken into account when calculating original cell density. This can be accomplished by including the dilution factor in the denominator (see equation below). (For an explanation of dilutions and dilution factors, see "Viable Count" Figure 6-1).

$$\text{Original cell density} = \frac{\text{Cells/small square}}{\text{(Volume) (Dilution factor)}}$$

For example, if an average of 16 cells/small square is counted in a sample with a dilution factor of 10^{-2}, the cell density in the original sample would be calculated as follows.

$$\text{Original cell density} = \frac{\text{Cells/small square}}{\text{(Volume) (Dilution factor)}}$$

$$\text{Original cell density} = \frac{16 \text{ cells}}{(5 \times 10^{-8} \text{ mL})(10^{-2})}$$

$$\text{Original cell density} = 3.2 \times 10^{10} \text{ cells/mL}$$

The advantages of direct counting are that it is fast, easy to do and is relatively inexpensive. The major disadvantage is that both living and dead cells are counted.

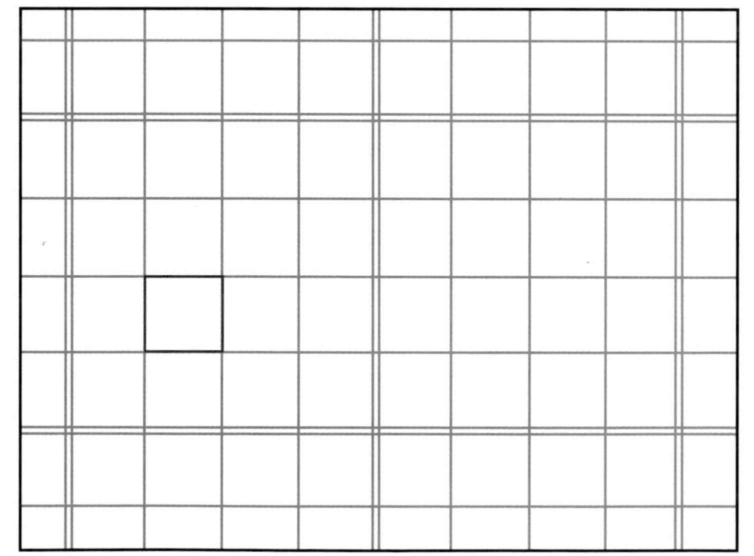

FIGURE 6-5 A portion of the Petroff-Hausser counting chamber grid with a small square highlighted in red. The volume above a small square is 5×10^{-8} mL. (Each small square is 0.05 mm $\times$ 0.05 mm giving an area of 2.5×10^{-3} mm^2. Since the well is 0.02 mm deep, the volume above a small square is 0.02 mm $\times$ 2.5×10^{-3} mm^2 = 5×10^{-8} cm^3. One cubic centimeter is the same as one milliliter, so the final units are: 5×10^{-8} mL.)

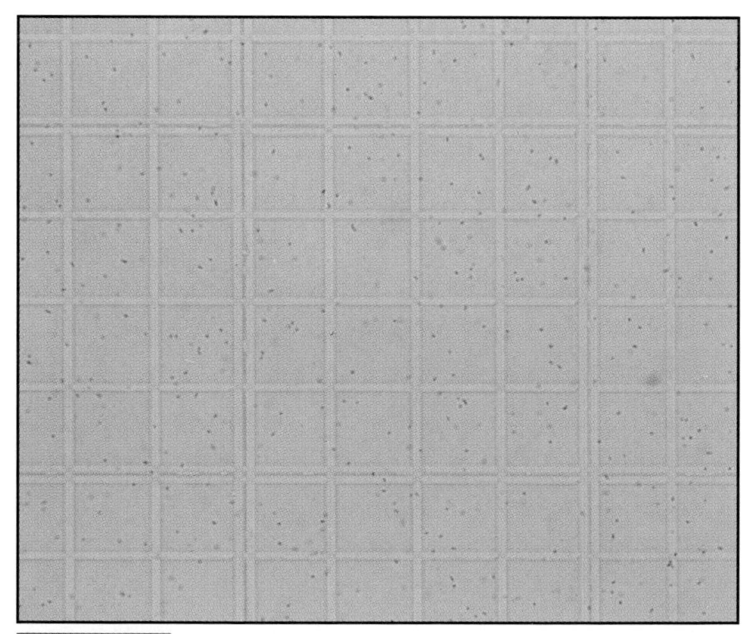

FIGURE 6-6 The Petroff-Hausser counting chamber. A 10^{-4} dilution of *Vibrio natriegens* has been stained with crystal violet and is visible on the grid. Pick five squares at random and determine the original cell density. Choose five different squares and compare your answers.

PLAQUE ASSAY FOR DETERMINATION OF PHAGE TITER

PURPOSE This technique is used to determine the concentration of viral particles in a sample. Samples taken over a period of time can be used to construct a viral growth curve.

PRINCIPLE Viruses that attack bacteria are called *bacteriophages* or simply *phages*. Some viruses attach to the bacterial cell wall and inject viral DNA into the bacterial cytoplasm. The viral genome then commands the cell to produce more viral DNA and viral proteins which are used for the assembly of more phages. Once assembly is complete, the cell lyses and releases the phages which then attack other bacterial cells and begin the replicative cycle all over again. This process, called the *lytic cycle*, is shown in Figure 10-2.

Lysis of bacterial cells growing on an agar plate and the subsequent attack of other cells in the immediate vicinity produces a clearing that can be viewed with the naked eye. These clearings are called *plaques*. Plaque assay uses this phenomenon as a means of calculating the phage concentration in a given sample.

In this procedure, as in the viable count (discussed earlier in this section), a serial dilution is used to reduce the concentration of viral particles in the original sample. For a description of serial dilution, see Figure 6-1. In contrast to the spread plate technique used in the viable count, the plaque assay uses a *pour plate technique*.

The pour plate technique adds a step to the conventional serial dilution. Instead of plating the dilution directly, it is added to a warm emulsion of dilute nutrient agar and living bacteria, then poured onto nutrient agar plates to produce an *agar overlay*. After incubation, the bacterial growth covers the entire plate but individual cellular movement has been restricted by the soft agar overlay. Conversely, the smaller bacteriophages are able to freely diffuse through the soft agar and infect the immobilized bacterial cells.

If between 30 and 300 plaques are formed on the plate, it is countable (Figs. 6-7 through 6-9). In addition, each plaque formed is assumed to have originated from a single virus or PFU (*plaque forming unit*). The number of plaques counted is then used to calculate the PFU titer (density) in the original sample using the following formula:

$$\text{Original phage density} = \frac{\text{\# Plaques counted}}{\text{(Volume plated) (Dilution factor)}}$$

For example, if 150 plaques are counted on a plate which received 0.1 mL of a 10^{-4} phage dilution, calculations would be as follows (substituting "PFU" for "plaques counted"):

$$\text{Original phage density} = \frac{\text{\# PFU}}{\text{(Volume plated) (Dilution factor)}}$$

$$\text{Original phage density} = \frac{150\ \text{PFU}}{(0.1\ \text{mL})\ (10^{-4})}$$

$$\text{Original phage density} = 1.5 \times 10^{7}\ \text{PFU/mL}$$

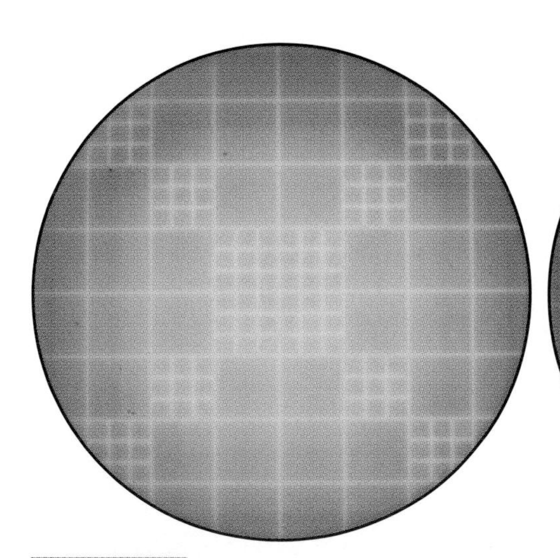

FIGURE 6-7 An uncountable plaque assay plate has less than 30 plaques and is scored as TFTC ("too few to count"). Less than 30 plaques is not a statistically sound sample size.

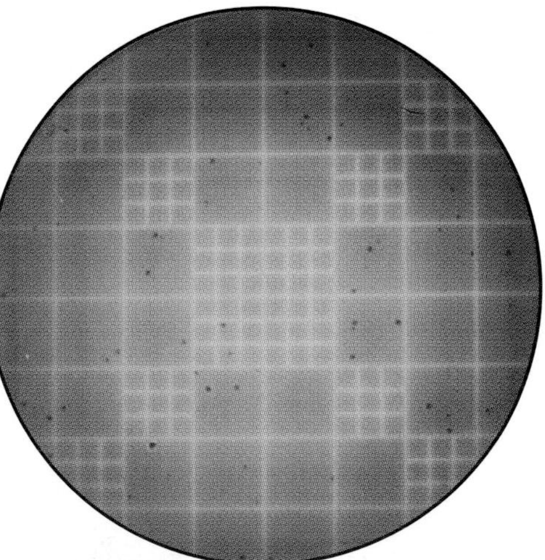

FIGURE 6-8 This plaque assay plate is countable, since it has between 30 and 300 plaques.

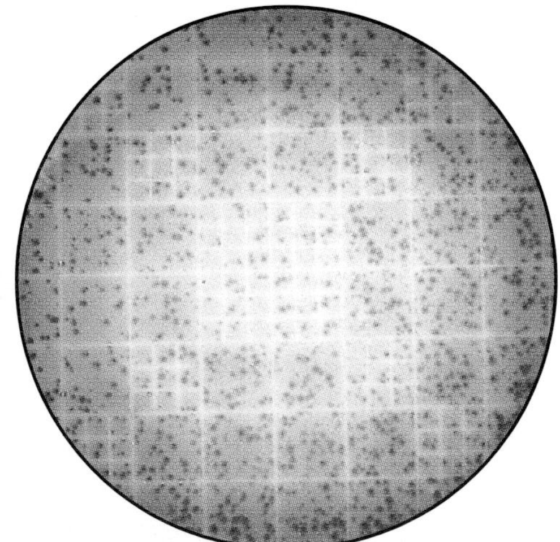

FIGURE 6-9 A plaque assay plate with greater than 300 plaques is TMTC, ("too many to count). More than 300 plaques is difficult to count accurately.

SEMIQUANTITATIVE STREAK OF A URINE SPECIMEN

PURPOSE The urine streak plate method is a fast and easy method for semiquantitatively determining the concentration of bacteria in a urine sample. Based on their differing growth characteristics the number of different species can also be determined.

PRINCIPLE The approximate concentration of bacteria in a urine sample can be determined by streaking a known volume of urine onto an agar plate, incubating, and counting the colonies that appear. This is routinely done in clinical settings using a volumetric loop. A volumetric loop is an inoculating loop calibrated to hold a specific volume of liquid. Both 0.001 mL and 0.01 mL sizes are available.

In this procedure a loopful of urine is carefully transferred to the agar plate by making a single streak across its diameter. The plate is then turned 90° and (without flaming the loop) streaked across the original line in order to evenly disperse the bacteria over the entire plate (Fig. 6-10). (An alternate method, shown in Figure 6-10, adds a third multiple streak at right angles to the second streak.) Following a period of incubation the resulting colonies are counted.

Bacterial concentration, recorded in colony forming units per milliliter (CFU/mL), can be calculated simply by counting the colonies on the plate and multiplying by 1000 if using a 0.001 mL loop or by 100 if using a 0.01 mL loop. The plate pictured in Figure 6-11 was inoculated with a 0.01 mL volumetric loop and it contains approximately 75 colonies. Therefore, the CFU concentration is 7.5×10^3 CFU/mL.

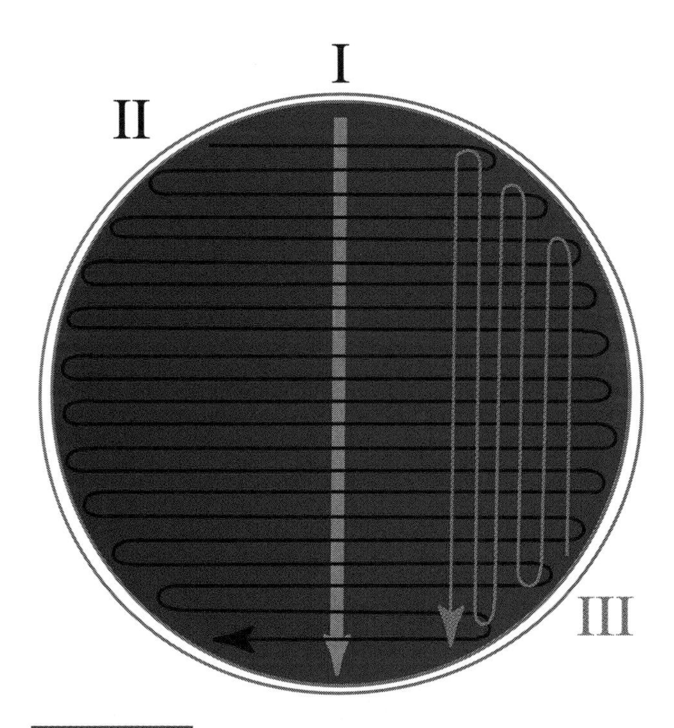

FIGURE 6-10 Semiquantitative streak method. Streak 1 is a simple streak line across the diameter of the plate. Streak 2 is a multiple streak at right angles to the first streak. As an alternate method, a third multiple streak may be added at right angles to the second streak.

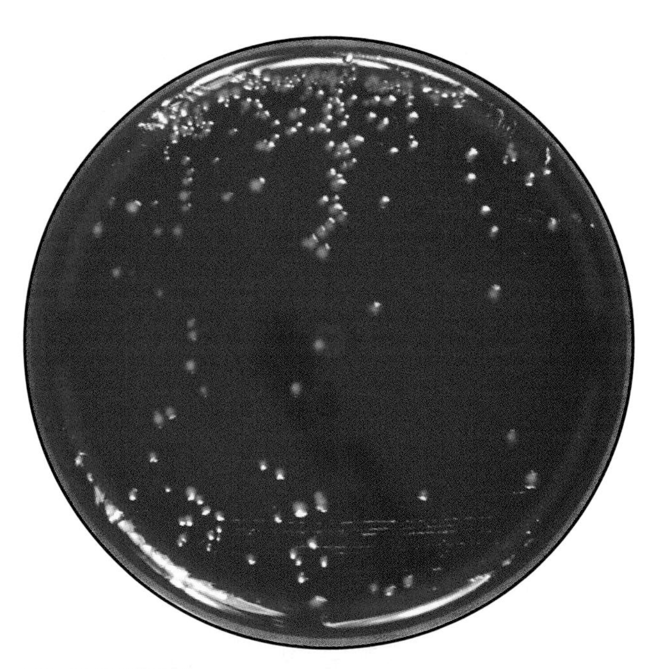

FIGURE 6-11 Urine streak on a sheep blood agar plate inoculated with a 0.01 mL volumetric loop.

Medical, Food, and Environmental Microbiology

7

AMES TEST

PURPOSE Many substances that are mutagenic to bacteria are also carcinogenic to higher animals. The Ames test is a rapid, inexpensive means of using specific bacteria to evaluate the mutagenic properties of potential carcinogens.

PRINCIPLE Bacteria that are able to enzymatically synthesize all of their necessary growth factors are called *prototrophs*. Bacteria that require the addition of growth factors (*e.g.*, vitamins, minerals, amino acids, purines, or pyrimidines) are called *auxotrophs*. Auxotrophs are typically created when mutations occur in a prototroph's genetic material.

The Ames test employs mutant strains of *Salmonella typhimurium* which have lost their ability to synthesize histidine. Some histidine auxotrophs are *frameshift* mutants. That is, they are missing one or have an extra nucleotide in the DNA sequence which would otherwise code for an enzyme necessary for histidine production. They are also missing the DNA repair mechanism which could correct the genetic error. Other strains are *substitution mutants,* in which one nucleotide in the histidine gene has been replaced, resulting in a faulty gene product.

The Ames test determines the ability of chemical agents to cause a reversal (*back-mutation*) of these mutant conditions to the original prototrophic state. In this test a small amount of the histidine auxotroph is spread onto *minimal agar plates* which contain only glucose, a few salts, and a trace of histidine. The organism typically exhibits faint growth, but does not develop into full-size colonies due to the rapid exhaustion of the histidine.

When a filter paper disc saturated with a suspected mutagen is placed in the middle of the inoculated minimal agar plate, the substance will diffuse outward into the medium. If it is mutagenic, it will cause back-mutation in some cells (converting them to histidine prototrophs) which freely grow into full-size colonies. (Note: histidine is initially included in the medium to allow the auxotrophs to grow for several generations and expose them to the effects of the mutagen.)

Several variations of the Ames test are possible. This example uses two minimal agar plates and two complete agar plates. The complete medium contains all of the nutrients necessary for unrestricted growth. Each plate is inoculated with histidine auxotroph. One minimal plate and one complete agar plate each receive a filter paper disk saturated with the test substance. The second minimal agar plate and complete agar plate each receive a filter paper disk saturated with Dimethyl Sulfoxide, DMSO (a substance known to be nonmutagenic).

The minimal agar plate containing the test substance determines the mutagenicity (by the appearance of back-mutations) of the test substance. Depending on the strain used, the type of mutation (either frameshift or base substitution) may also be determined. The minimal agar plate with DMSO serves as a control for the minimal agar/test substance plate by measuring the *spontaneous* back-mutations (natural mutations which occur without the presence of a mutagen). The purpose of the complete agar plate containing the test substance is a control to evaluate the toxicity of the substance. The creation of a *zone of inhibition* around the disk indicates toxicity. The more toxic the substance is, the larger the zone will be. If the substance is toxic, there may be no indication of mutagenicity because the cells are killed before they can back mutate. Finally, the complete agar plate containing DMSO serves as a control for comparison to the growth and zone of inhibition on the complete agar/test substance plate (Figs. 7-1 through 7-4).

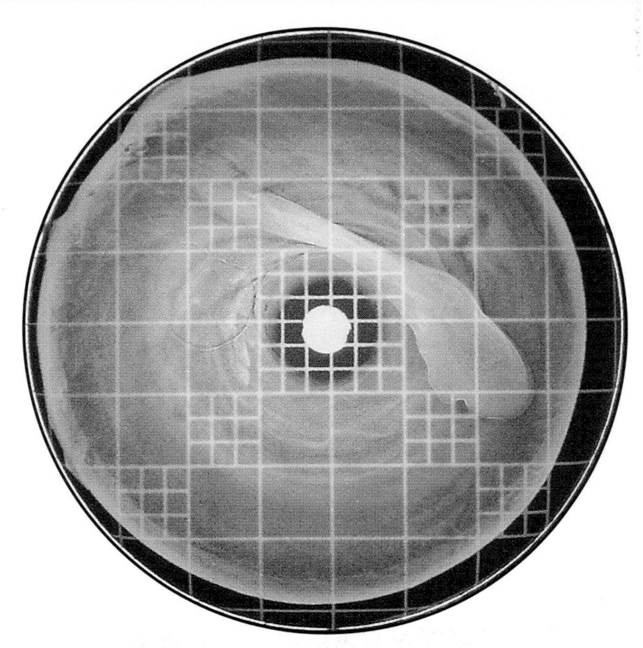

FIGURE 7-1 *Salmonella* histidine auxotroph grown on complete agar (containing histidine) and exposed to the test substance. The zone of inhibition around the paper disk indicates toxicity of the substance.

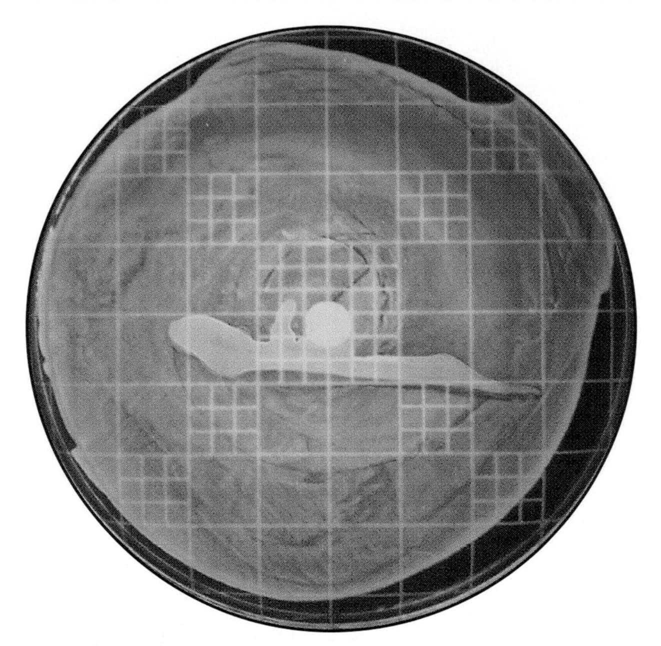

FIGURE 7-2 *Salmonella* histidine auxotroph grown on complete agar (containing histidine) and exposed to an inert substance (DMSO). This plate serves as a control for comparison of zone size with the plate in Figure 7-1. The absence of a zone around the disk indicates that neither the paper disk nor the DMSO are toxic or inhibitory.

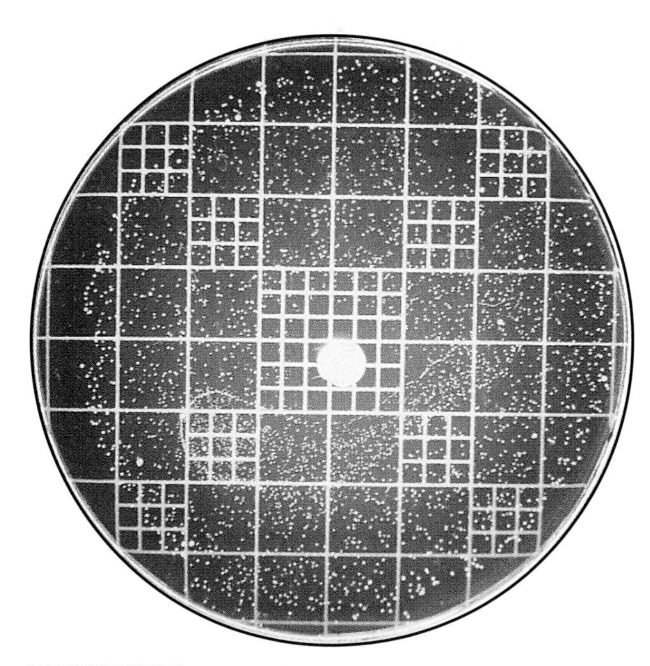

FIGURE 7-3 *Salmonella* histidine auxotroph grown on histidine minimal agar and exposed to a suspected mutagen. A zone of inhibition is visible on this plate, but is not as well-defined as on complete agar because the growth is not as dense. The number of colonies suggests that back-mutations have occurred, however, it is not known at this time whether or not they are due to the effects of the test substance. Therefore, comparison must be made with the plate in Figure 7-4.

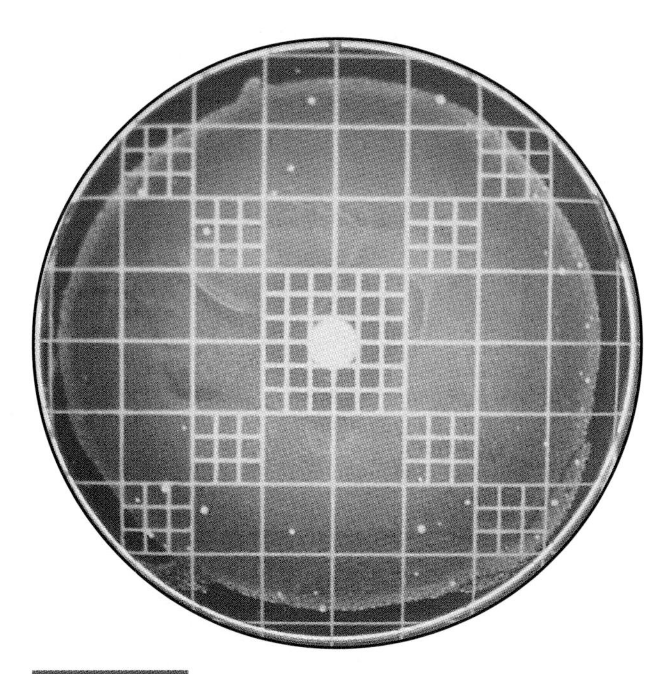

FIGURE 7-4 *Salmonella* histidine auxotroph grown on histidine minimal agar and exposed to a known nonmutagen (DMSO). This plate is a control to demonstrate spontaneous mutation for comparison with an identical plate containing test substance in Figure 7-3. The faint background growth is due to the small amount of histidine in the medium necessary to promote initial growth.

ULTRAVIOLET RADIATION: ITS CHARACTERISTICS AND EFFECTS ON BACTERIAL CELLS

PURPOSE This is a lab demonstration to illustrate the effects of ultraviolet radiation on bacterial cells. It also allows observation of UV radiation's ability to penetrate cells and other matter. Finally, it illustrates the ability of bacterial cells to repair UV damage.

PRINCIPLE Ultraviolet radiation is part of the electromagnetic spectrum, but with shorter, higher energy wavelengths than visible light. Because DNA absorbs UV radiation, prolonged exposure can be lethal to cells. Studies have shown that when DNA absorbs UV radiation, the energy is used to break the bonds between the strands and form new covalent bonds between adjacent pyrimidines: either cytosine-cytosine, cytosine-thymine, or more typically, thymine-thymine (*thymine dimers*).

Thymine dimers (Fig. 7-5) distort the DNA molecule and are incapable of base-pairing during DNA replication. The probable result (15/16 or 94% probability) is a complementary (second generation) strand that has nucleotides other than the two correct adenines. The third generation strand (which should be identical to the original) is going to be complementary to the second generation strand — including the randomly inserted nucleotides. This means a 94% probability of *not* having two thymines in the same position as the dimer of the original strand. *Both* strands end up different from the original where the dimer formed after only two replications. If enough genes possess thymine dimers and the subsequent mistakes, faulty gene products will be made and the result can be death for the cell.

Many bacteria have mechanisms to repair such DNA damage. Some have what is called *light repair,* in which the repair enzymes are activated by light (which includes visible light and UV radiation). The repair consists of removing the thymine dimer and replacing it with two other thymines. A second repair mechanism, *excision repair* or *dark repair,* involves a number of enzymes (Fig. 7-6). The thymine dimer distorts the sugar-phosphate backbone of the strand. This is detected by an endonuclease that breaks a bond in the sugar-phosphate backbone several nucleotides away from the damage. An exonuclease removes nucleotides (including the dimer), leaving a single strand for that DNA segment. DNA polymerase inserts the appropriate complementary nucleotides in a 5' to 3' direction to make the molecule double stranded again. Finally, DNA ligase closes the gap between the last nucleotide of the new segment and the first nucleotide of the old DNA, and the repair is complete.

This demonstration involves four plates inoculated with *Serratia marcescens* by the spread plate technique. Two are exposed to the UV for 30 seconds, without their lids but covered by a tag board cutout (mask). The other two plates are exposed for 3 minutes, one with its lid on, the other with its lid off (Fig. 7-7). One of the 30 second plates and the two 3 minute plates are incubated for 24 to 48 hours where sunlight can reach them. These three plates demonstrate the ability (or inability) of UV radiation to penetrate tag board and the plastic lid, and the consequences of different exposure times for the bacteria not covered by the mask. The other 30 second plate is incubated in the dark for the same time period.

The results are shown in Figure 7-7. After incubation, you should see clearing in the center of the plates shielded by a mask but without a lid. Those parts of the plate covered by the mask should show confluent growth. Any colonies growing in the cleared zone have undergone light or dark repair and have survived the exposure. The plate with the lid and the mask should show confluent growth over the entire surface since UV doesn't penetrate plastic well. The plate incubated in the dark should have a pattern similar to the other two uncovered plates, but with colonies which survived only because of dark repair.

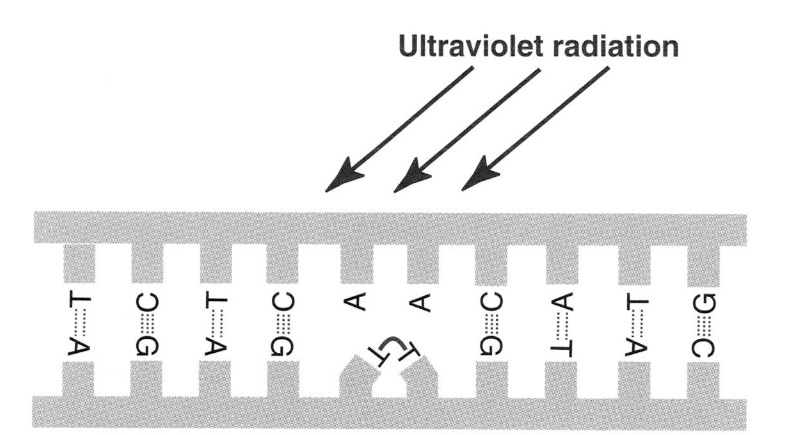

FIGURE 7-5 A thymine dimer in one strand of DNA. During replication, the other strand is not likely to receive the correct adenines, so it now has a mutation. When that mutated strand acts as a template during replication, base pairing will proceed normally, and the complementary bases to whatever nucleotides are present will be inserted. This mutation of both strands may result in a nonfunctional gene product, and if enough thymine dimers occur, death for the cell.

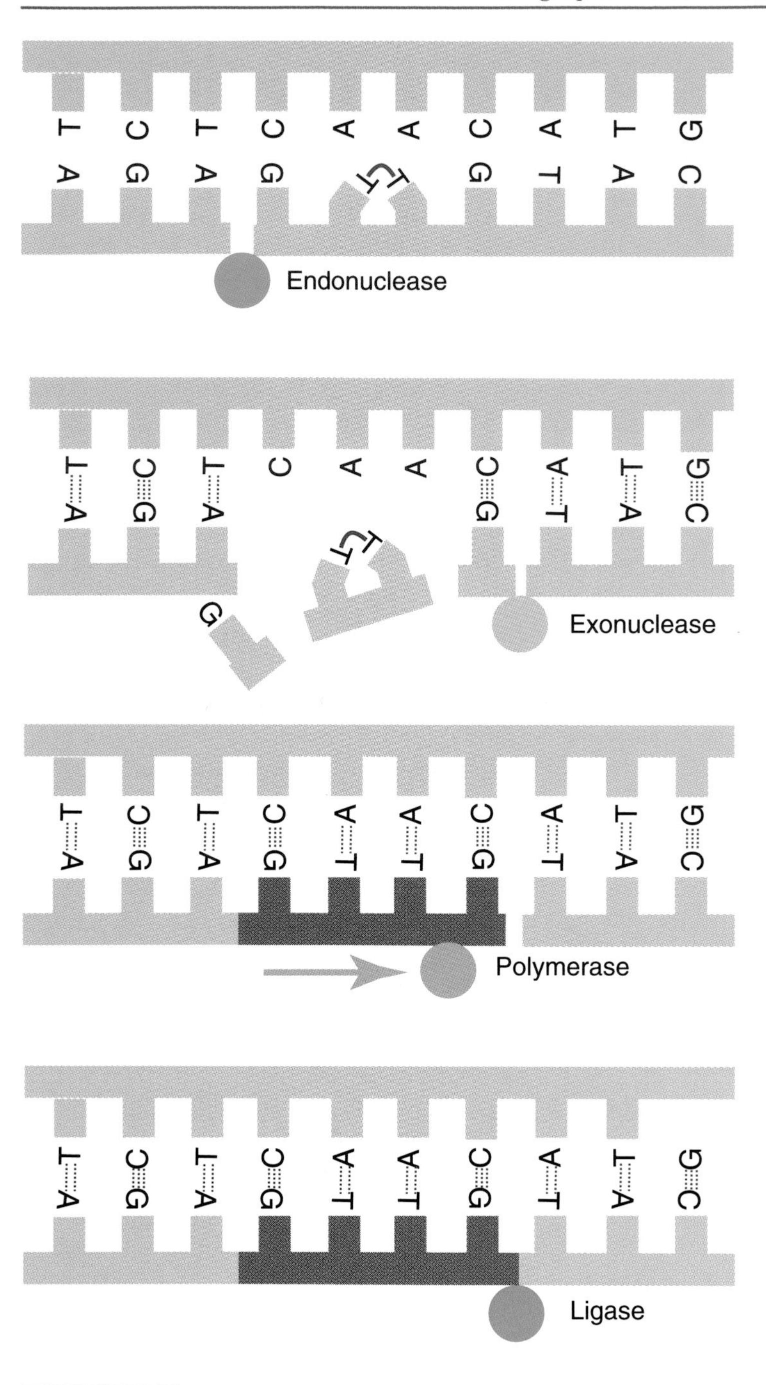

FIGURE 7-6 Excision or dark repair. In the sequence, four enzymes are used: An endonuclease to break a covalent bond in the sugar-phosphate backbone of the damaged strand, an exonuclease to remove the nucleotides in the damaged segment, DNA polymerase to synthesize a new strand, and ligase to form a covalent bond between the new and the original strands.

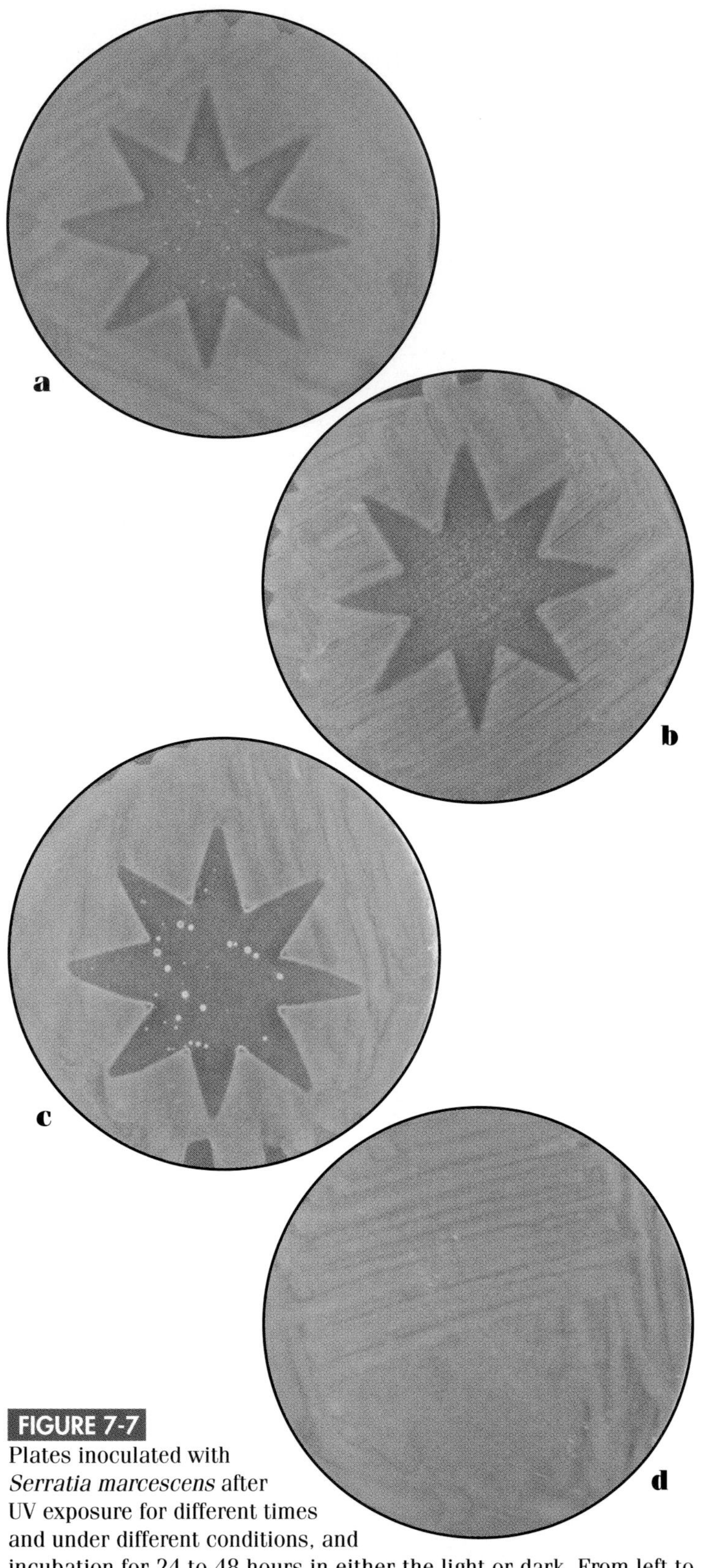

FIGURE 7-7 Plates inoculated with *Serratia marcescens* after UV exposure for different times and under different conditions, and incubation for 24 to 48 hours in either the light or dark. From left to right: (*a*) 3 minute exposure with a mask, no lid, and incubation in the light; (*b*) 30 second exposure under the same conditions; (*c*) 30 second exposure under the same conditions except for incubation in the dark; (*d*) 3 minute exposure with a mask and a lid, and incubation in the light. Colonies inside the pattern have undergone DNA repair. Regions of confluent growth were protected from irradiation by the mask or the mask and the plastic lid.

ANTIBIOTIC SENSITIVITY (KIRBY-BAUER METHOD)

PURPOSE Antibiotic sensitivity testing is used to determine the susceptibility of bacteria to various antibiotics. This standardized testing method is typically used to measure the effectiveness of a variety of antibiotics on a specific organism in order to prescribe the most suitable antibiotic therapy.

PRINCIPLE In the Kirby-Bauer test a series of antibiotic impregnated paper disks is placed on a plate inoculated to form a bacterial lawn (Fig. 7-8 and 7-9). The plates are incubated to allow growth of the bacteria and time for the antibiotics to diffuse into the agar. If an organism is susceptible to an antibiotic a clear zone will appear around the disk where the growth of that organism has been inhibited. The size of this *zone of inhibition* depends upon the sensitivity of the bacteria to the specific antibiotic and the antibiotic's ability to diffuse from the disk through the agar.

All aspects of the Kirby-Bauer procedure are standardized to ensure reliable results. Therefore, care must be taken in preparation to adhere to these standards. The medium used, Mueller-Hinton agar, is formulated to have a pH between 7.2 and 7.4 and is poured into petri dishes to a depth of 4 mm. Inoculation is made with a broth culture diluted to match a 0.5 McFarland turbidity standard (Fig. 7-10).

The antibiotic disks used for this test are also standardized to contain a specific amount of antibiotic (printed on the disk). The disks are dispensed onto the inoculated plate and incubated at 35°C. After 18 hours incubation the plates are removed and the clear zones measured (Figs. 7-8 and 7-9). Table 7-1 shows the standard interpretations of various antibiotic zones.

Normally, the zones on an agar plate will be distinct and separate halos around each antibiotic disk. On occasion, however, two zones will appear to join, producing an area of clearing between them extending beyond the perimeters of the otherwise circular zones (Fig. 7-11). This is due to the *synergistic effect* of the two antibiotics. In other words, this is the area where the concentration of each antibiotic is too low to be effective by itself, but in combination with the other antibiotic, has sufficient strength to kill the bacteria.

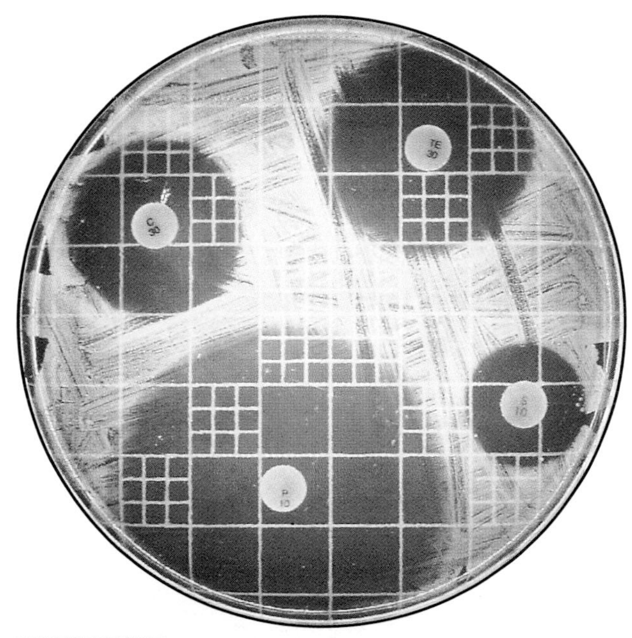

FIGURE 7-8 The Kirby-Bauer test illustrating the effect of (clockwise, from left) chloramphenicol (C30), tetracycline (TE30), streptomycin (S10), and penicillin (P10) on Gram-positive *Staphylococcus aureus*. The number on each disk indicates the amount of antibiotic in µg. As an example of how this test is used, the zone diameter for penicillin is approximately 50 mm. According to Table 7-1, the minimum zone for susceptibility is ≥29 mm. Penicillin would be an effective antibiotic for use against this organism.

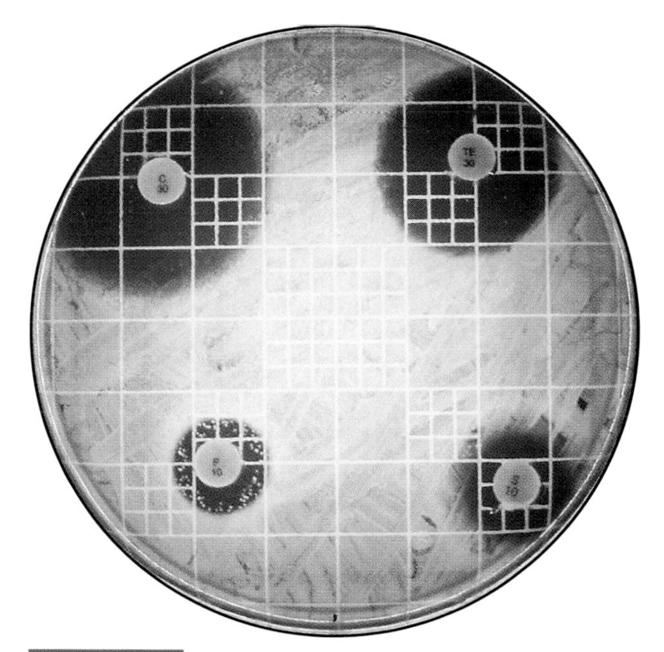

FIGURE 7-9 The Kirby-Bauer test illustrating the effect of the same antibiotics as in Figure 7-8 on Gram-negative *Escherichia coli*. Compare the zone produced by penicillin on this plate with that in Figure 7-8. Penicillin is generally ineffective against Gram-negative organisms. Note also the penicillin resistant colonies within the zone.

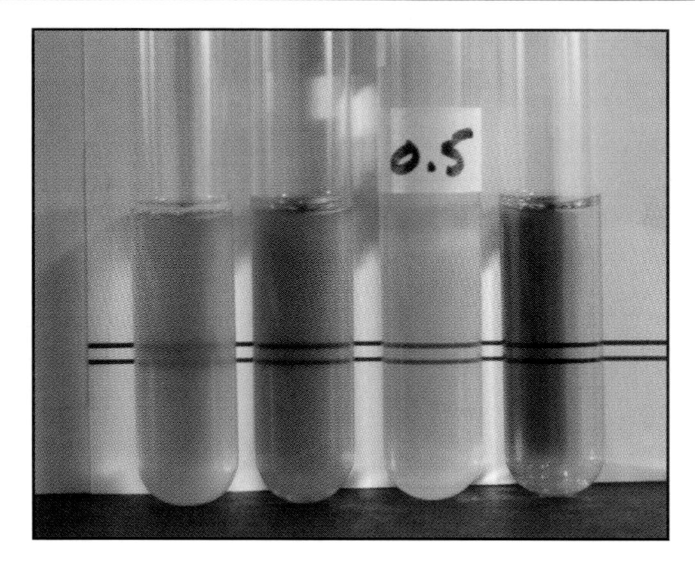

FIGURE 7-10 Comparison of a 0.5 (#5) McFarland turbidity standard to three broths with varying degrees of turbidity. (There are ten McFarland standards, each of which contains a specific mixture of barium sulfate and sulfuric acid. The standards are used as controls by which to compare actual broth cultures. In many procedures (as in the Kirby-Bauer method), the cultures to be inoculated are first diluted to match the turbidity of a specific McFarland standard. Visual comparison is made by placing a card with sharp black lines running horizontally behind the tubes.) As can be seen in the above photo, the clarity of the lines decreases as turbidity increases.

TABLE 7-1 **Zone Diameter Interpretive Chart for Selected Antibiotics Based on Data Provided by the National Committee for Clinical Laboratory Standards (NCCLS)**

Antimicrobial Agent	Code	Disc Potency	Zone Diameter Interpretive Standards (mm)		
			Resistant	Intermediate	Susceptible
Chloramphenicol .(for non-*Haemophilus* species)	C-30	30 µg	≤12	13–17	≥18
Penicillin (for staphylococci)	P-10	10 U	≤28		≥29
Penicillin (for enterococci)	P-10	10 U	≤14		≥15
Penicillin (for enterococcal streptococci)	P-10	10 U	≤19	20–27	≥28
Streptomycin	S-10	10 µg	≤11	12–14	≥15
Tetracycline (for most organisms)	TE-30	30 µg	≤14	15–18	≥19
Trimethoprim	TMP-5	5 µg	≤10	11–15	≥16

Permission to use portions of M100-S5 (Performance Standards for Antimicrobial Susceptibility Testing; Fifth Informational Supplement) has been granted by NCCLS. The interpretive data are valid only if the methodology in M2-A5 (Performance Standards for Antimicrobial Disk Susceptibility Tests — Fifth Edition; Approved Standard) is followed. NCCLS frequently updates the interpretive tables through new editions of the standard and supplements. Users should refer to the most recent editions. The current standard may be obtained from NCCLS, 940 West Valley Road, Suite 1400, Wayne, PA 19087.

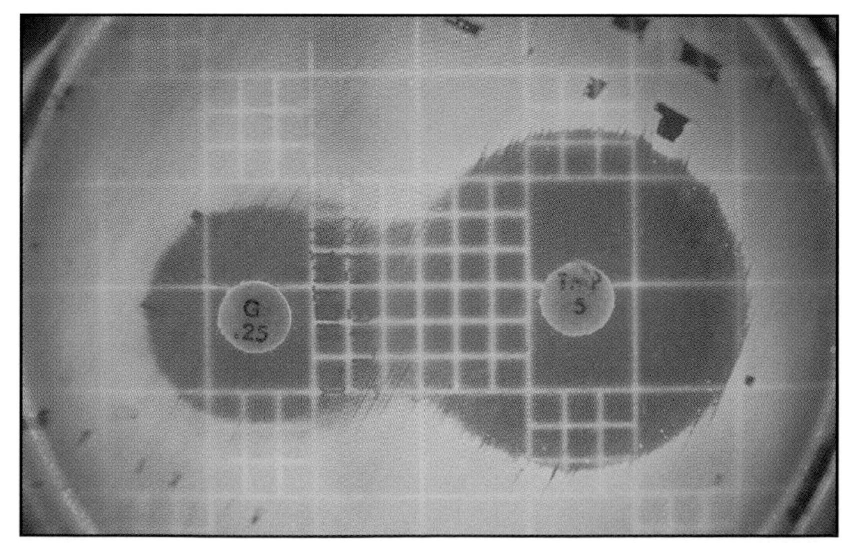

FIGURE 7-11 Synergism between the antibiotics Sulfisoxazole (G–.25) and Trimethoprim (TMP–5).

COLILERT®*METHOD FOR TESTING DRINKING WATER

PURPOSE The Colilert® test is a commercial preparation that examines drinking water for the presence of total coliforms generally and *Escherichia coli* specifically.

PRINCIPAL Colilert® reagent contains nutrients and salts which favor the growth of coliforms and inhibit the growth of noncoliforms. It also contains the indicator nutrients, o-Nitrophenyl-β-D-Galactopyranoside (ONPG) and 4-Methylumbelliferyl-β-D-Glucuronide (MUG). The test is conducted by adding Colilert® reagent to a nonfluorescing bottle containing a 100 mL sample of drinking water. The sample is incubated at 35°C for 24–28 hours and then compared to the control.

A positive ONPG result is seen as formation of a yellow color (Fig. 7-12). If the test bottle is as yellow or more yellow than the control, the first portion of the test is positive and reported as "total coliforms present." A positive MUG result produces fluorescence when viewed under a U.V. lamp (Fig. 7-13). If the test bottle fluorescence is equal to or greater than that of the control, the presence of *E. coli* has been confirmed. This portion of the test is reported as "*E. coli* present." A sample that is not as yellow and does not fluoresce is considered negative and is reported as "total coliforms absent" and "*E. coli* absent."

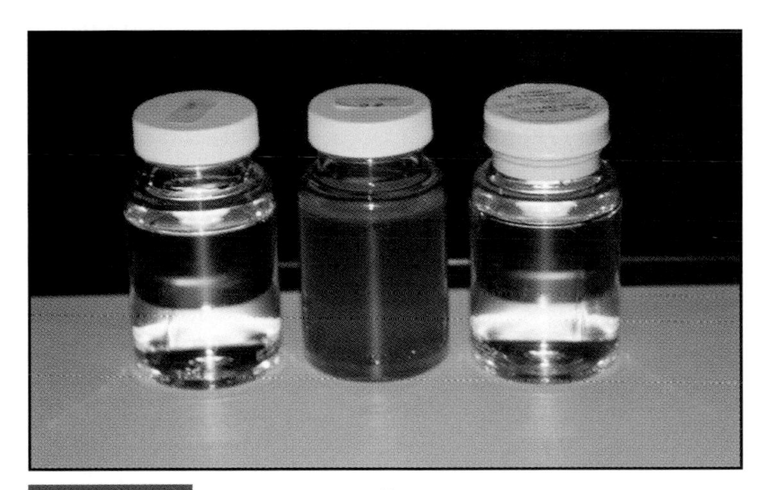

FIGURE 7-12 This is a Colilert® ONPG (o-Nitrophenyl-β-D-Galactopyranoside) test for total coliforms. The bottle on the left is negative; the bottle in the center is positive; the control (comparator) is on the right.

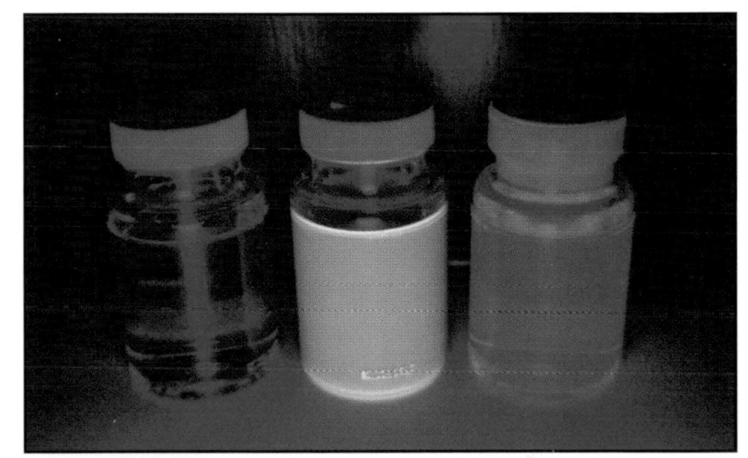

FIGURE 7-13 This is a Colilert® MUG (4-Methylumbelliferyl-β-D-Glucuronide) test for *E. coli* seen under a UV Lamp. The bottle on the left is negative; the bottle in the center is positive; the control (comparator) is on the right.

*Colilert® is a registered trademark of IDEXX Laboratories, Inc.

MEMBRANE FILTER TECHNIQUE

PURPOSE In this country, public drinking water is tested and treated daily to maintain potability and assure public safety. The membrane filter technique is commonly used in combination with other tests to identify the presence of fecal coliforms.

PRINCIPLE Fecal contamination is a common pollutant in open water and a potential source of serious disease causing organisms. Because these pathogens are usually in very low concentration in water and can be fairly short-lived, the membrane filter technique tests for the presence of the much more abundant coliform bacteria. The coliforms are defined as any member of *Enterobacteriaceae* which produces acid and gas from the fermentation of lactose within 48 hours at 35°C. The presence of coliform bacteria is viewed as an indication of the possible presence of fecally transmitted pathogens.

Eosin methylene blue (EMB) agar contains peptone, lactose, and the dyes eosin Y and methylene blue. The purpose of the dyes is twofold: 1) they inhibit the growth of Gram-positive organisms and, 2) they react with each other under acidic conditions to form a dark purple complex usually accompanied by a green metallic sheen. This green metallic sheen serves as an indicator of the vigorous lactose fermentation typical of fecal coliforms (Fig. 2-7).

In the EMB membrane filter technique, a water sample suspected of containing contaminants is drawn through a sterile membrane filter. The filter, a special micropore membrane designed to trap any microorganisms larger than 0.45 mm, is then placed onto an EMB plate. After a 24 hour incubation at 35°C the plate is removed and the colonies which produce the characteristic green metallic sheen are counted (Fig. 7-14). An acceptable plate contains between 20 and 80 coliform colonies with a total colony count no larger than 200. To fall within this range it is sometimes necessary to filter larger volumes of water, or to dilute heavily polluted samples.

The number of colonies counted is converted to "coliorm colonies/100mL" using the following formula:

$$\frac{\text{coliform colonies}}{100\text{ mL}} = \frac{\text{coliform colonies} \times 100\text{ mL}}{\text{mL } \textit{original} \text{ sample filtered}}$$

In the above formula, it is important to divide by the volume of *original* sample filtered to account for any dilutions performed. More precise results can be achieved by performing this test several times and taking an average of the counts. Water must contain less than 1 coliform/100 mL to be *potable*.

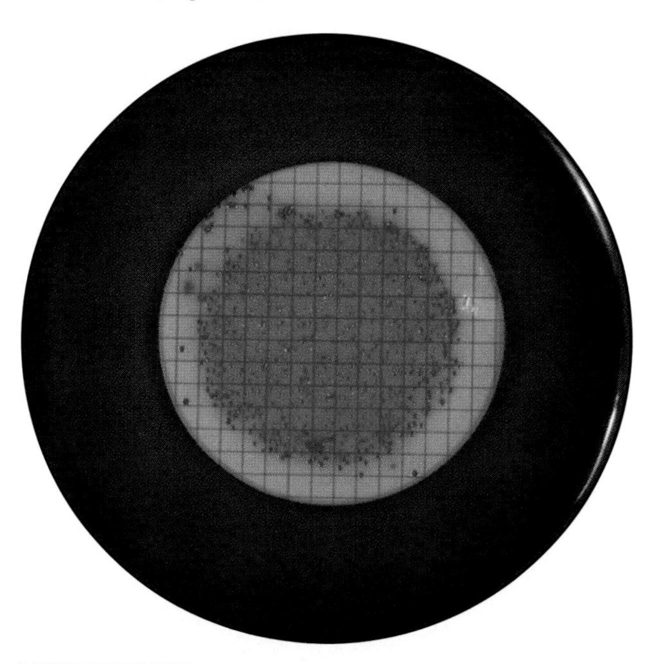

FIGURE 7-14 A membrane filter on eosin methylene blue (EMB) medium. Note the characteristic dark colonies with green metallic sheen which indicate that this water sample has been contaminated with fecal coliforms. Potable water has fewer than one coliform per 100 mL of sample tested.

MPN (MOST PROBABLE NUMBER) METHOD FOR TOTAL COLIFORM DETERMINATION

PURPOSE This standardized test, also called the "Multiple Tube Fermentation Technique," published by the American Public Health Association is used to measure coliform density in water. It may be used to calculate the density of all coliforms present (*total coliforms*) or to calculate the density of *Escherichia coli* specifically. Although a sea water sample suspected of sewage contamination is used in this example, water from any source can be tested using the MPN method.

PRINCIPAL The presence of coliform bacteria in a water sample suggests possible presence of fecally transmitted pathogens. Therefore, most water testing techniques are designed to measure the presence of fecal coliforms. A coliform is defined as any member of *Enterobacteriaceae* which produces acid and gas from lactose fermentation within 48 hours at 35°C.

Three media are used in this procedure — lauryl tryptose broth (LTB), brilliant green lactose bile (BGLB) broth, and EC (*E. coli*) broth. LTB, which includes lactose and lauryl sulfate, is selective for the coliform group. Because it does not screen out *all* noncoliforms it is used to *presumptively* determine the presence or absence of total coliforms. BGLB broth, which includes lactose and 2% bile, inhibits noncoliforms and is used to confirm the presence of total coliforms. EC broth, which includes lactose and bile salts, is selective for *E. coli* when incubated at 45.5°C.

Each tube used in this procedure contains 10 mL of broth and an inverted Durham tube to trap any gas produced by fermentation. The LTB tubes are arranged in six groups of five tubes as shown in Figure 7-15. (Less polluted water samples may be tested with only three groups of tubes.) Each tube in the first set of five receives 1.0 mL of the original sample. Each tube in the remaining five groups receive 1.0 mL of a specific dilution of the sample. The dilutions, from 10^{-1} through 10^{-5}, are added to groups 1 through 5 respectively (*i.e.*, each tube in group one receives 1.0 mL of 10^{-1}, each tube in group two receives 1.0 mL of 10^{-2}, group three 10^{-3}, *etc.*). For an explanation of dilutions see "Viable Count" in Section 6.

The LTB tubes are incubated at 35˚C for up to 48 hours and examined for gas production (Fig. 7-16). Any positive LTB tubes are then subcultured to BGLB tubes using an inoculating loop or applicator stick. (*The BGLB tubes are clearly labeled with the LTB dilution from which they are inoculated.* This information is later used to calculate the total coliform count.) Again, the cultures are allowed to grow for up to 48 hours at 35°C and examined for gas production (Fig. 7-17). Positive BGLB cultures are then transferred to EC broth (again clearly labeled) and incubated at 45.5°C for up to 48 hours (Fig. 7-18). EC tubes which show gas production are counted and the MPN calculated. (Note: The preceding three fermentation tests may be run in sequence. However, to save time once the presumptive (LTB) phase is determined to be positive, the BGLB and EC tests may be run simultaneously.) A negative result in the LTB or BGLB tests suggest the water is not contaminated by coliforms and are cause for terminating the procedure.

In Table 7-2 shows three examples of possible combinations of positive and negative tubes. In each case only three dilutions are used for calculation — the highest dilution (most dilute sample) in which *all five* tubes are positive and the next two higher dilutions (sample #1). When no group has five positive tubes, the group with the highest number of positives and the next two higher dilutions are used (sample #2). However, as in the preceding example, once the three dilutions have been chosen, all remaining positives in higher dilutions are counted and added to the third number of positives in the series. If, as in sample #3, only one group is found to contain positives, the three dilutions considered are the one containing the positives, the dilution preceding it and the dilution following it.

The MPN is calculated using the following formula:

$$\text{MPN/100 mL} = \frac{\text{number of positive tubes} \times 100}{\sqrt{\text{mL sample in negative tubes} \times \text{mL sample in all tubes}}}$$

For example, MPN calculation for sample #1 in Table 7-2 would be:

$$\text{MPN/100 mL} = \frac{\text{number of positive tubes} \times 100}{\sqrt{\text{mL sample in negative tubes} \times \text{mL sample in all tubes}}}$$

$$\text{MPN/100 mL} = \frac{9 \times 10}{\sqrt{(.024 \times .555}} = \frac{900}{\sqrt{.013}} = \frac{900}{.114} = 7895$$

It is customary to calculate and report *both* total coliform *and E. coli* densities. Total coliform MPN is calculated using BGLB broth results and *E. coli* MPN is based on EC broth results.

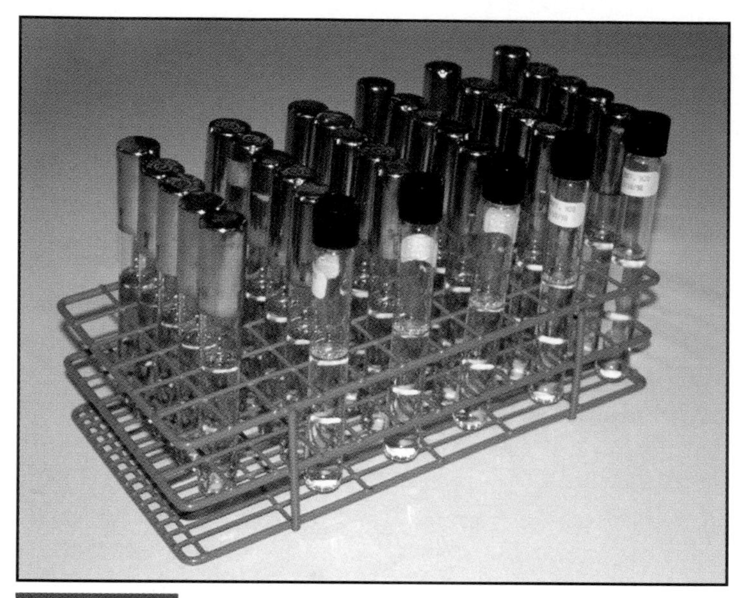

FIGURE 7-15 This is a multiple tube fermentation of a sea water sample contaminated by sewage. The tubes contain lauryl tryptose broth and a measured volume of water sample as described in the text. Following incubation, each positive broth (based on gas production) will be used to inoculate a BGLB broth.

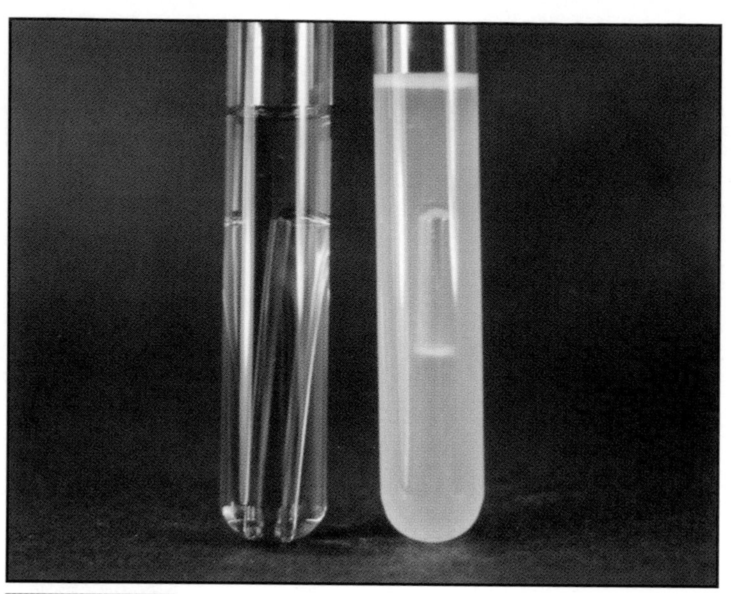

FIGURE 7-16 Lauryl tryptose broth tubes. The bubble in the Durham tube on the right is presumptive evidence of coliform contamination. The tube on the left is negative.

FIGURE 7-17 Brilliant green lactose bile broth tubes. The bubble in the Durham tube on the right is seen as confirmation of coliform contamination. The tube on the left is negative.

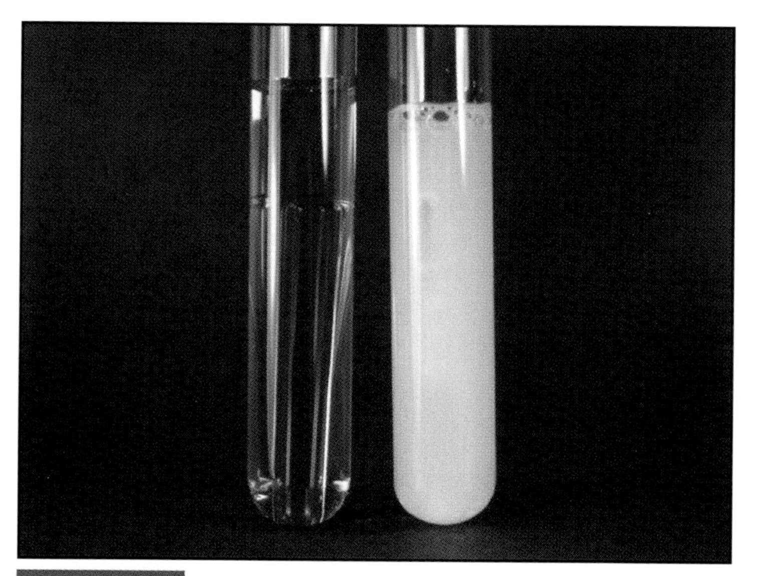

FIGURE 7-18 EC broth tubes. The bubble in the Durham tube on the right is seen as confirmation of *E. coli* contamination. The tube on the left is negative.

TABLE 7-2 Three Examples of Possible Combinations of Positive and Negative Tubes

Sample	1.0 mL (10^0)	0.1 mL (10^{-1})	0.01 mL (10^{-2})	0.001 mL (10^{-3})	0.0001 mL (10^{-4})	0.00001 mL (10^{-5})	Positive tubes in three groups
1	5/5	**5/5**	**3/5**	**1/5**	0/5	0/5	**5–3–1**
2	**4/5**	**1/5**	**1/5**	**1/5**	0/5	0/5	**4–1–2**
3	**0/5**	**1/5**	**0/5**	0/5	0/5	0/5	**0–1–0**

The amounts in the upper row represent the actual amount of *original* sample being added to each tube in the group if one milliliter of the dilution (*i.e.*, 10^{-1}, 10^{-2}, 10^{-3}, etc.) is transferred. As indicated by bold type, only three groups per sample are selected for MPN calculation. The numerator in the fractions under each dilution represents the number of positives in that group; the denominator represents the number of tubes in that group.

METHYLENE BLUE REDUCTASE TEST

PURPOSE This test is helpful in differentiating the enterococci from other members of *Streptococcus*. It also tests for the presence of coliforms in raw milk.

PRINCIPLE Methylene blue is a dye which is blue when oxidized and colorless when reduced. It can be reduced by a reductase enzyme either aerobically or anaerobically. Aerobically, methylene blue can be substituted for any substrate in the electron transport system. It is reduced by cytochromes to leuco-methylene blue which is colorless, but is quickly reoxidized by oxygen. In this reaction hydrogen peroxide is produced but no color change occurs. Anaerobically, methylene blue is converted by nucleotides such as nicotinamide adenine dinucleotide (NADH) or diphosphopyridine nucleotide (DPNH) to leuco-methylene blue and, in the absence of an oxidizing substance (*i.e.*, oxygen), remains colorless.

The reduction of methylene blue may be used as an indicator of milk quality. In the methylene blue reductase test a small quantity of a dilute methylene blue solution is added to a sterilized test tube containing raw milk. The tube is then tightly sealed and incubated in a 35°C water bath. The time it takes the milk to turn from blue to white (due to methylene blue reduction) is a qualitative indicator of the number of microorganisms living in the milk (Fig. 7-19). Good quality milk takes greater than 6 hours to convert the methylene blue. The sample used in Figure 7-19 was purchased from a local health food store and, we are pleased to say, took approximately 20 hours to change from blue to white.

FIGURE 7-19 The methylene blue reductase test. The tube on the left is a control to illustrate the original color of the medium. The tube on the right indicates bacterial reduction of methylene blue after 20 hours. The speed of reduction is related to the concentration of microorganisms present in the milk.

SNYDER TEST

PURPOSE The Snyder test is used to detect the presence of *Lactobacillus* in saliva.

MEDICAL APPLICATION Tooth decay is a phenomenon that practically everyone is familiar with. Although several microorganisms are involved in the process, only species in the genus *Lactobacillus* seem to be capable of lowering the pH enough to dissolve tooth enamel. The Snyder test measures dental *caries* susceptibility by detecting the presence of Lactobacilli in saliva.

PRINCIPLE Snyder Test medium is designed to favor growth of Lactobacilli and discourage growth of most other species. This is accomplished by adjusting the pH of the medium to 4.8 and by adding glucose, a carbohydrate easily fermented by the bacteria. Lactobacilli thrive in the low pH and ferment the glucose, producing more acid which reduces the pH even more. The medium includes the pH indicator, bromcresol green, which is green at pH 4.8 and above, and yellow below pH 4.8.

The medium is autoclaved for sterilization, cooled to just over 40 degrees and maintained in a warm water bath until needed. The still molten agar is then inoculated with 0.2 ml of saliva, mixed well, incubated, and allowed to grow for up to 72 hours. The agar tubes are checked at 24 hour intervals for any change in color. Yellow color indicates that fermentation has taken place and is a positive result (Fig. 7-20). High susceptibility to dental caries is indicated if the medium turns yellow within 24 hours. Moderate and slight susceptibility are indicted by a change within 48 and 72 hours, respectively. No change within 72 hours is considered a negative result.

FIGURE 7-20 The Snyder test for dental caries susceptibility. A positive result is on the left, a negative on the right. The broad fissure in the agar on the left is due to gas production.

Host Defenses

8

DIFFERENTIAL BLOOD CELL COUNT

PURPOSE A differential blood cell count is done to determine approximate numbers of the various *leukocytes* (white blood cells). An excess or a deficiency of all or a particular group is indicative of certain disease states.

PRINCIPLE White blood cells (WBCs) are divided into two groups: *granulocytes* (which have prominent cytoplasmic granules) and *agranulocytes* (which lack these granules). There are three basic types of *polymorphonuclear granulocytes* or *PMNs*: *neutrophils*, *basophils*, and *eosinophils*. The two types of agranulocytes are *monocytes* and *lymphocytes*.

A sample of blood is observed under the microscope and at least 100 white blood cells are counted and tallied (this task is automated now). Approximate normal percentages for each leukocyte are as follows: 55–65% neutrophils (mostly segs–see below), 25–33% lymphocytes, 3–7% monocytes, 1–3% eosinophils and 0.5–1% basophils.

Neutrophils (Figs. 8-1 through 8-3) are the most abundant WBCs in blood. They leave the blood and enter tissues to phagocytize foreign material. An increase in neutrophils is indicative of a systemic bacterial infection. Mature neutrophils are sometimes referred to as *segs* because of their segmented nucleus, usually in two to five lobes. Immature neutrophils lack this segmentation and are referred to as *bands* (Fig. 8-4). This distinction is useful, since a patient with an active infection will be producing more neutrophils, so a higher percentage will be of the band (immature) type. Neutrophils are 12–15μm in diameter.

Eosinophils are phagocytic, and their numbers increase during allergic reactions and parasitic infections (Fig. 8-5). They are 12–15μm in diameter (about twice the size of an RBC) and generally have 2 lobes in their nucleus.

Basophils (Fig. 8-6) are the least abundant WBCs in normal blood. They are structurally similar to tissue mast cells and produce some of the same chemicals (histamine and heparin), but are derived from different stem cells in bone marrow. They are 12–15μm in diameter and have two lobes in their nucleus or an unlobed nucleus.

Agranulocytes include monocytes and lymphocytes. Monocytes (Fig. 8-7) are the blood form of macrophages. They are the largest of leukocytes, being two to three times the size of RBCs (12–20μm). Their nucleus is horseshoe shaped. Lymphocytes (Fig. 8-8) are cells of the immune system. Two functional types of lymphocytes are the T-cell, involved in cell-mediated immunity, and the B-cell, which converts to a *plasma cell* when activated and produces antibodies. Lymphocytes are approximately the same size as RBCs or up to twice their size. The nucleus is usually spherical and takes up most of the cell.

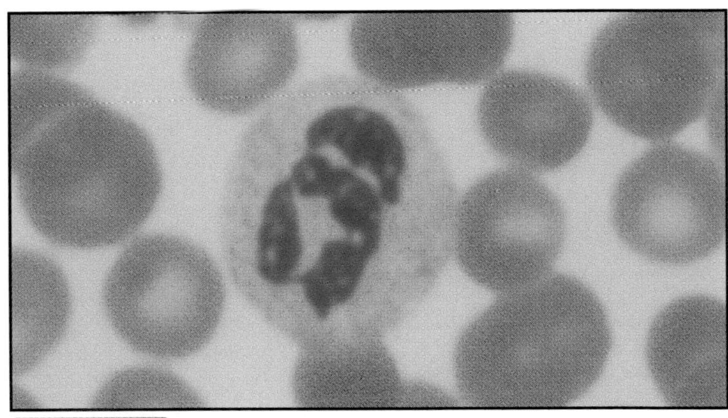

FIGURE 8-1 A segmented neutrophil (Wright's stain, X2376). Note its size relative to the RBCs and the lobed nucleus.

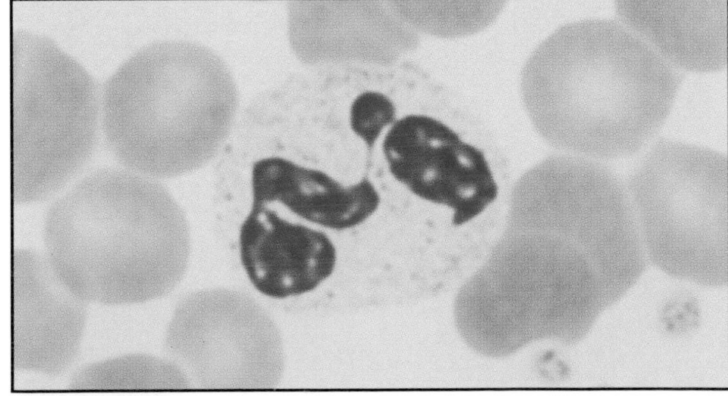

FIGURE 8-2 A segmented neutrophil with four nuclear lobes (Wright's stain, X2640).

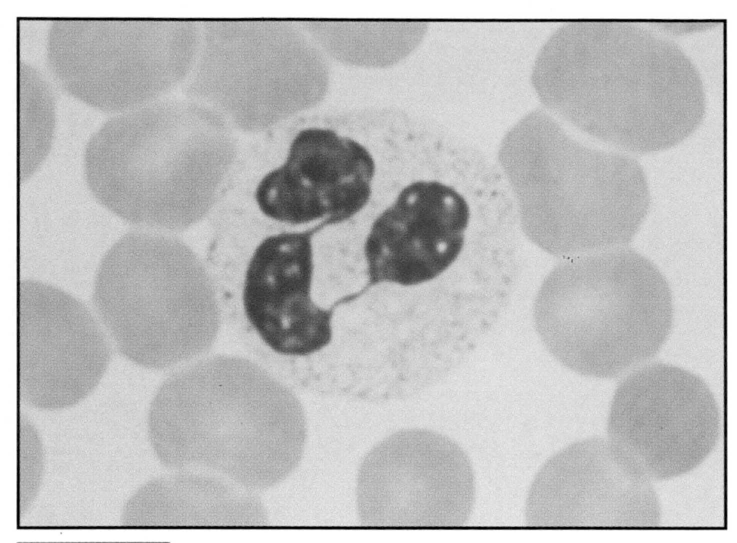

FIGURE 8-3 A segmented neutrophil with three nuclear lobes (Wright's stain, X2640).

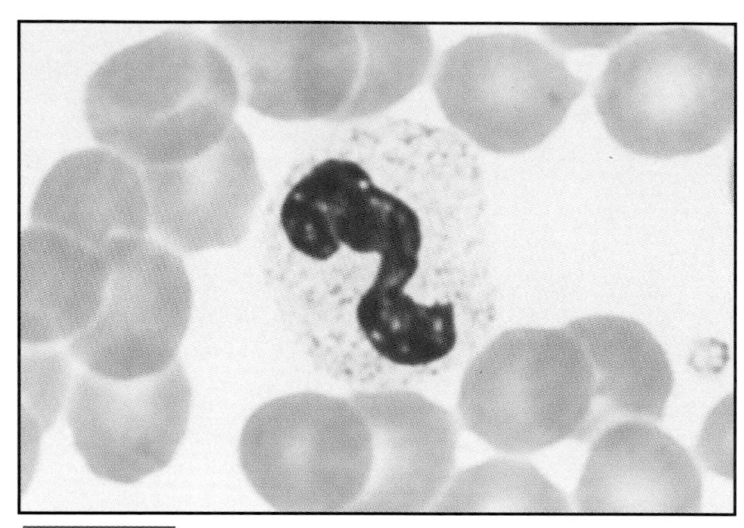

FIGURE 8-4 A band neutrophil (Wright's stain, X2640). Note the lack of nuclear segmentation.

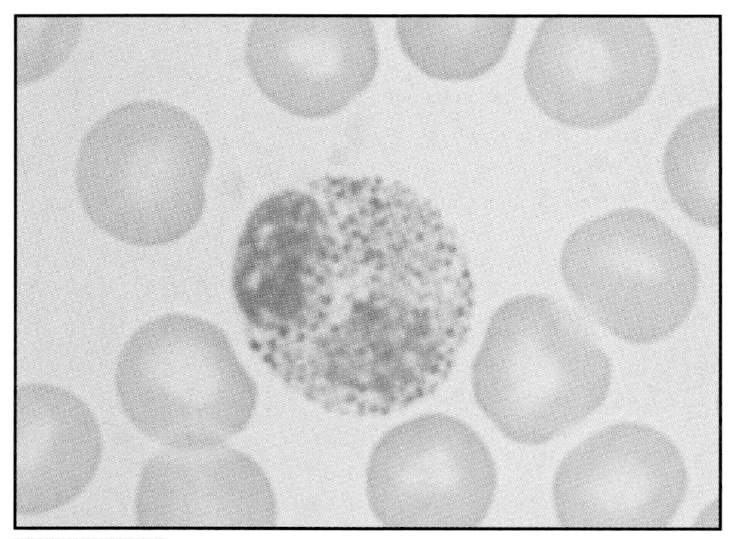

FIGURE 8-5 An eosinophil (Giemsa stain, X2640). Note the cell's size relative to the RBCs and its red cytoplasmic granules.

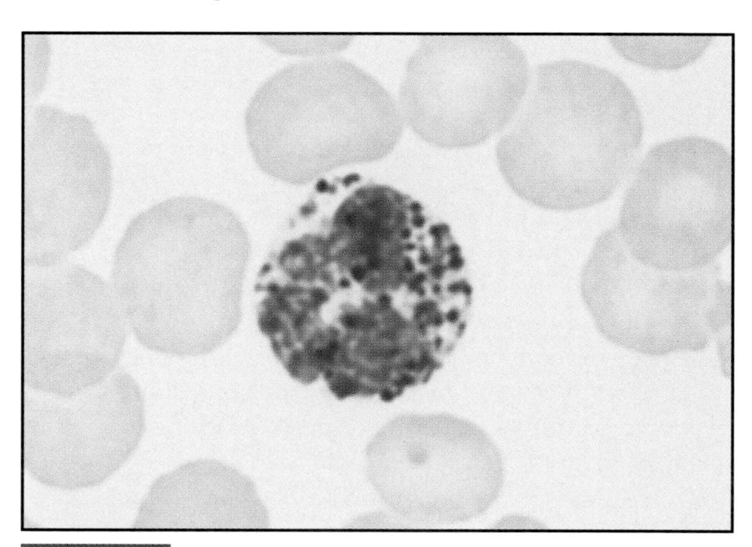

FIGURE 8-6 A basophil (Giemsa stain, X2640). Note the abundant purple cytoplasmic granules that partially obscure the nucleus.

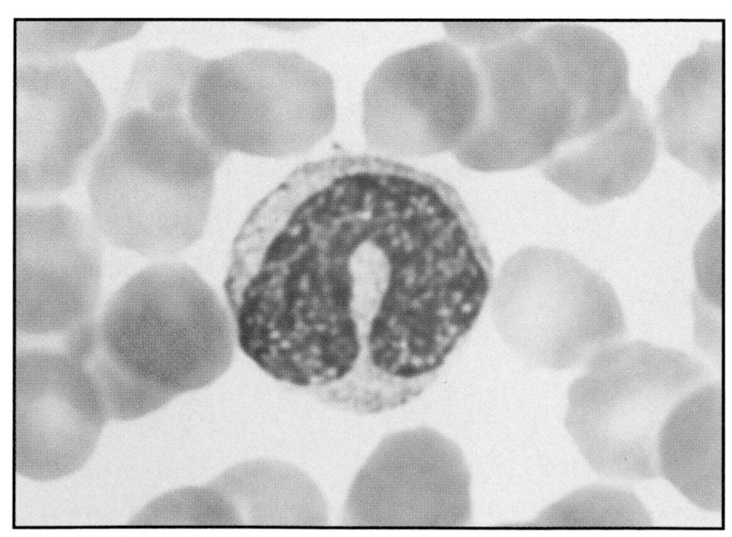

FIGURE 8-7 A monocyte (Wright's stain, X2640). Note its large size and horseshoe-shaped nucleus.

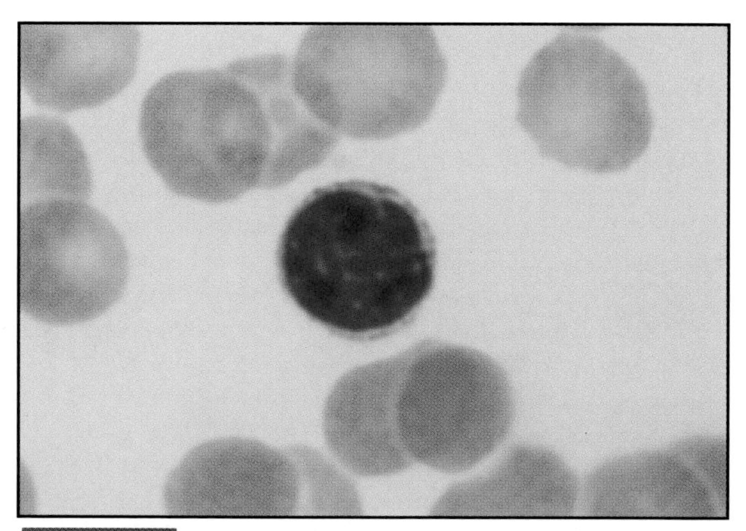

FIGURE 8-8 A lymphocyte (Wright's stain, X2640). Note its size compared to the RBCs in the field and the size of its nucleus relative to cell size. A small percentage of lymphocytes attain a size of up to 18μm.

OTHER IMMUNE CELLS AND ORGANS

Cells of the reticuloendothelial system and immune system are involved in defense of the body and are found scattered throughout the body. Examples are shown in the following photomicrographs (Figs. 8-9 through 8-18).

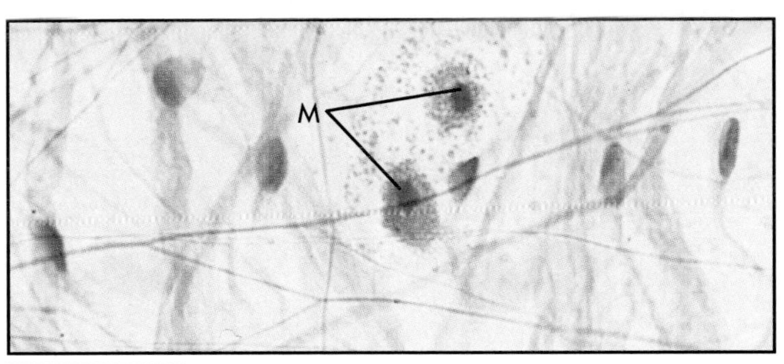

FIGURE 8-9 Two mast cells (M) of loose connective tissue (X370). The cytoplasmic granules contain histamine and other chemicals involved in inflammation. When there is tissue damage, the mast cells degranulate and release those chemicals. The mast cells may also be coated with IgE antibodies which cause degranulation when they bind antigen — as in Type I hypersensitivity (allergic) reactions.

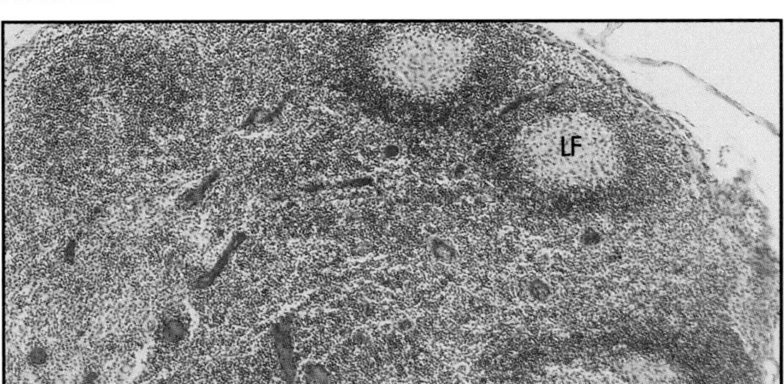

FIGURE 8-11 A section through a lymph node (X73). The large spherical purple objects are masses of lymphocytes called *lymph nodules* or *lymph follicles* (LF). As lymph passes through the sinuses (channels) within the node, antigens may contact a lymphocyte with the ability to produce antibodies against it. This provides the stimulus for cloning of the lymphocyte and conversion of some clones into antibody-secreting plasma cells. Other clones become memory cells and reside in lymphatic tissue around the body. Macrophages are also found in the sinuses.

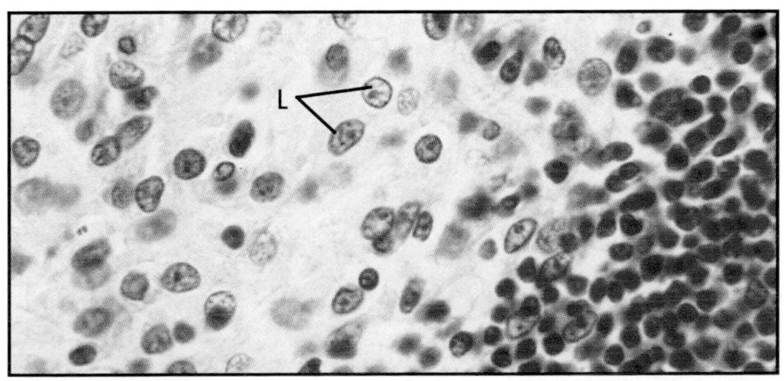

FIGURE 8-13 The germinal center of a lymph follicle (X554). Medium-sized lymphocytes (L) are abundant in the lighter region, whereas mature lymphocytes predominate in the darker region. Their differences in size and nuclear straining are apparent.

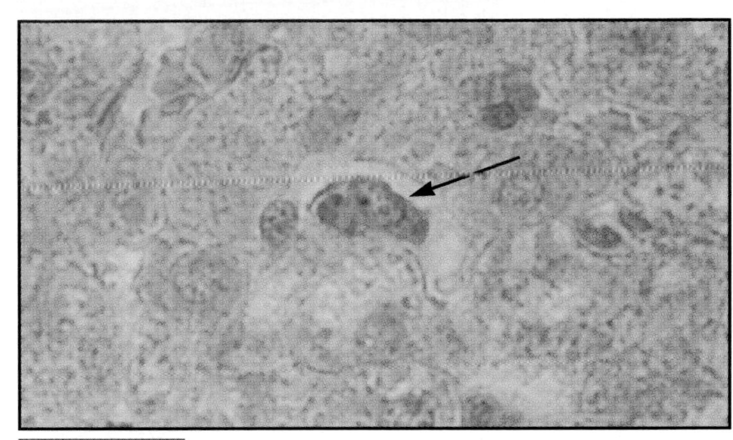

FIGURE 8-10 A section of the liver showing a Kuppfer (arrow) cell (X1280). This example of a fixed macrophage has ingested a dye. Fixed macrophages may also be found in the lungs.

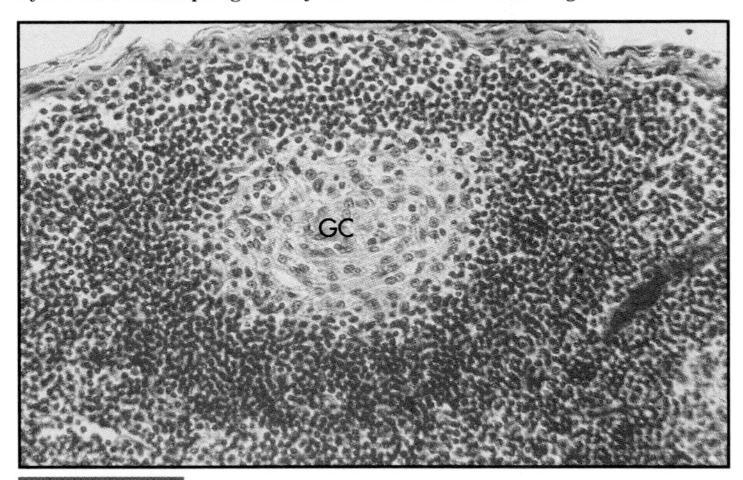

FIGURE 8-12 A single lymph follicle (X2005). The germinal center (GC) is occupied by lymphoblasts and medium-sized lymphocytes. More mature lymphocytes are found in the darker, outer portion of the nodule.

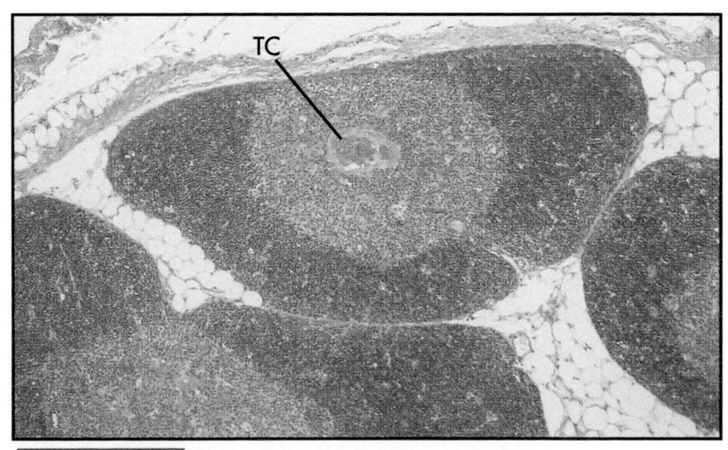

FIGURE 8-14 A section through the thymus (X78). One lobule (composed of lymphocytes) with its thymic corpuscle (TC) is shown. The thymus is the site of T-lymphocyte maturation.

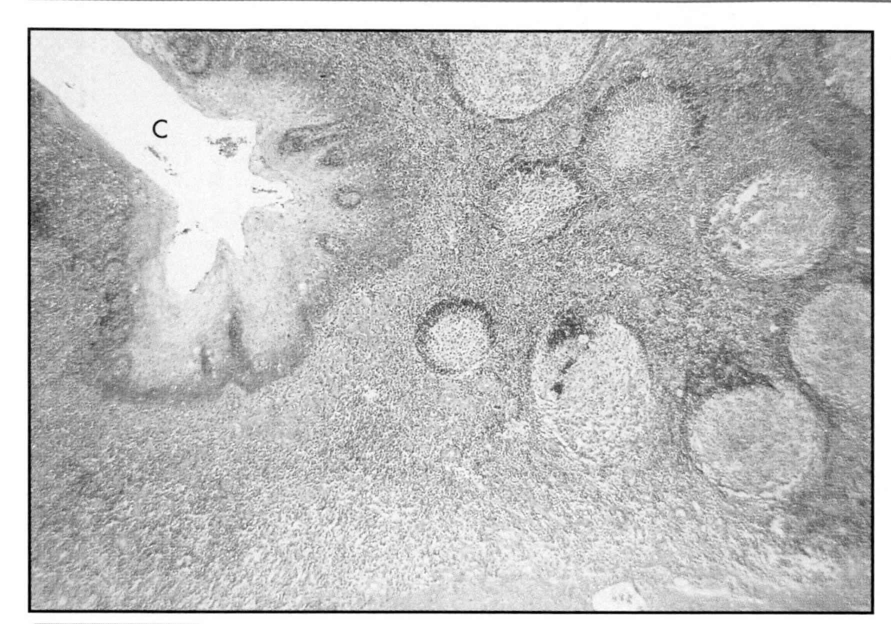

FIGURE 8-15 A section through a palatine tonsil (X58). Palatine tonsils are found on either side of the opening of the oral cavity to the throat. Two other sets of tonsils are also found in the throat: the pharyngeal tonsils (adenoids) are posterior to the nasal cavity and the lingual tonsils are in the base of the tongue. Tonsils act much like lymph nodes in that they are composed of lymph follicles (evident in the right half of the field) that contact fluids passing through them. A tonsilar crypt (C) is also visible.

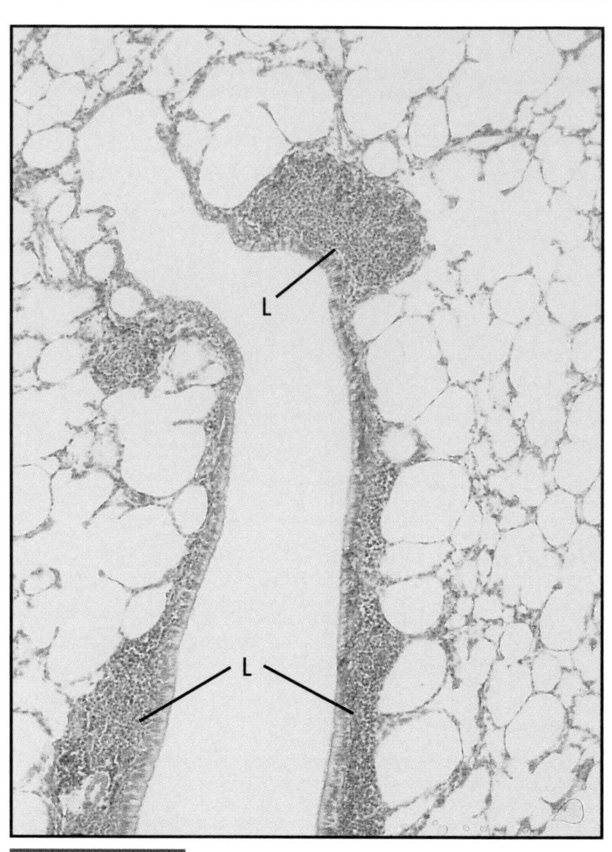

FIGURE 8-16 Lymphatic tissue is found in many places besides the lymph nodes. Shown is a section through the lung with lymphatic tissue (L) in the wall of a bronchiole (X100).

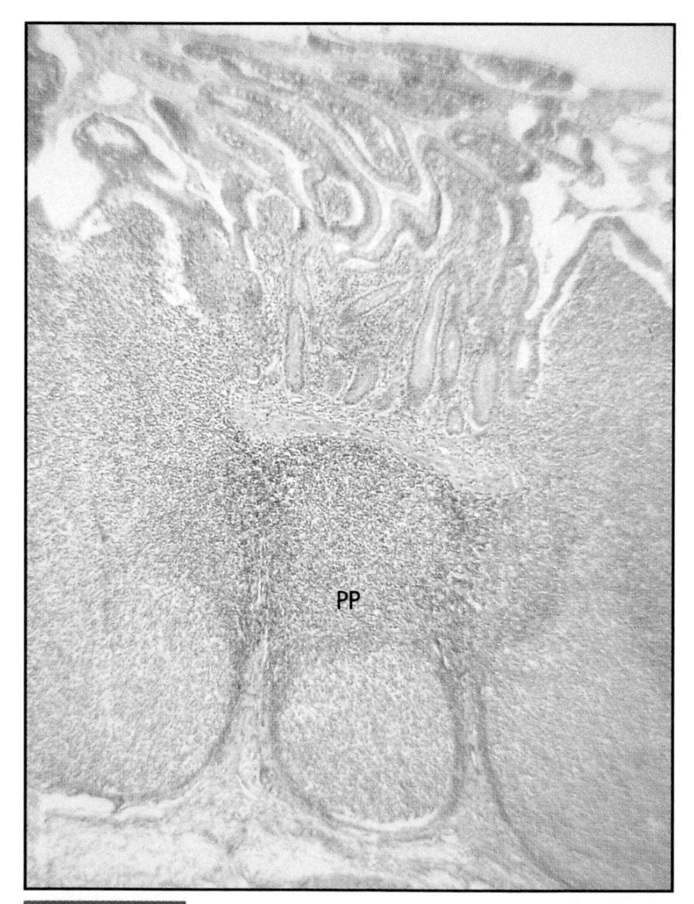

FIGURE 8-17 A section through the ileum showing a Peyer's patch (PP) of lymph follicles (X50). Peyer's patches may consist of 10 to 70 follicles separated from the intestinal lumen only by a single layer of epithelial cells.

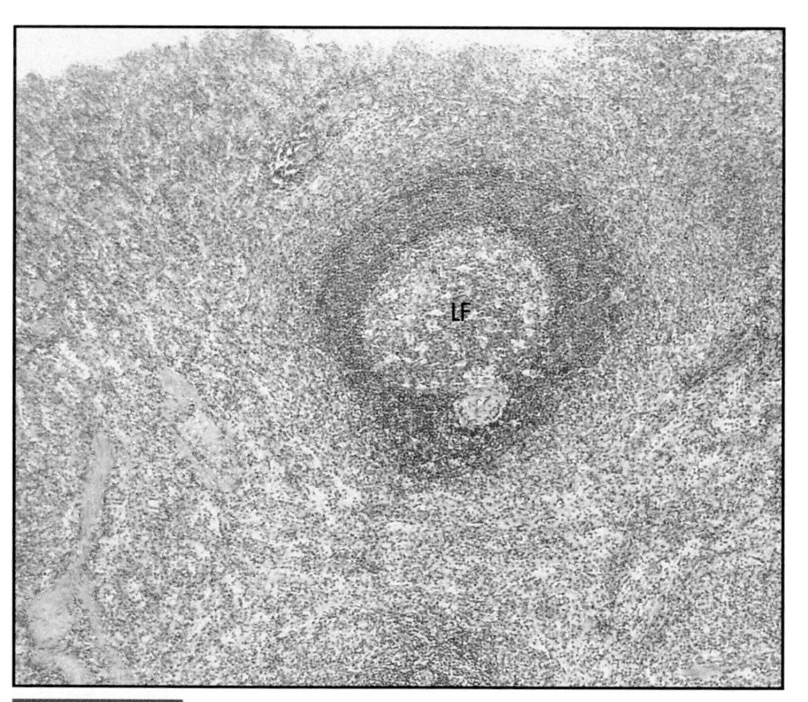

FIGURE 8-18 A section through a spleen (X80). In addition to being a blood filter and reservoir, the spleen contains lymph follicles collectively referred to as *white pulp*. Lymphocytes of white pulp respond to antigens in the blood. The portion of the spleen devoted to blood-vascular functions is referred to as *red pulp*. A single follicle (LF) with its central artery is shown surrounded by the red pulp.

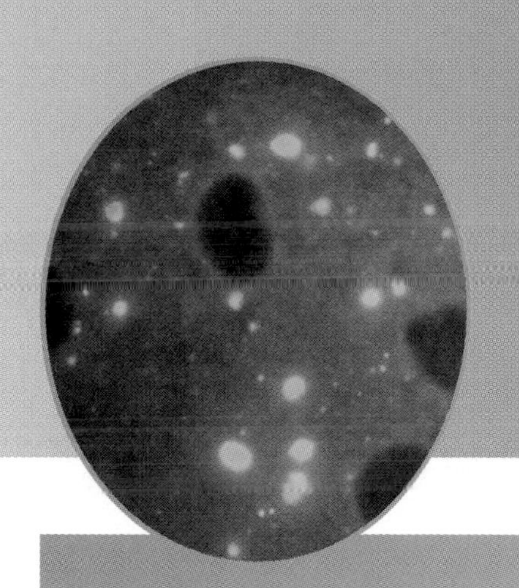

Molecular Techniques, Immunology, and Serology

9

ELECTROPHORESIS

PURPOSE Electrophoresis is a technique in which molecules are separated by size and electrical charge in a gel. Once separated, the gel may be used in a number of ways. Protein or nucleic acid patterns may be compared for taxonomic or identification purposes. Separated molecules may be removed from the gel and used in biochemical studies. Other techniques, such as DNA fingerprinting and Southern, Western, and Northern Blotting, begin with electrophoresis.

PRINCIPLE The gels used in this procedure are typically prepared from agarose or polyacrylamide. In preparation for the procedure the gel is cast as a thin slab containing tiny wells at one end and placed in a buffered solution to maintain proper electrolyte balance. Samples to be examined (either nucleic acid or protein) are loaded into the different wells and electrodes are attached to create an electrical field in the gel.

Under the influence of the electrical field the molecules in the samples migrate through the gel. They travel different distances due to differences in size and/or electrical charge. At the end of the run, the gel is stained with an appropriate chemical to show the location of the separated molecules as bands in each lane. Coomassie blue is commonly used for protein (Fig. 9-1), whereas ethidium bromide, a fluorescent dye, may be used for nucleic acids. If the nucleic acid has been radioactively labeled, *autoradiography* may be done. The gel can be placed on X-ray film and the radioactivity from the bands expose the film, as in Figure 9-2.

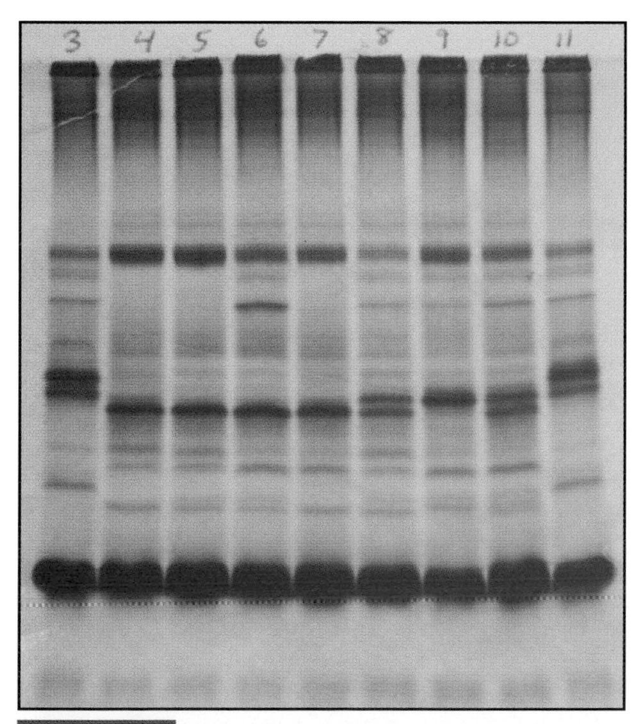

FIGURE 9-1 A polyacrylamide gel of serum proteins from several lemur species to illustrate staining with Coomassie blue. Migration was from top to bottom.

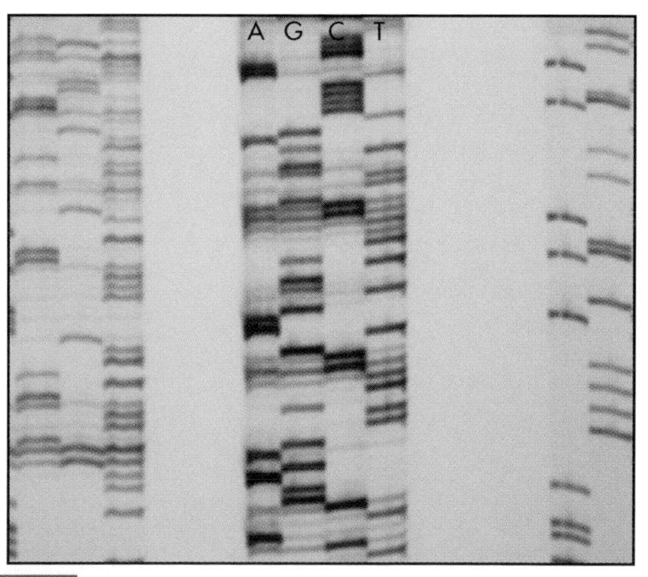

FIGURE 9-2 A portion of an autoradiograph produced by an electrophoretic gel used in DNA sequencing. The radioactively labeled DNA fragments were separated by size under the influence of an electrical field. Migration of the samples was from top to bottom. The bands were visualized by placing the gel on X-ray film and allowing the radioactivity to expose the film. The smaller, faster molecules are at the bottom of the autoradiograph. Each set of four lanes allows determination of nucleotide sequence, with the lanes representing A, G, C, and T nucleotides from left to right.

PRECIPITATION REACTIONS

PURPOSE Precipitation reactions may be used to detect either the presence of antigen or antibody in a sample. They have mostly been replaced by more sensitive serological techniques for diagnosis, but are still useful to simply demonstrate serological reactions. Double-gel immunodiffusion may be used in identification of antibodies formed in autoimmune diseases.

PRINCIPLE *Soluble antigens* may combine with *homologous antibodies* to produce a visible *precipitate*. Precipitate formation thus serves as evidence of antigen-antibody reaction and is considered a positive result.

Precipitation is produced because each antibody has (at least) two antigen binding sites and many antigens have multiple sites for antibody binding. This results in the formation of a complex lattice of antibodies and antigens and produces the visible precipitate — a positive result. As shown in Figure 9-3, if either antibody or antigen is found in too high a concentration relative to the other, no visible precipitate will be formed even though both are present. *Optimum proportions* of antibody and antigen are necessary for precipitate formation and occur in the *zone of equivalence*.

Several styles of precipitation tests are used. The precipitin ring test is performed in a small test tube or capillary tube. Antiserum is placed in the bottom of the tube. The sample with the suspected antigen is layered on the surface of the antiserum in such a way that the two solutions have a sharp interface. As the two fluids diffuse into each other, precipitation occurs where optimum proportions of antibody and antigen are found (Fig. 9-4). This test may also be run to test for antibody in a sample.

Double-gel immunodiffusion tests are used to check samples for identical, related or unrelated antigens. Wells are formed in a gel and a mixture of antibodies is placed in the center well. Samples with unknown antigens are placed in the surrounding wells (Fig. 9-5). As antigens and antibodies diffuse radially from their respective wells,

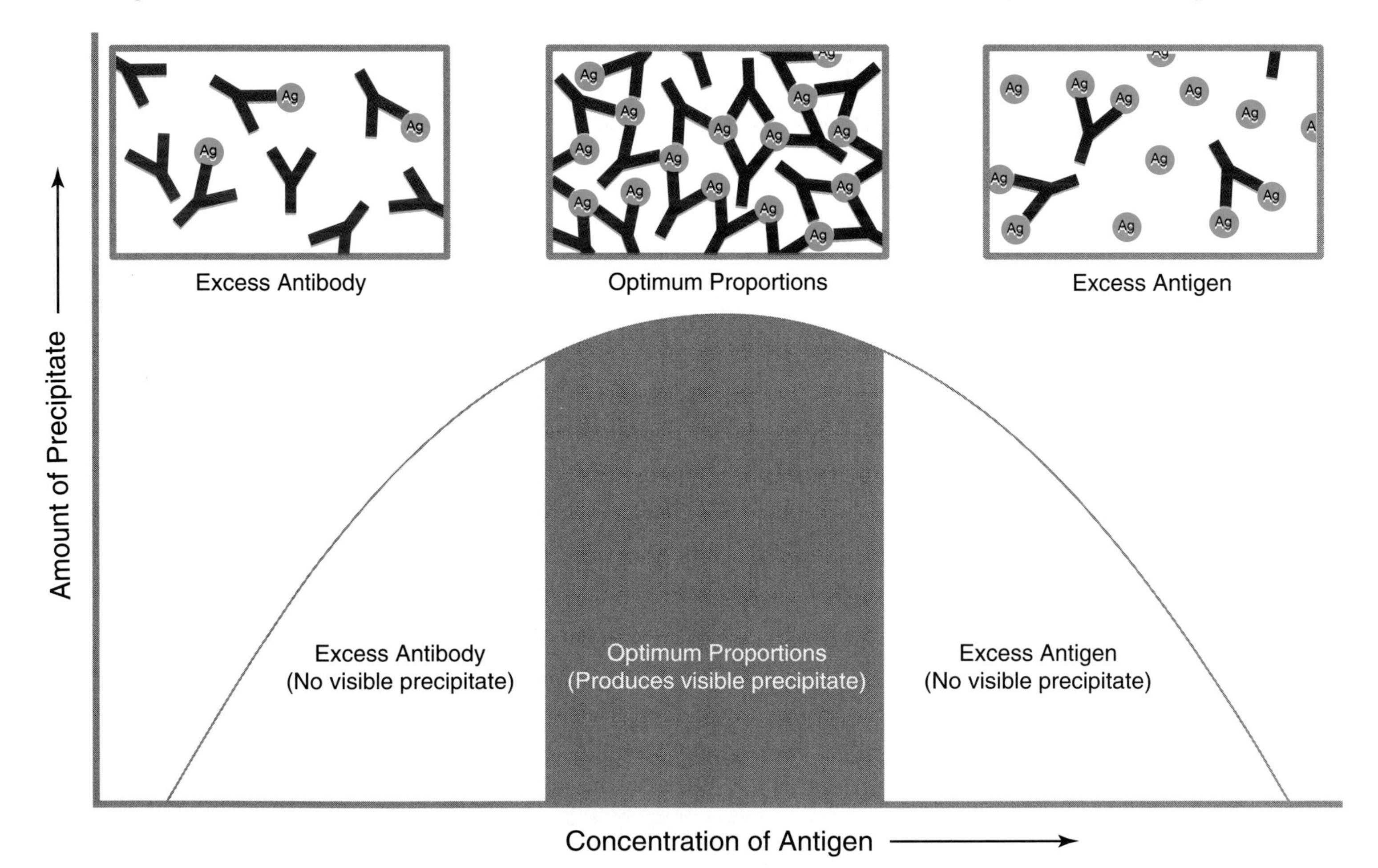

FIGURE 9-3 Precipitation occurs between soluble antigens and homologous antibodies where they are found in optimal proportions to produce a cross-linked lattice. Excess antigen or excess antibody prevent substantial cross-linking, so no lattice is formed and no visible precipitate is seen — even though both antigen and antibody are present.

precipitation lines occur where optimal proportions occur. The precipitation line pattern is indicative of antigen relatedness: a single, smooth curved line indicates the two antigens in neighboring wells are identical ("identity"); a line with a spur indicates the antigens are related, but not identical ("partial identity"); two spurs indicate unrelated antigens ("nonidentity").

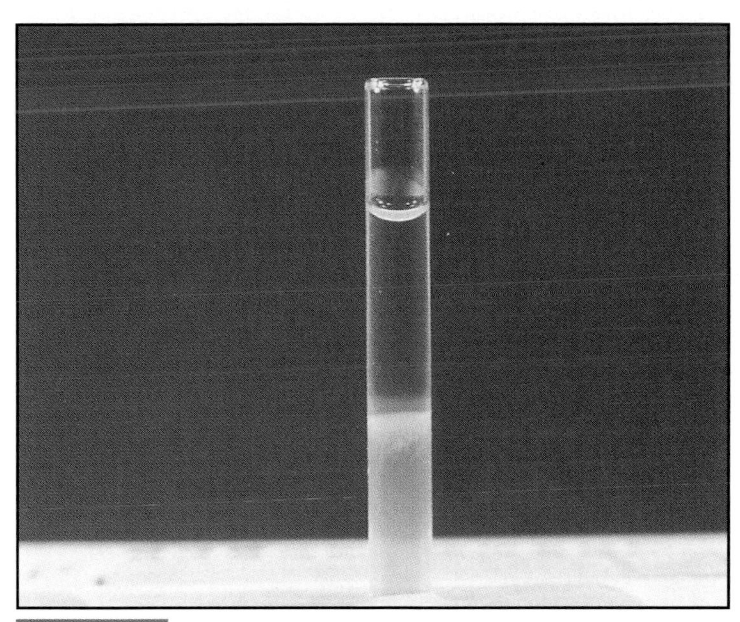

FIGURE 9-4 A positive precipitin ring test. A sample of antigen has been layered over an antiserum. The white precipitation ring formed at the site of optimal proportions of antibodies and antigens.

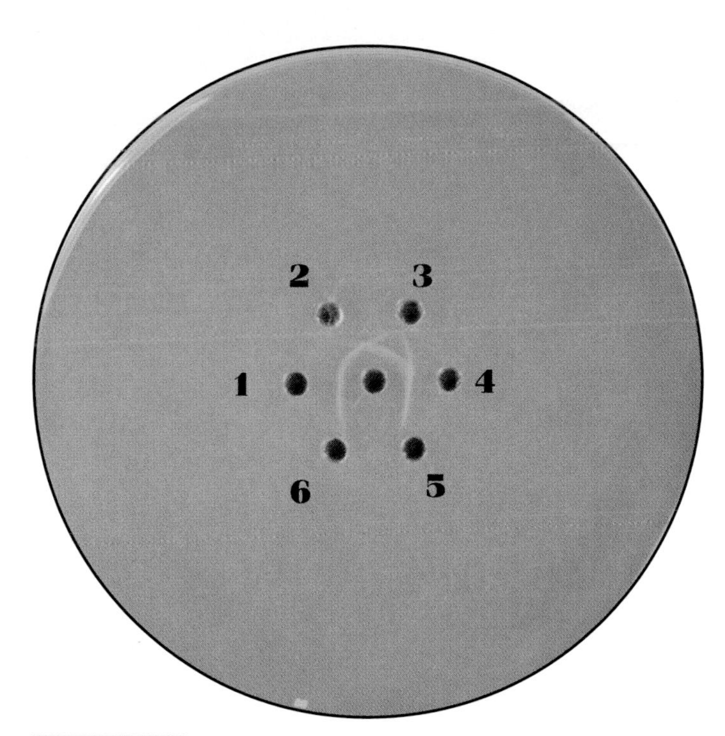

FIGURE 9-5 Double-gel immunodiffusion with antibodies in the center well and antigens in the outer wells. Precipitation lines formed between the central well and wells 1 and 2 illustrate identity of the antigens in wells 1 and 2. Nonidentity is illustrated by the antigens in wells 2 and 3. Partial identity is illustrated by the lines formed between the central well and wells 3 and 4. No reaction occurred between the antibodies and antigens in wells 5 and 6.

AGGLUTINATION REACTIONS

PURPOSE Agglutination reactions may be used to detect either the presence of antigen or antibody in a sample. Direct agglutination reactions are used to diagnose some diseases, determine if a patient has been exposed to a particular pathogen, and are involved in blood typing. Indirect agglutination is used in some pregnancy tests as well as in disease diagnosis.

PRINCIPLE Particulate antigens (such as whole cells) may combine with homologous antibodies to form visible clumps called *agglutinates*. *Agglutination* thus serves as evidence of antigen-antibody reaction and is considered a positive result.

There are many variations of agglutination tests (Fig. 9-6). *Direct agglutination* relies on the combination of antibodies and naturally particulate antigens. *Indirect agglutination* relies on artificially constructed systems in which agglutination will occur. These involve coating particles (such as RBCs or latex microspheres) with either antibody or antigen, depending on what is being looked for in the sample. Addition of the appropriate antigen or antibody will then result in clumping of the artificially constructed particles.

Slide agglutination (Fig. 9-7) is an example of a direct agglutination test. Samples of antigen and antiserum are mixed on a microscope slide and allowed to react. Visible aggregates indicate a positive result.

Hemagglutination is a general term applied to any agglutination test in which clumping of red blood cells indicates a positive reaction. Blood tests as well as a number of indirect diagnostic serological tests are hemagglutinations.

The most common form of blood typing detects the presence of A and B antigens on the surface of red blood cells. An individual with type A blood has RBCs with the A antigen and produces anti-B antibodies. Likewise, an individual with type B blood has RBCs with the B antigen and produces anti-A antibodies. People with type AB blood have *both* A and B antigens on their RBCs and lack anti-A and anti-B antibodies. Type O individuals lack A and B antigens but produce *both* anti-A and anti-B antibodies.

ABO blood type is determined by adding a patient's blood to anti-A and anti-B antiserum and observing any signs of agglutination (Fig. 9-8). Agglutination with anti-A antiserum indicates the presence of the A antigen and type A blood. Agglutination with anti-B antiserum indicates the presence of the B antigen and type B blood. If both agglutinate, the individual has type AB blood; lack of agglutination occurs in individuals with type O blood.

A similar test is used to determine the presence or absence of the *Rh factor* (antigen). If clumping of the patient's blood occurs when mixed with anti-Rh (anti-D) antiserum, the patient is Rh positive (Fig. 9-9).

Indirect hemagglutination may be used to detect the presence of either antigens or antibodies in a sample. In the example shown in Figure 9-10, sheep RBCs coated with *Treponema pallidum* (the causative agent of syphilis) antigen represent the particulate antigen. When added to antiserum containing anti-*Treponema pallidum* antibodies, agglutination occurs.

Another application of indirect agglutination is the Rapid Plasma Reagin Test. It also is used for diagnosis of syphilis (Fig. 9-11).

Direct agglutination occurs with naturally particulate antigens. Either antigen or antibody may be detected in a sample using this style of agglutination test.

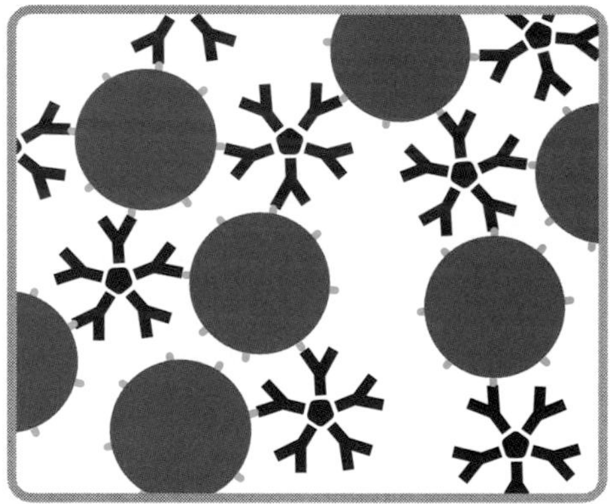

Detection of antibody in a sample by indirect agglutination. A test solution is prepared by artificially attaching homologous antigen (blue) to a particle (red) such as red blood cells or latex beads and mixing with the sample suspected of containing the antibody (purple).

Another style of indirect agglutination detects antigen (light blue) in a sample. Antibodies (purple) are artificially attached to particles (dark blue).

FIGURE 9-6 Direct and indirect agglutination tests. Direct agglutinations involve either naturally particulate antigens or antibodies. Indirect agglutination relies on attaching either the antigen or antibody to a particle, such as a latex bead or RBC.

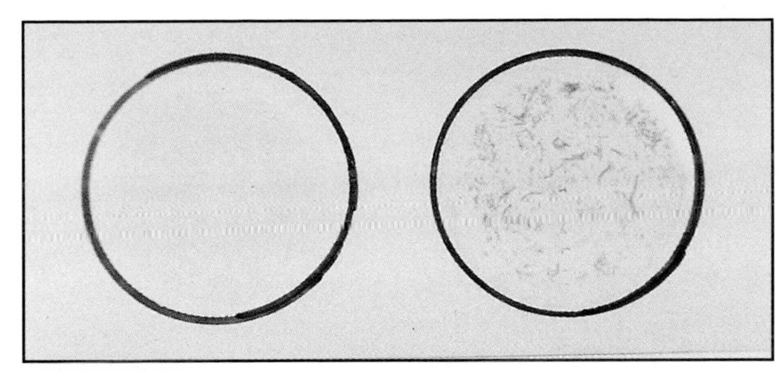

FIGURE 9-7 A positive slide agglutination of *Salmonella* H antigen is shown on the right. Anti-H antiserum mixed with *Salmonella* O antigen is shown on the left. Serological variation of H (flagellar) and O (somatic lipopolysaccharide) antigens is an important diagnostic feature of *Salmonella* serovars.

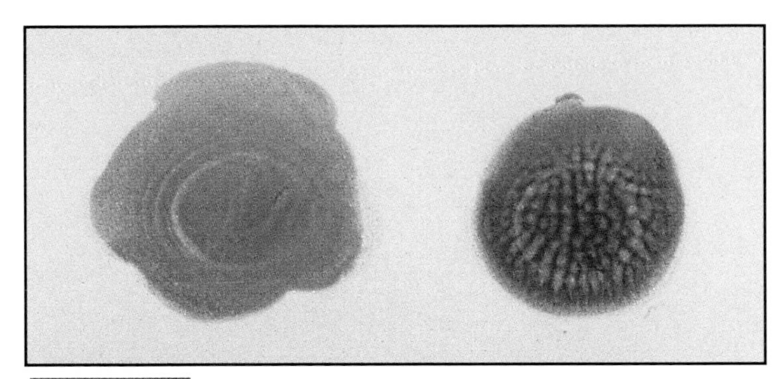

FIGURE 9-9 Rh blood type is determined by agglutination. Rh-negative blood is on the left, Rh-positive is on the right.

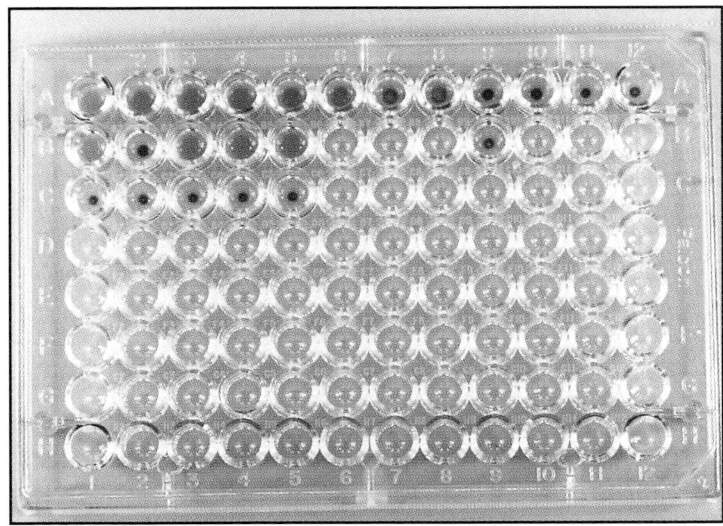

FIGURE 9-10 Coating RBCs with *Treponema pallidum* antigens is the basis for this hemagglutination test. Since the antigens do not naturally cause agglutination, this test is an example of an indirect agglutination. The top row consists of serially diluted standards to act as positive (reactive) controls. A positive result is evidenced by a smooth mat of cells in the well (as in well A1). A negative result is a button of cells (as in well A12). Patient samples are in rows B and C. Patients B1, B3, B4, and B5 test positive for syphilis antibodies; patient B2 is negative. Samples in row C correspond to samples in row B and are negative (nonreactive) controls. In each case, the patient's serum is mixed with unsensitized RBCs to assure that agglutination is actually due to reaction with *Treponema* antibodies.

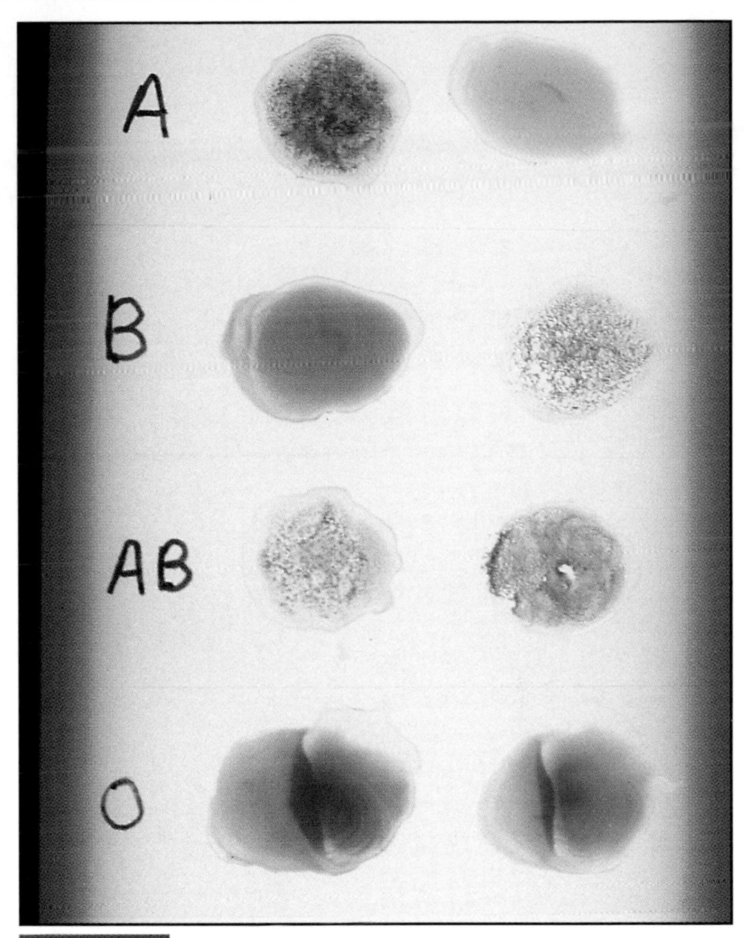

FIGURE 9-8 Blood typing relies on agglutination of RBCs by Anti-A and/or Anti-B antisera. The blood types are as shown.

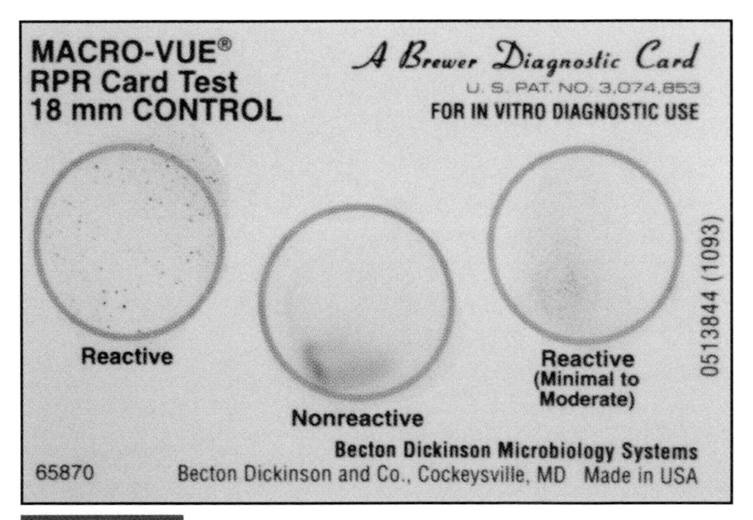

FIGURE 9-11 The Rapid Plasma Reagin (RPR) test for syphilis relies on antigen-coated particles agglutinating reagin (an antibody-like substance) present in infected individual's plasma. Shown here is a control card that must be run with every set of tests.

ENZYME-LINKED IMMUNOSORBENT ASSAY (ELISA)

PURPOSE ELISA may be used to detect the presence and amount of either antigen or antibody in a sample. The indirect ELISA is used to screen patients for the presence of HIV antibodies, rubella virus antibodies and others. The direct ELISA may be used to detect hormones (such as HCG in some pregnancy tests), drugs, and viral and bacterial antigens.

PRINCIPLE As with other serological tests, ELISAs can be used to detect antigen in a sample or antibody in a sample (Fig. 9-12). All rely on a second antibody with an attached (*conjugated*) enzyme as an indicator of antigen-antibody reaction.

The direct ELISA detects the presence of antigen in a sample. A microtiter well is coated with *homologous antibody* specific to the antigen being looked for. The sample being assayed is added to the well. If the antigen is present, it will react with the antibody coating the well; if none is present, no reaction occurs. In this form of ELISA, the enzyme-linked antibody is also specific for the antigen being tested for. When added to the well, it binds to the antigen, if present. After allowing time for the enzyme-linked antibody to react with the antigen, the well is washed to remove any unbound enzyme-linked antibody (which would produce a false positive when the substrate is added). Upon addition of substrate, a color change is evidence of the enzymatic conversion of substrate to its product which indicates presence of the antigen.

The indirect ELISA detects the presence of antibody in a sample. In this style of test, the microtiter well is lined with antigen specific to the antibody being looked for. The sample being assayed is added to the well and, if the antibody is present, it will react with the antigen coating the well; if none is present, no reaction occurs. In this case, the enzyme-linked antibody is an *antihuman immunoglobin antibody* — its antigen is actually an antibody! When added to the well, it binds to the antibody in the sample, if present. As with the direct ELISA, unbound enzyme-linked antibody must be washed away to prevent a false positive result. Enzyme

Direct ELISA Method

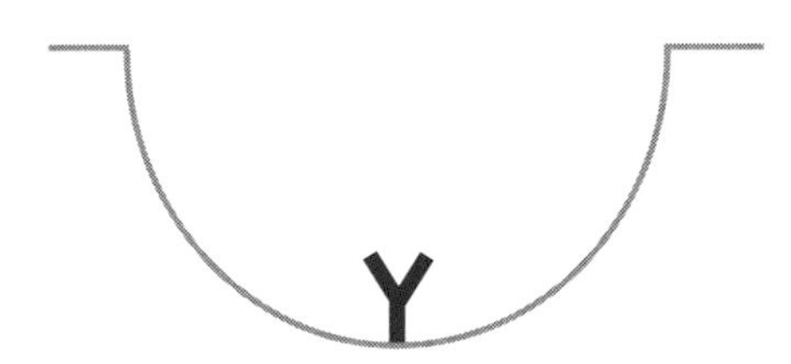

This ELISA detects the presecnce of *antigen* in a sample. Antibody (antiserum) specific for that antigen is coated (adsorbed) onto the wall of a microtiter well.

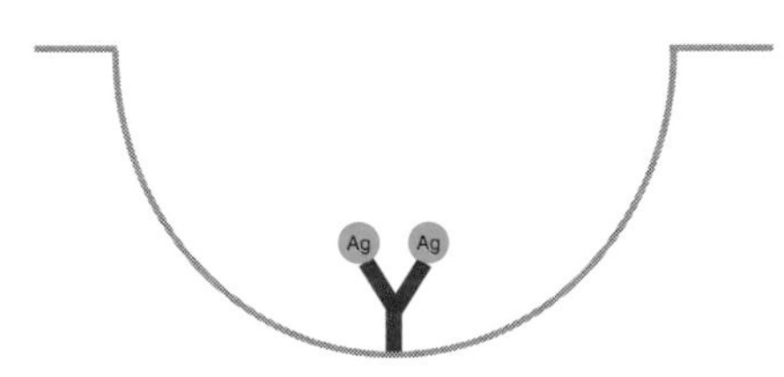

The sample is added. If the *antigen* is present, it will bind to the antibody coating the well.

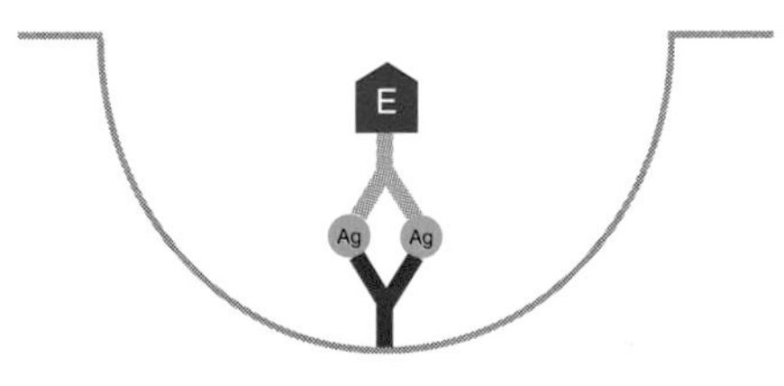

A second antibody with an attached enzyme specific for the same antigen is added. Unbound enzyme-linked antibody is washed away.

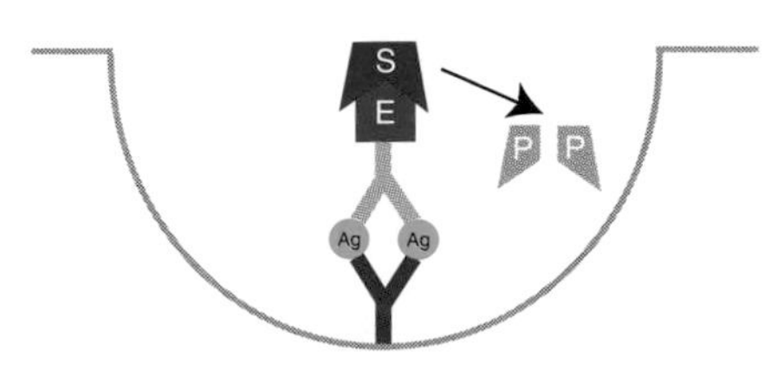

Substrate for the enzyme is added. Conversion of substrate to product is evidenced by a color change. A color change means the sample has the antigen; no color change is a negative result.

Indirect ELISA Method

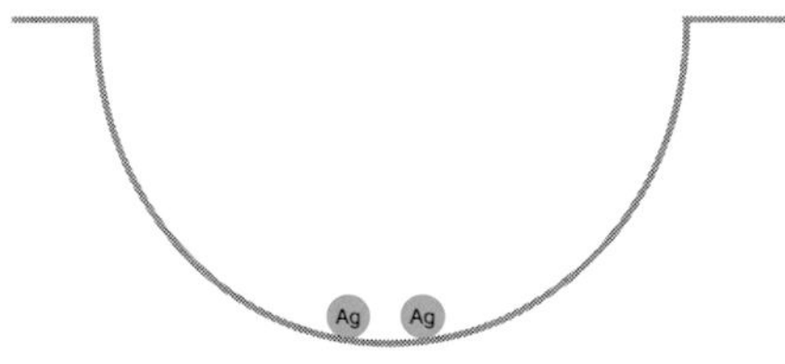

This ELISA detects the presecnce of *antibody* in a sample. The antigen specific for that antibody is coated (adsorbed) onto the wall of a microtiter well.

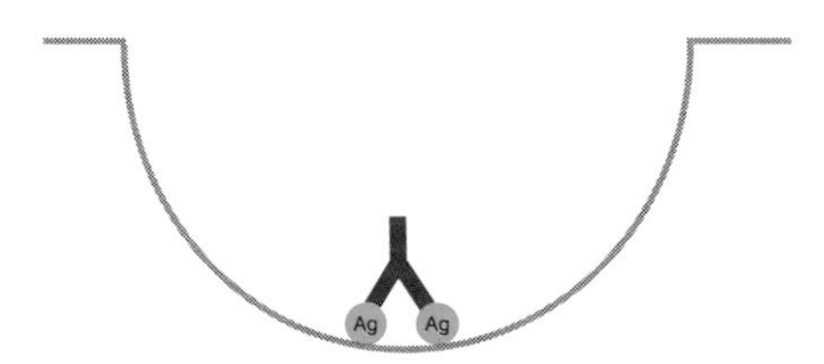

The sample is added. If the *antibody* is present, it will bind to the antigen coating the well.

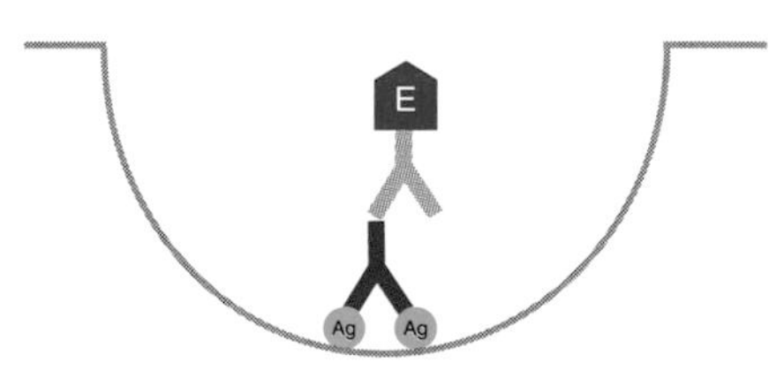

An antihuman immunoglobin antibody with an attached enzyme is added. Unbound enzyme-linked antibody is washed away.

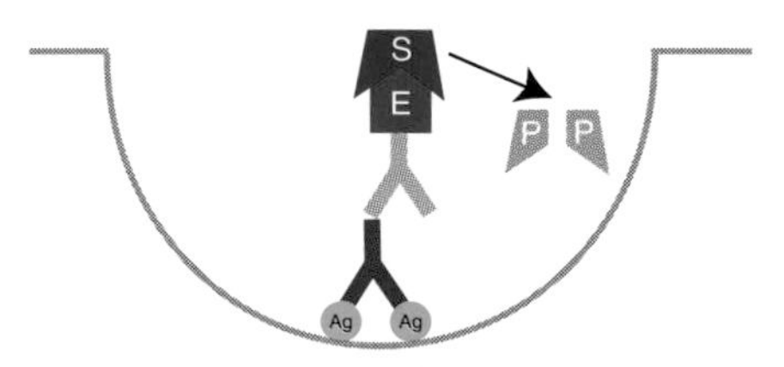

Substrate for the enzyme is added. Conversion of substrate to product is evidenced by a color change. A color change means the sample has the antibody; no color change is a negative result.

FIGURE 9-12 Direct and indirect ELISAs.

substrate is added and a color change indicates a positive test.

Figures 9-13 and 9-14 illustrate a form of ELISA which is quantitative, *i.e.*, it is used to determine both the presence of antibody in a sample *and* the amount. Figure 9-15 illustrates another practical use of the ELISA technique — a home pregnancy test — that screens for the presence of *human chorionic gonadotropin* found in pregnant women.

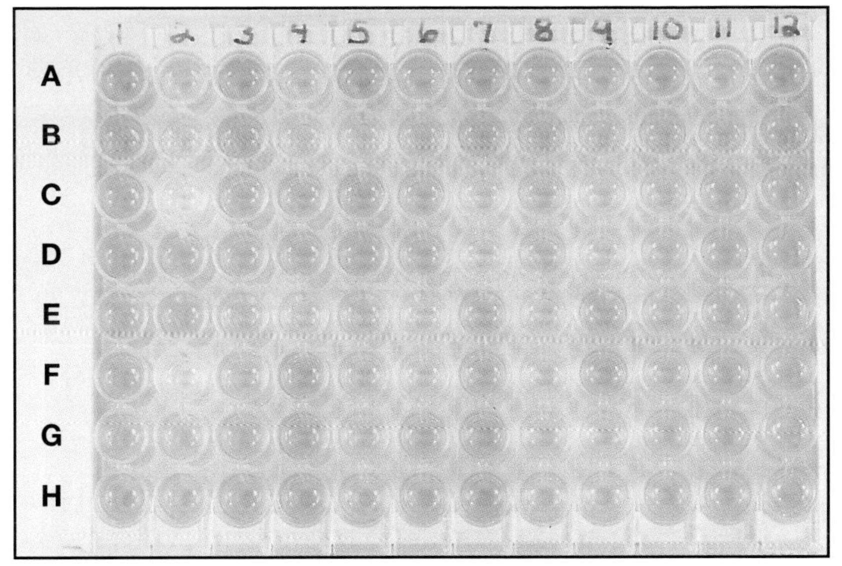

FIGURE 9-13 One example of an ELISA used to determine antibody titer. In this ELISA, a dark yellow color indicates a negative reaction. The lighter the color, the higher the antibody titer. Color is read by a photometer (Fig. 9-14) and results are fed into a computer. A variety of controls is also used. Serially diluted antibody samples of known concentration are in column 1, rows A through H and column 2, rows A through C. Absorbance values from these are used to develop a standard curve correlating antibody titer with absorbance. Patient samples are in the other wells. Each patient's antibody titer may be determined by comparison with the standard curve.

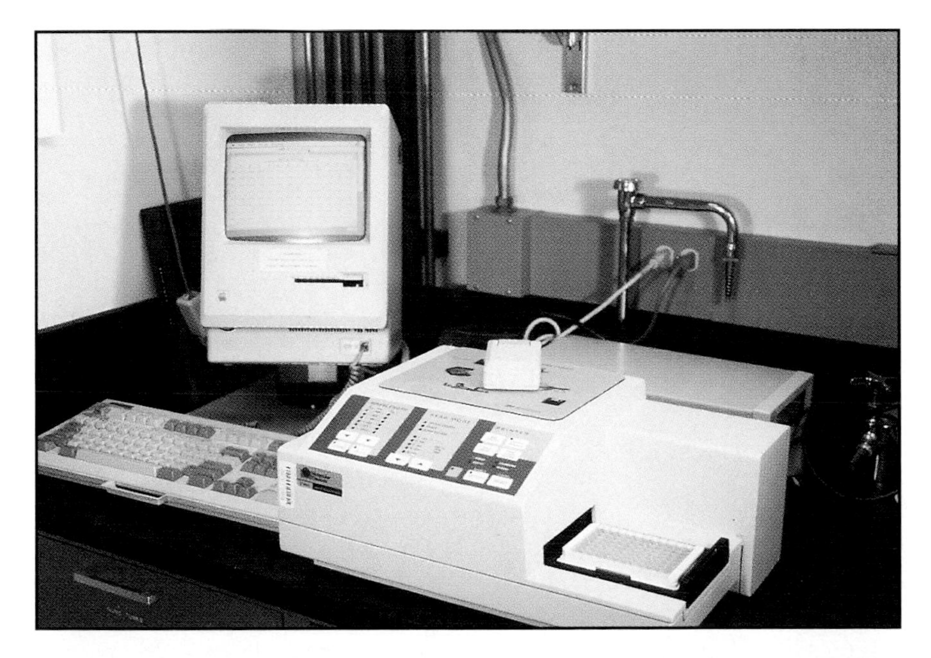

FIGURE 9-14 Set-up for the ELISA illustrated in Figure 9-13. The photometer is in the foreground. The microtiter plate is visible on the photometer at the right. The computer is visible in the background.

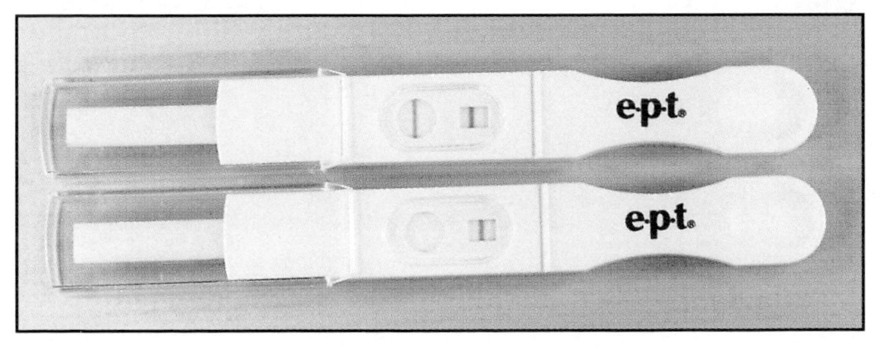

FIGURE 9-15 Another use of ELISA technique is home pregnancy tests. Human chorionic gonadotropin (HCG) is an antigen present only in pregnant women. A sample of urine is applied to the wick on the left of the test system. Two pink lines indicate a reaction of HCG with antibodies in the test system. A single line indicates absence of HCG in the urine and is interpreted as a negative result.

FLUORESCENT ANTIBODY (FA) TECHNIQUE

PURPOSE Like most serological tests, the fluorescent antibody technique may be used to identify the presence of either antigen or antibody in a sample. Direct tests (DFA) identify the presence of antigens; indirect tests (IFA) detect the presence of antibody in a sample. FAs are useful in diagnosing many viral infections as well as certain parasitic diseases.

PRINCIPLE Fluorescent antibodies are labeled with fluorescein isothiocyanate (FITC) dye which fluoresces when illuminated with UV light. In a DFA (Fig. 9-16), the sample containing the suspected antigen is fixed to a microscope slide. The fluorescent antibody is added and allowed to react with the antigen. After rinsing to remove unbound antibody, the slide is viewed with a fluorescent microscope with a UV light source. If the suspected antigen is present, the labeled antibodies will have bound to it and will emit an apple green color (Fig. 9-17).

IFAs (Fig. 9-16) are used to detect antibodies in a sample. In this form of the test, the specific antigen is fixed to a microscope slide. Dilutions of the patient's sample are added to several slides and given time to react with the antigen. The FITC-labeled antibody is an anti-gamma globulin antibody, so if there is patient antibody bound to antigen on the slide, the fluorescent antibody will bind to it. After rinsing to remove any unbound antibodies, the slide is viewed under a fluorescent microscope with a UV light source. If the suspected antibody is present, the labeled antibodies will fluoresce and appear apple green (Fig. 9-18).

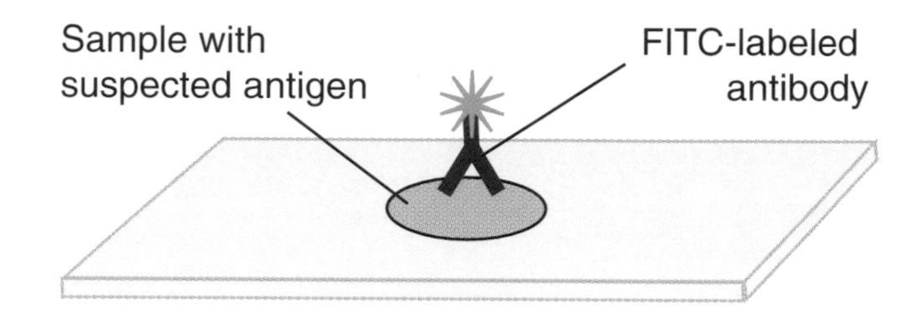

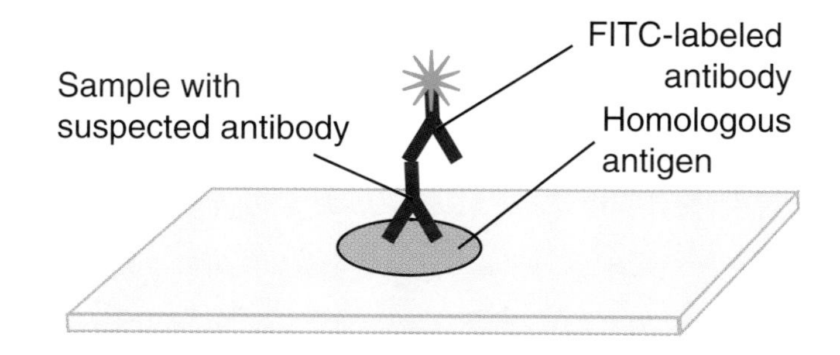

FIGURE 9-16 Direct and indirect fluorescent antibody techniques. The DFA and IFA are used to identify antigens and antibodies in a sample, respectively.

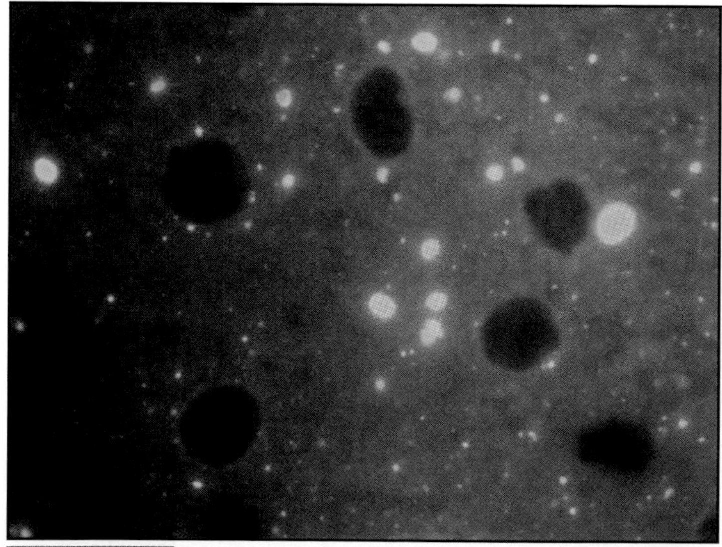

FIGURE 9-17 A fluorescent micrograph of a positive DFA for rabies virus (X576). The fluorescent green color is due to anti-rabies virus antibodies labeled with FITC dye bound to the virus in the specimen.

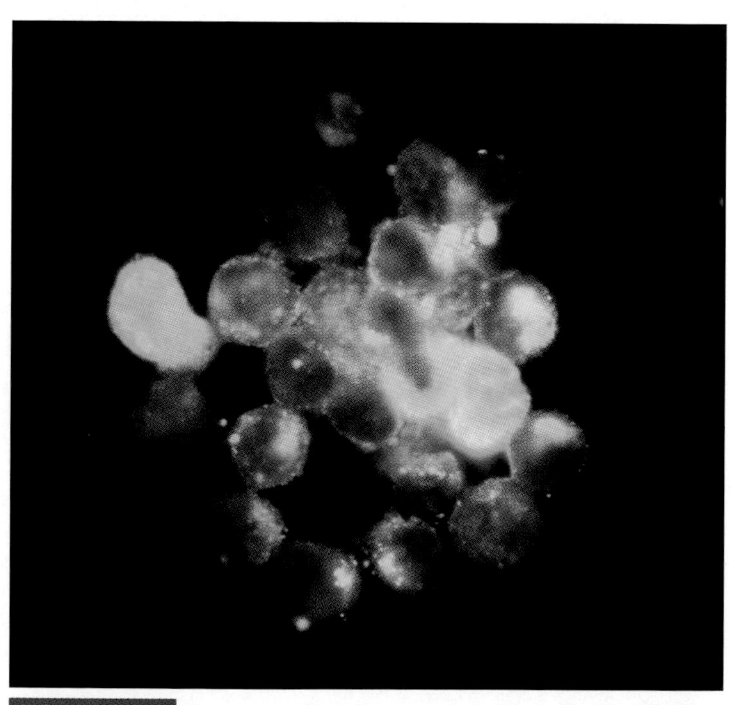

FIGURE 9-18 A positive IFA for influenza B virus (X624). Infected cells fluoresce an apple green color. Uninfected cells appear reddish due to a second stain (Evans blue) in the preparation.

WESTERN BLOT TECHNIQUE

PURPOSE Western blots are used to identify proteins in a sample. (Similar techniques called the Southern blot and Northern blot are used for identifying DNA and RNA in samples, respectively.) This technique has applications in research, especially with viruses, but its main clinical use is to identify HIV antibodies in a patient's serum or plasma. Since it is more expensive and requires greater skill than other serological tests, it is only used to confirm the results from samples that are repeatedly positive for HIV antibodies in preliminary ELISA screening.

PRINCIPLE This test is used to check for specific antibodies in a sample, so known antigens (proteins) are used. The procedure begins with polyacrylamide gel electrophoresis (see "Electrophoresis" in this section) of the proteins to separate them by size and charge. Then the bands of protein are transferred to a nitrocellulose membrane by the blotting technique. Simply, the gel is sandwiched together with the membrane between layers of absorbent material. The proteins are moved into the membrane and remain fixed in the same position as the original bands (but are not visible). The nitrocellulose is then cut into strips which are ready for use. (The electrophoresis and blotting steps may be done by the manufacturer if a commercial kit is used.)

A tray with troughs is used to hold the strips so that several tests may be run simultaneously. After careful preparation, each sample (suspected of containing antibody) is applied to a strip and allowed to react for the prescribed time (usually 12 or 24 hours). This incubation allows any antibody in the sample to bind to the known antigens in the strip. After rinsing to remove unbound antibody, the strips are exposed to an anti-immunoglobin antibody which is labeled with an enzyme or a fluorescent chemical. The strips are rinsed again to remove any unbound antibody.

Visualization of the bands depends on the type of labeled antibody. If an enzyme is used, then substrate is added. If a fluorescent chemical is used, then the strip is observed under UV light. Comparing band patterns on each strip to positive and negative controls allows determination of whether the sample contains the relevant antibodies. Figures 9-19 and 9-20 show Western blot strips used in HIV screening. The HIV antigens on the strip are shown in Figure 9-21.

The ELISA test identifies the presence or absence of antibodies in a sample. Although a Western blot is more expensive to run and requires more skill than an ELISA, it has the advantage of allowing identification of the specific antibodies in the patient's serum.

HIV-1 WESTERN BLOT WORKSHEET

Organon Teknika Corporation • 100 AKZO Avenue • Durham, NC 27712

FACILITY: DATE PERFORMED: 8-25-97
READ BY: REVIEWED BY: DATE: 8/26/97

KIT / TESTING INFORMATION

KIT MASTER LOT NO. M121601	NEGATIVE CONTROL 1115602 11/98	STRIP LOT NO. M091360I 3/98	CONJUGATE 1104602 11/97
KIT EXPIRATION DATE 11-97	LOW POSITIVE CONTROL 1011602 11/98	SAMPLE DILUENT 1127605 5/98	CHROMOGEN 1112602 11/97
PERFORMING TECHNOLOGIST	HIGH POSITIVE CONTROL 1101603 11/98	POWDERED MILK 1024601 4/98	

BAND EVALUATION: A - Absent I - Indeterminate P - Present
RESULTS: P - Positive I - Indeterminate N - Negative

(MOUNT THE WESTERN BLOT STRIPS HERE)

No.	SAMPLE ID	p18	p24	p31	gp41	p51	p55	p65	gp120	gp160	NOTES	RESULT
1	Hi Pos	P	P	P	P	P	I	P	P	P		P
2	Low Pos	P	P	P	P	P	A	P	I	P		P
3	Neg	A	A	A	A	A	A	A	A	A		N
4		A	P	P	P	A	A	P	I	P		P
5		A	P	A	A	A	A	A	A	A		I
6		A	P	P	P	A	A	P	I	P		P
7		A	P	A	I	A	A	A	A	A		I
8		A	P	A	I	A	A	A	A	A		I
9												
10												
11												
12												
13												
14												
15												
16												
17												
18												
19												
20												

ORIGIN LINE

FORM VI-105-96

FIGURE 9-19 A sample lab sheet with the blotted nitrocellulose strips attached. This Western blot was used to detect anti-HIV antibodies in several patients' serum (identified only by a code number which we blacked out). Strips 1 and 2 are high positive and low positive controls, respectively. Strip 3 is a negative control. To be considered positive, the patient's strip must react equal to or greater than the low positive control. Of the strips shown, samples 4 and 6 were interpreted as positive (P), and samples 5, 7 and 8 were negative (N). The test may also produce indeterminate results (I). Unfortunately, the criteria for interpreting a positive differ slightly between different agencies, but all involve the presence of two or more bands (usually p24, gp41, and/or gp120/160. See Figure 9-21.)

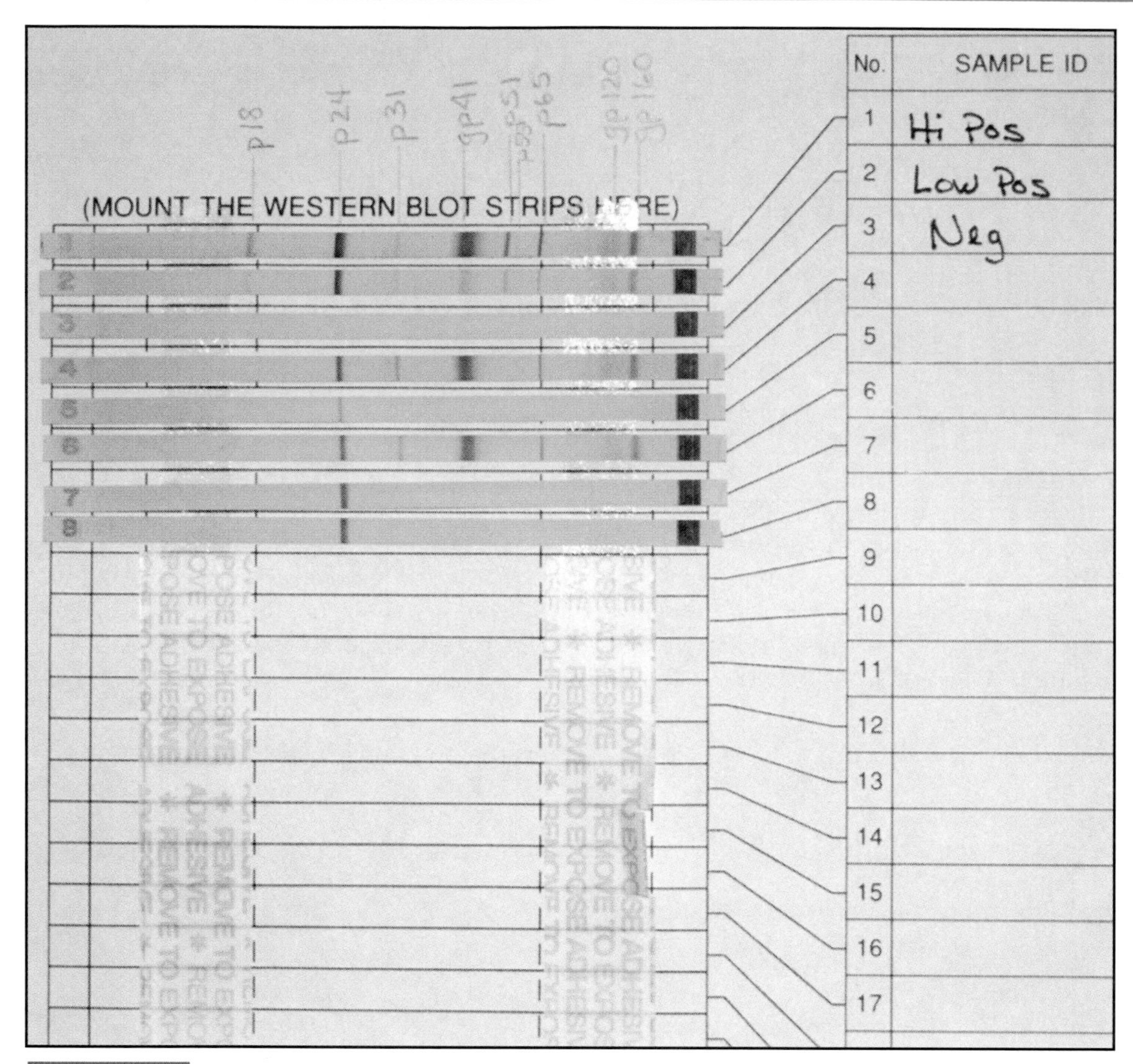

FIGURE 9-20 These are the same strips as in Figure 9-20 to show the banding pattern more clearly. Above the strips are handwritten codes identifying each HIV protein blotted. The "p" and "gp" in each refers to protein and glycoprotein, respectively. The numbers are the molecular weight of each protein. Notice that the smaller proteins are at the left. During electrophoresis, smaller proteins migrate faster than bigger ones. Based on that, these proteins migrated from right to left.

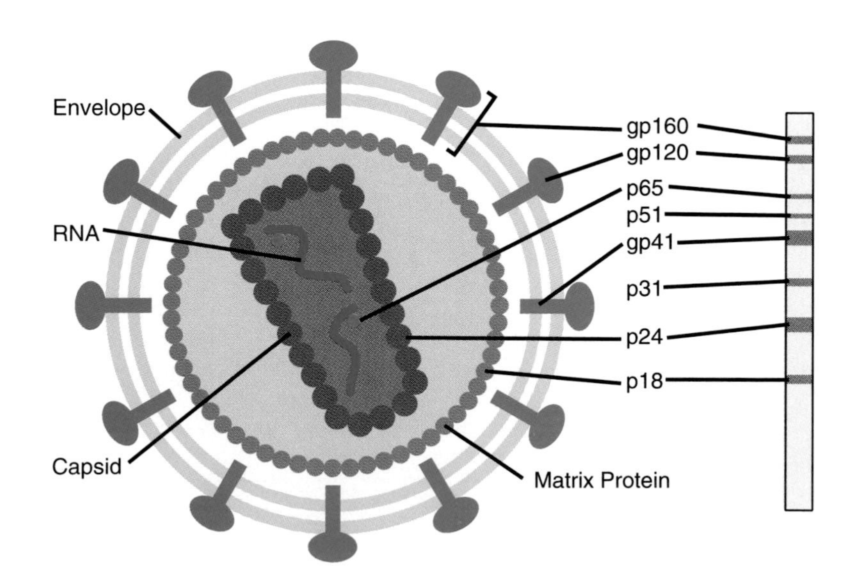

FIGURE 9-21 HIV is an enveloped retrovirus with virus-specific glycoproteins extending from it as spikes. The spikes are used for attachment to cells with the CD4 protein (such as Helper-T cells). There are also matrix and capsid proteins. Inside the capsid are two copies of its RNA with each attached to a reverse transcriptase enzyme. To the right is a drawing of a Western blot strip with each band labeled to indicate the HIV antigen to which it corresponds. The location of each band protein in the virus is also shown.

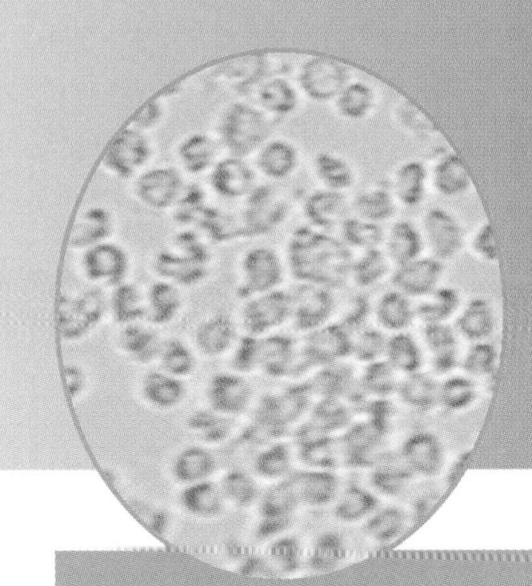

Viruses

10

T4 VIRUS

Viruses are not cellular organisms, having no cell membrane or metabolism of their own. Rather, they are infectious particles minimally consisting of an outer protein coat covering a nucleic acid (either RNA or DNA, not both — another significant difference between viruses and cellular life) as shown in Figure 10-1. In order to replicate (Fig. 10-2), the virus must infect a living cell which provides the raw materials and organelles necessary for viral replication. Thus, viruses are referred to as "obligate *intracellular* parasites." Viruses that have bacterial hosts are referred to as *bacteriophages* or *phages*.

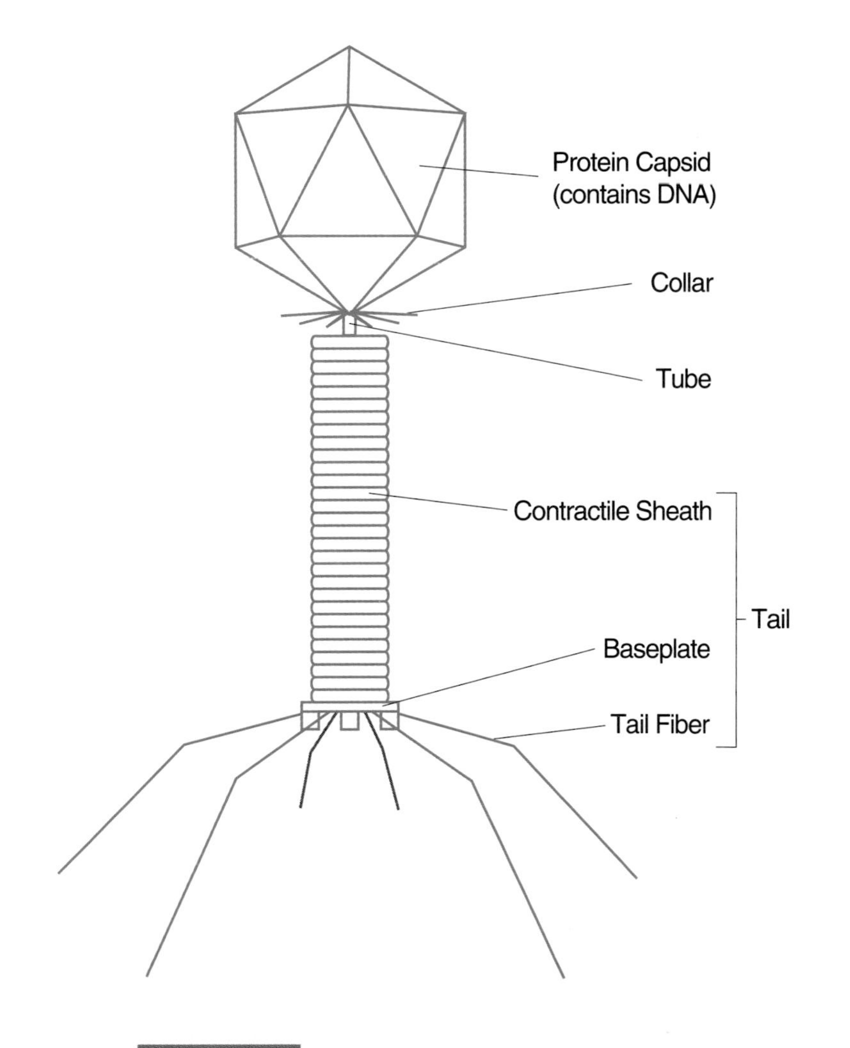

FIGURE 10-1 A diagram of the T4 virus.

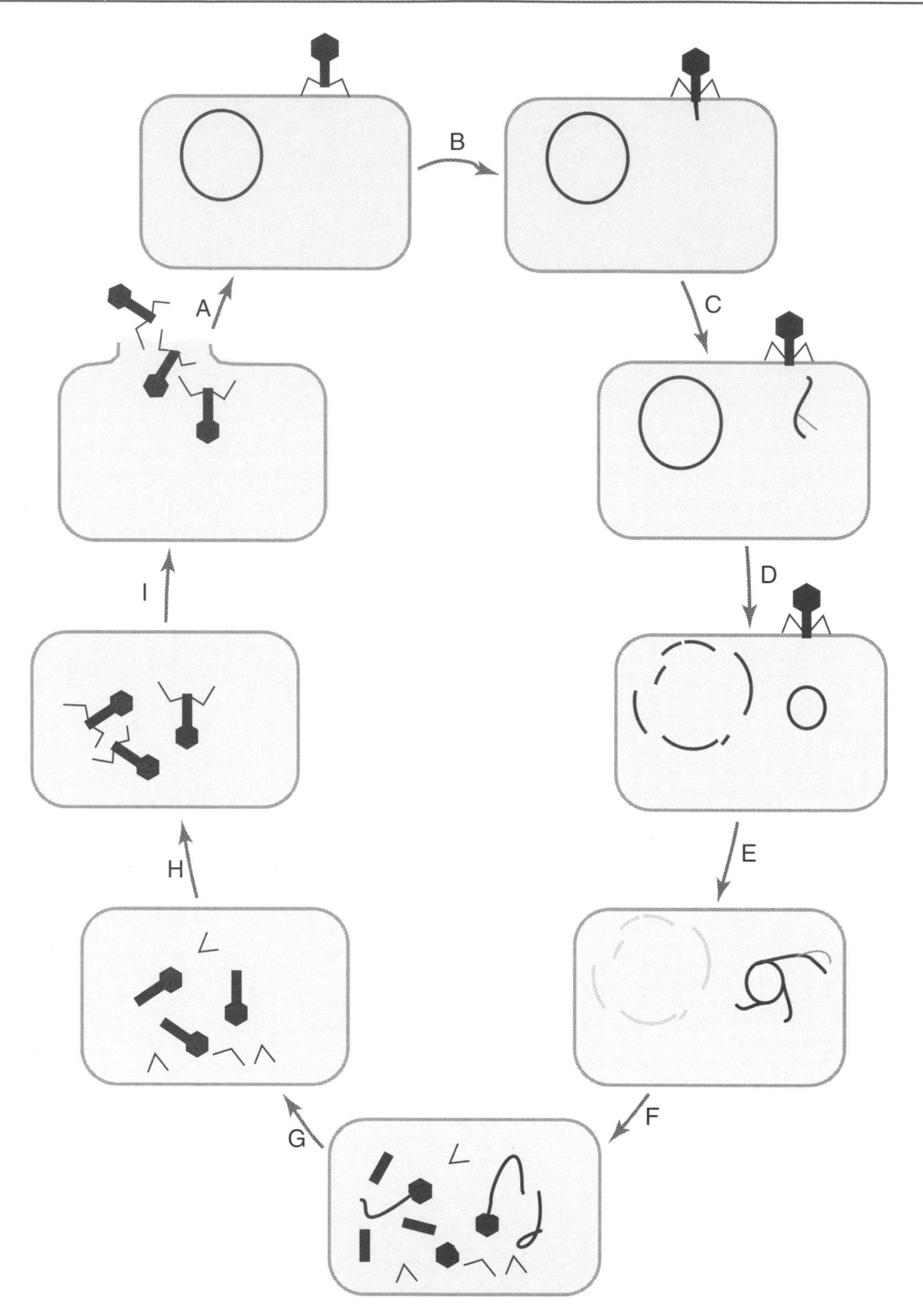

FIGURE 10-2 A schematic diagram of the T4 replicative cycle. *a.* The virus particle (shown in blue) attaches to the host cell. Each virus will infect only a specific host and this specificity is based on the ability of the virus to make attachment with viral receptors on the host. *b.* The virus particle acts like a syringe and injects its DNA (shown in pink) into the host cell. *c.* Viral DNA is not transcribed all at once. Rather, the genes necessary for the early events in replication are transcribed first, with other genes being transcribed as the appropriate time arises. In this diagram, *early* and *middle genes* are being transcribed into mRNA (shown in green). *d.* The viral DNA circularizes and the host DNA is broken apart. *e.* Viral DNA is replicated by a *rolling circle* mechanism. Host DNA is further degraded into nucleotides which are used for viral DNA replication. *Late genes* are also transcribed. *f.* Capsid, tail and tail fiber proteins are made and assemble into their respective components in separate assembly lines. DNA enters the capsids. g & *h.* Assembly continues as first tails and then tail fibers attach to form a complete virus particle (*virion*). *i.* The host cell is lysed and releases the completed virus particles, each capable of infecting another host cell. The typical *burst size* for T4 is about 300 viruses. The whole process from attachment to host lysis takes less than 25 minutes!

HUMAN IMMUNODEFICIENCY VIRUS (HIV)

HIV is the causative agent of AIDS (*A*cquired *I*mmune *D*eficiency *S*yndrome). At first, only a single type of HIV was known, but in 1985 a second HIV was isolated. The two forms are now referred to as HIV-1 and HIV-2, respectively.

HIV (Figs. 9-21, 10-3 and 10-4) is a *retrovirus*. It has a spherical capsid enclosed by a phospholipid envelope derived from the host cell membrane. Within it is a protein core surrounding two single stranded RNA molecules, each of which is associated with a molecule of *reverse transcriptase*. The HIV genome consists of only 9 genes. Glycoprotein spikes emerging from the envelope are involved in attachment to the host cell.

HIV infects cells with CD4 membrane receptors, normally used for antigen recognition, but used by HIV for attachment. A subpopulation of T cells, the T_4 helper cells, are most commonly infected and die as a result. Other cells, such as dendritic cells — a class of antigen presenting cells, macrophages and monocytes may also be infected. After infection, the reverse transcriptase catalyzes the production of viral DNA (provirus) which integrates into the host chromosome and is the template for production of viral RNA and proteins. After assembly, new virions emerge from the host cell by budding and infect other cells. Latent infection, in which no new virus is made, is also a possible outcome of infection.

T_4 helper cells are essential to the normal operation of the immune system since they promote development of immune cells in both humoral and cell-mediated responses. Depletion of T_4 cells cripples the immune system and the patient becomes susceptible to infections by organisms not typically pathogenic to humans. Thus, AIDS is not a single disease, but rather a *syndrome* of diseases characteristic of patients with HIV infection.

HIV is transmitted via body fluids such as blood, breast milk, semen, and vaginal secretions. Infection can occur as a result of sexual intercourse with or blood transfusion from an infected individual. Infection may also occur across the placenta during pregnancy or via contaminated needles used for injection of intravenous drugs. It also may be transmitted to a newborn during delivery or nursing by an infected mother. Casual social contact does not appear to be a route of infection.

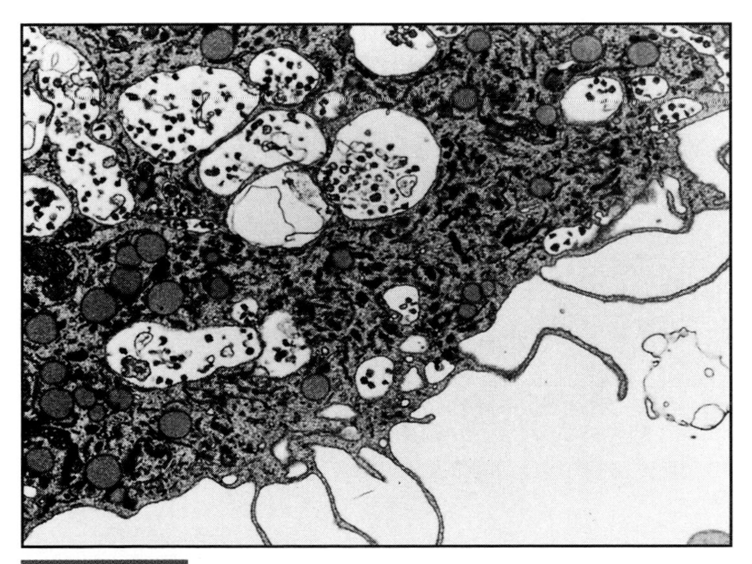

FIGURE 10-3 Electron micrograph of a macrophage from an HIV infected person (X7200). Vacuoles contain viral particles. (Photo courtesy of Dr. Rachel Schrier and Dr. Clayton Wiley.)

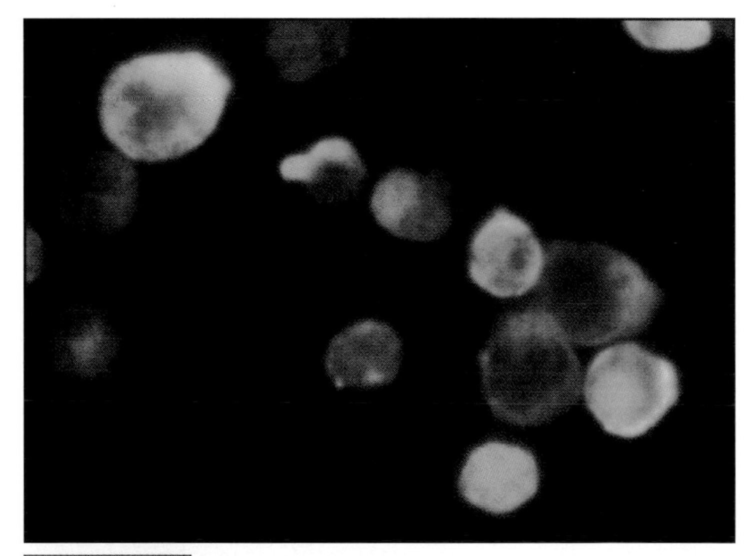

FIGURE 10-4 An indirect fluorescent antibody (IFA) test positive for HIV (X960). The test system consists of HIV-infected cells attached to the glass slide. When patient serum or plasma is added, any anti-HIV antibodies bind to the infected cells. The bright apple green color occurs because fluorescent anti-human immunoglobin antibodies bind to patient antibodies attached to HIV-positive cells.

VIRAL CYTOPATHIC EFFECTS IN CELL CULTURE

PURPOSE This procedure is used for *in vitro* identification of viruses. By determining the cell line(s) in which a virus replicates, how quickly it causes damage, and the type of cytopathic effect it produces, a presumptive identification may be made of a virus in a specimen. Confirmation may be made using a specific serological test.

PRINCIPLE Supplied with the appropriate nutrients and environment, viable virus host cells may be grown in a tube or flat flask. This is a *cell culture*. Incubation is done with the tube on a small angle, which promotes growth only on one side. The cells divide and produce a characteristic monolayer on the container's inside surface.

Different media are used for cell culture at different stages. These are: growth medium and maintenance medium. Growth medium is used to begin a cell culture. When the cell layer is confluent or nearly so, the growth medium is replaced with maintenance medium. Both media are supplemented with amino acids, vitamins and calf serum. To ensure that the serum is free of viral antibodies and certain infectious agents, only fetal, neonatal or agammaglobulinemic calf serum is used. Antibiotics are also included to inhibit bacterial growth. The accumulation of carbon dioxide and the associated acidification of the medium is counteracted by a buffer. A pH indicator (such as phenol red) is used to monitor the effectiveness of the buffer.

A sample suspected of containing virus is introduced into the cell culture. As they replicate, viruses inflict damage upon the host cells called *cytopathic effect* (CPE). Depending on the virus and the host cell, CPEs will be evident after as little as 4 days to as much as four weeks. Most of the time, they start as small spots (*foci*) in the cell layer, then spread outwards. Common damage to the cells includes rounding (either small or large), a change in texture (either granular or hyaline) or formation of a *syncytium* (fusion of infected cells). Figures 10-5, 10-9 and 10-11 illustrate three cell types used in cell culture. Figures 10-6 through 10-8, 10-10, and 10-12 through 10-14 illustrate various CPEs in these three host cells.

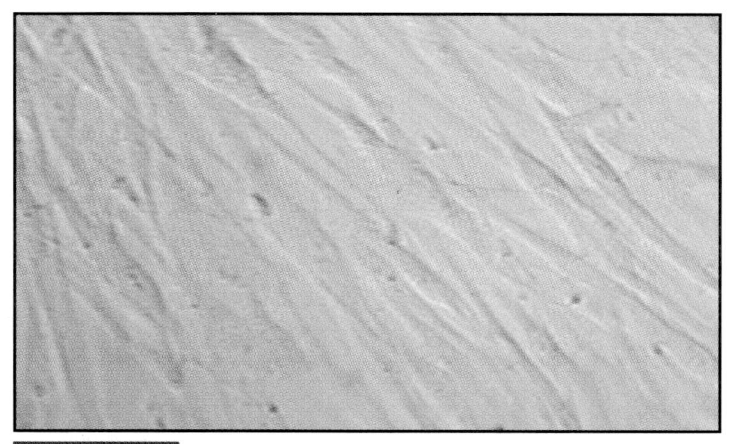

FIGURE 10-5 Normal MRC-5, a cell line of human diploid fibroblasts (X480). A cell line has a limited number (about 50) of passages (transfers) before it is no longer useful.

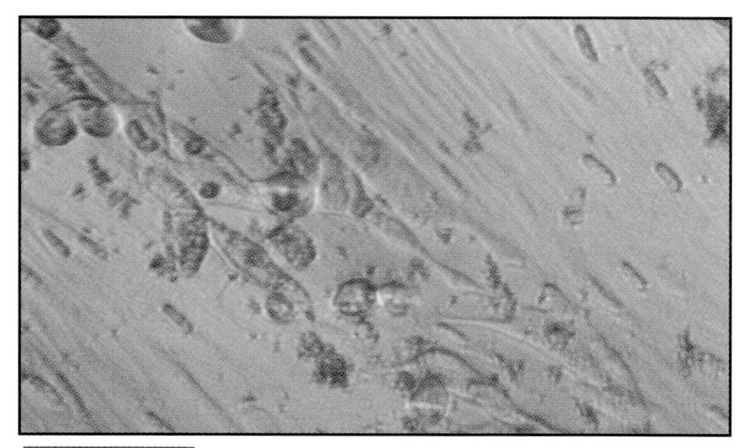

FIGURE 10-6 Cytomegalovirus (CMV) CPE in a cell culture of MRC-5 (X480). The infected fibroblasts form a row and are rounded and hyaline.

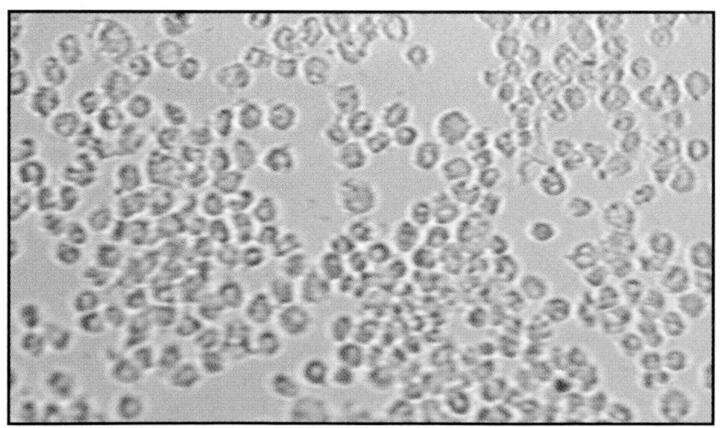

FIGURE 10-7 Enterovirus CPE in a cell culture of MRC-5 (X480). The infected cells become small (pycnotic) and round.

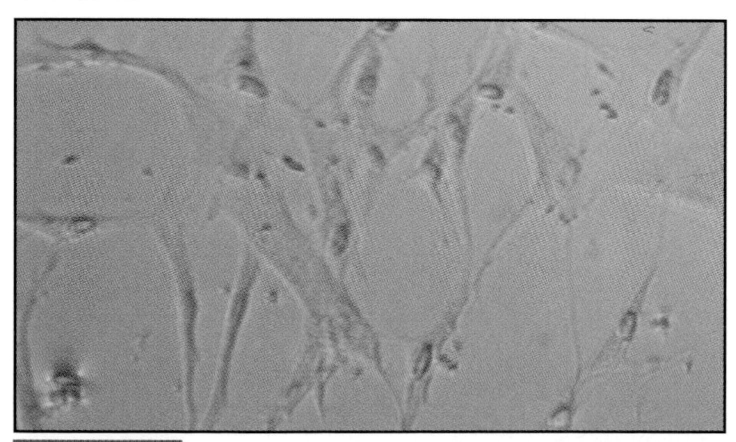

FIGURE 10-8 Varicella-zoster (VZ) CPE in a cell culture of MRC-5 (X480). Enlarged cells with odd shapes characterize the VZ CPE.

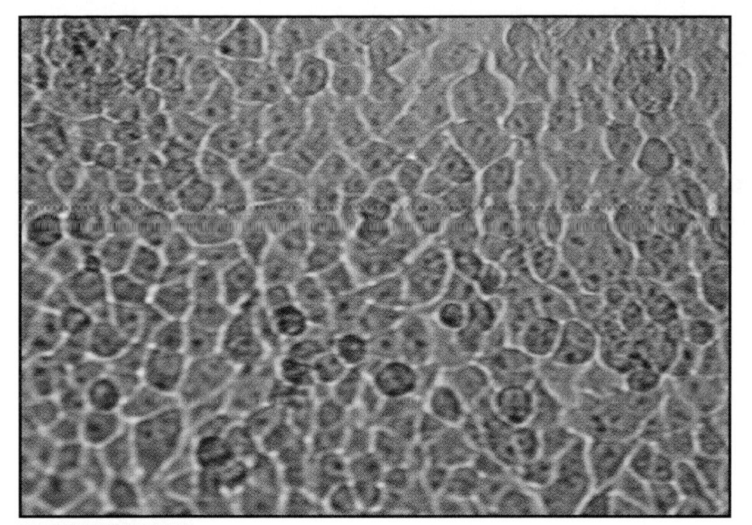

FIGURE 10-9 Normal HeLa cells in culture (X480). HeLa is an established cell line; that is, it has lasted more than 70 passages.

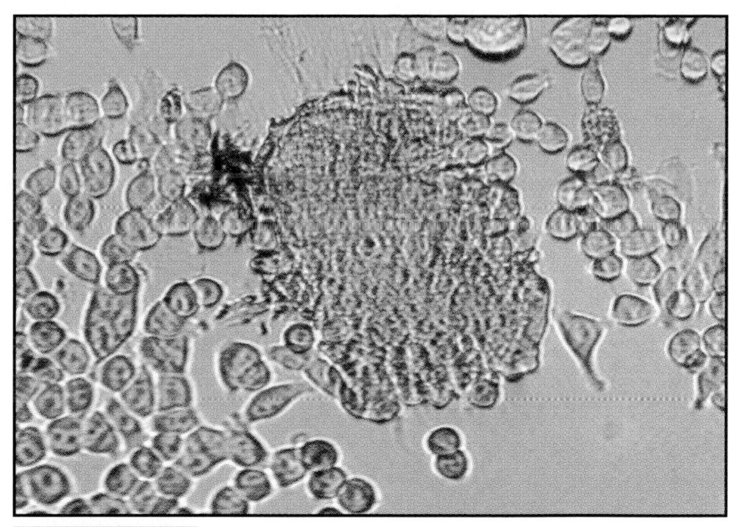

FIGURE 10-10 Respiratory Syncytial Virus (RSV) CPE in HeLa cell culture (X480) produces characteristic syncytia.

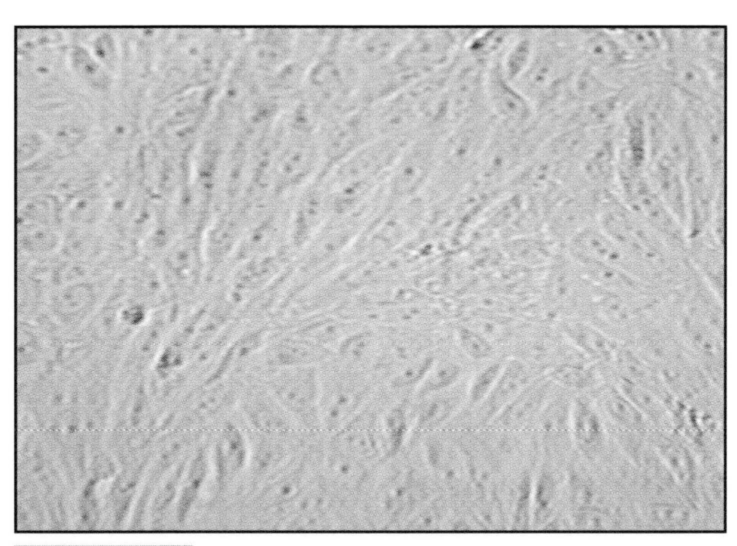

FIGURE 10-11 Normal African Green Monkey Kidney (AGMK) cell culture (X480).

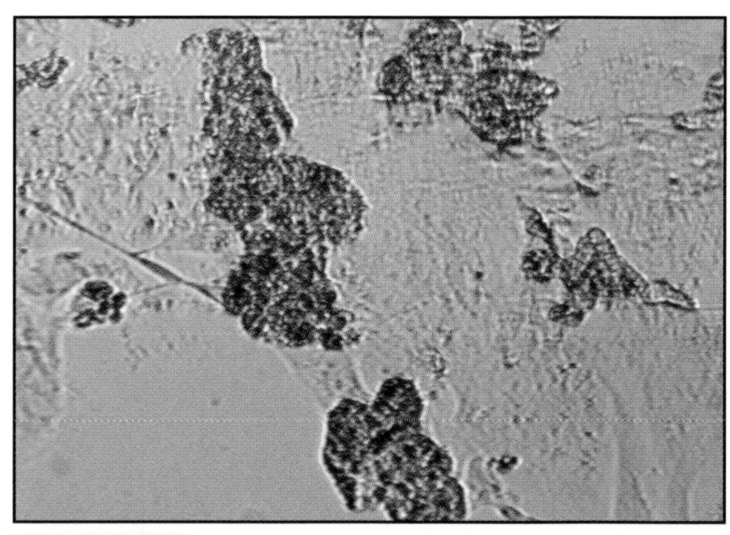

FIGURE 10-12 Measles virus CPE in AGMK cell culture (X480). Syncytia formation is the typical CPE for measles virus.

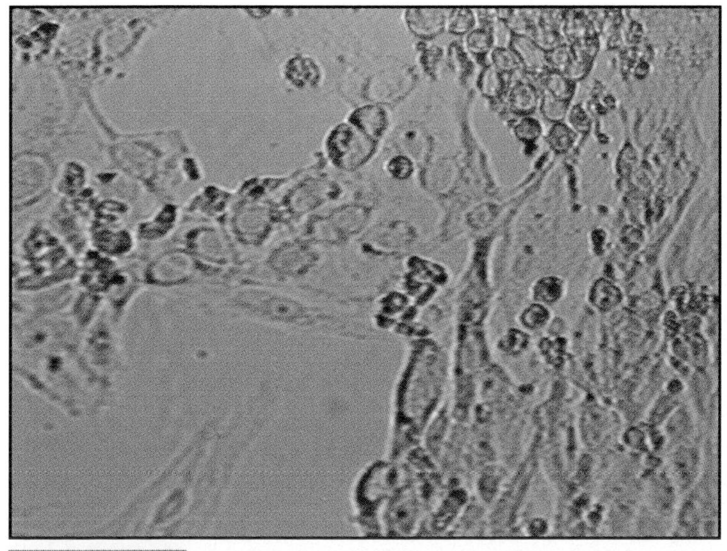

FIGURE 10-13 Influenza A virus CPE in AGMK cell culture (X480). Cell degeneration and syncytia formation characterize the CPEs of influenza viruses.

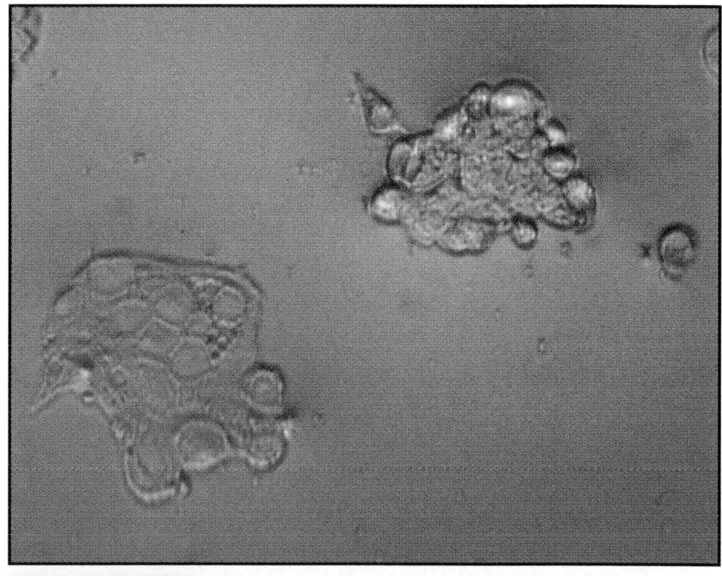

FIGURE 10-14 Adenovirus CPE in AGMK (X480). The typical CPE consists of rounded cells clustered like grapes.

HEMADSORPTION IN CELL CULTURE

PURPOSE Hemadsorption is used for the presumptive identification of influenza, parainfluenza, and sometimes mumps virus, since these viruses do not produce much cytopathic effect (CPE) in cell culture. Confirmation is accomplished by serological tests.

PRINCIPLE Infection with influenza, parainfluenza, or mumps virus results in viral glycoproteins being present in the infected cell's membrane. When these viruses emerge from the host, they carry the glycoproteins in their envelope (which is actually host cell membrane) and use them for attachment to and penetration of a new host. They also have the ability to adsorb (attach) to guinea pig RBCs. This property is exploited in the hemadsorption test.

Because influenza, parainfluenza, and mumps viruses often produce no CPE in cell culture, a hemadsorption test can be a useful diagnostic tool. After incubation of the cell culture inoculated with the patient's sample, guinea pig RBCs may be added to the medium. If the virus is present, the infected cells will have viral glycoproteins in their membranes, and the RBCs will adsorb to them (Fig. 10-15).

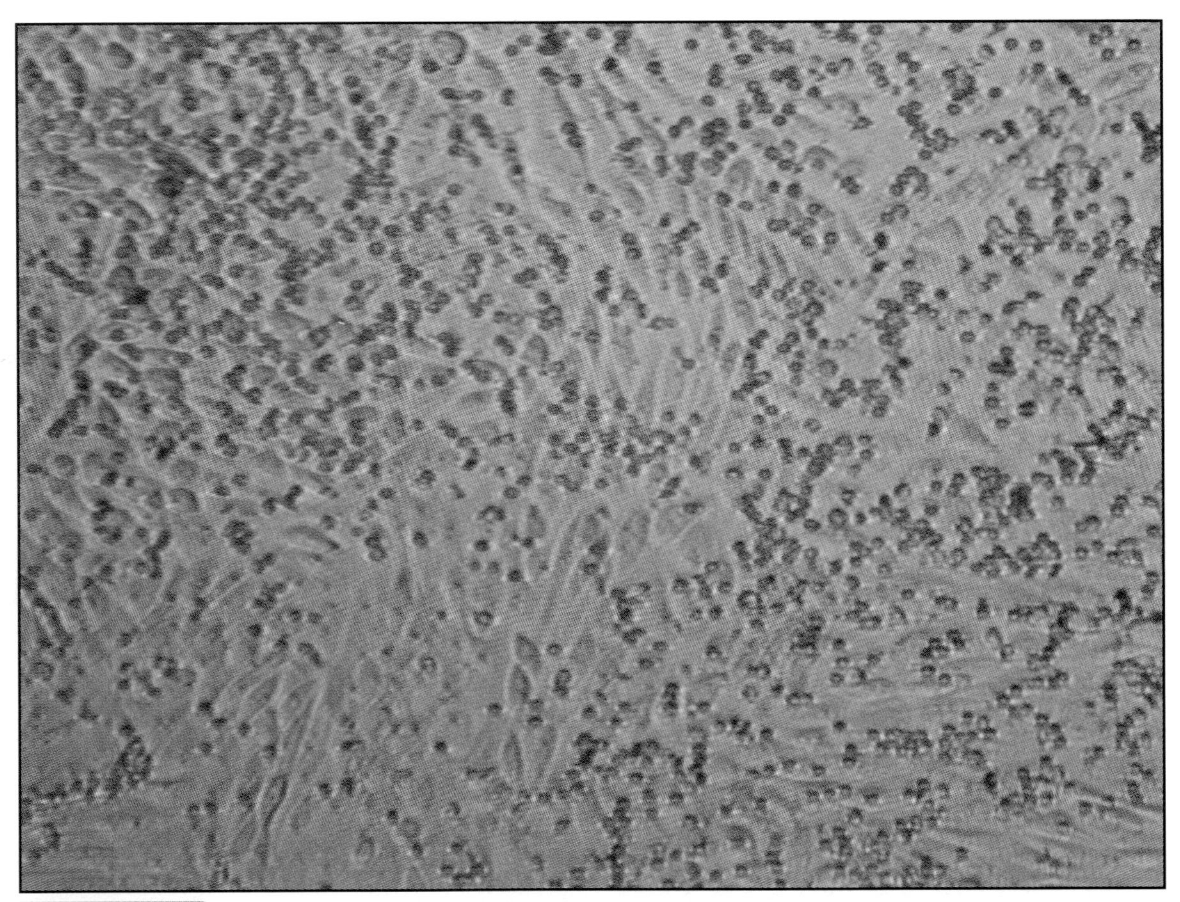

FIGURE 10-15 Hemadsorption of guinea pig RBCs by human diploid fibroblast cells infected with parainfluenza virus (X480). Notice that the RBCs are always associated with the cells.

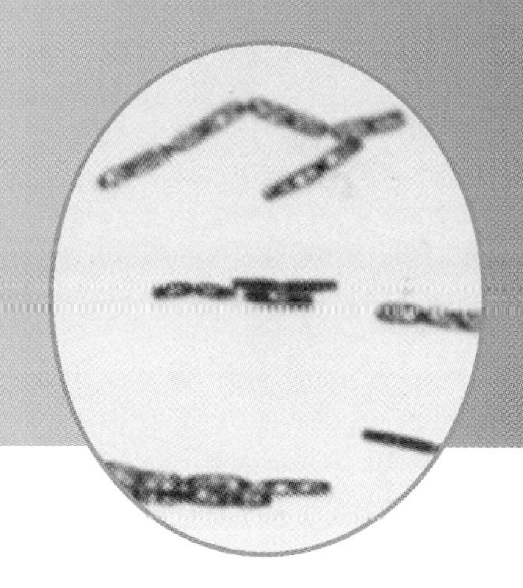

Bacterial Pathogens 11

Aeromonas hydrophila

Aeromonas hydrophila is a motile, facultatively anaerobic, Gram-negative rod found in sink drains, water faucets, and aquatic habitats around the world. It causes infections in both warm blooded animals and cold blooded aquatic animals. The organism enters the body by colonizing open wounds or as a contaminant in food or water. When ingested, it adheres to the intestinal mucous membrane and produces the cholera-like *Asao toxin* which causes watery diarrhea. Untreated infections can lead to septicemia, endocarditis, and meningeal infection.

Differential characteristics *A. hydrophila* ferments sucrose and mannitol, is positive for ONPG, indole, lysine decarboxylase and esculin hydrolysis, and negative for ornithine decarboxylase.

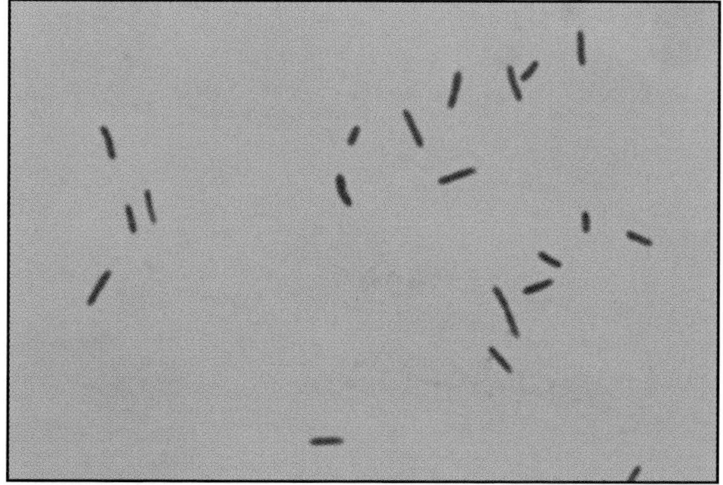

FIGURE 11-1 Simple crystal violet stain of *Aeromonas hydrophila* (X1600).

Bacillus anthracis

Bacillus anthracis is a nonmotile, nonhemolytic, encapsulated, facultatively anaerobic, endospore-forming, Gram-positive rod. It causes *anthrax* in animals and more rarely, in humans. The intestinal tract of asymptomatic large animals is its principal habitat, however, *B. anthracis* spores have been known to remain viable in the soil for more than twenty years. Three types of anthrax occur in humans: cutaneous anthrax, inhalation anthrax ("woolsorter's disease") and gastrointestinal anthrax. The cutaneous and inhalation forms are most frequently the result of contact with spores carried in wool or goat hair; gastrointestinal disease usually results from the consumption of contaminated meat. Two virulence factors account for the organism's pathogenicity: 1) a capsule which enables it to survive phagocytosis and disseminate via the bloodstream, and 2) secretion of a powerful tissue destroying exotoxin. Although all forms of the disease can be fatal, inhalation almost always results in death within hours of contact.

Differential characteristics *B. anthracis* will not grow on phenylethyl alcohol blood agar and is gelatinase-negative.

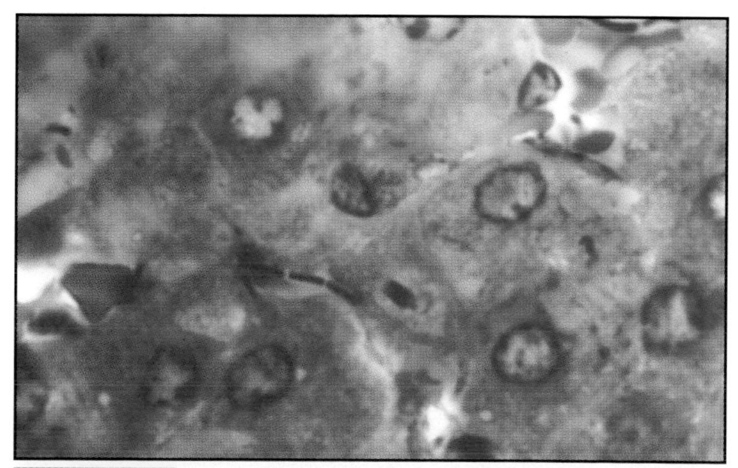

FIGURE 11-2 Gram stain of *Bacillus anthracis* in the capillaries of mouse liver tissue (X1600). Although not shown, spores are central and oval.

Bacillus cereus

Bacillus cereus is a motile, facultatively anaerobic, endospore-forming, Gram-positive rod. It is one of the two principal pathogens in the *Bacillus* genus, the other being *Bacillus anthracis* (shown pn page 121). *B. cereus* has become an increasingly serious cause of food poisoning in the United States due to the production of both heat-stable and heat-labile exotoxins. The heat-stable toxin, which causes vomiting, is a particular problem in Oriental style rice dishes. It is common practice in many restaurants to cook large amounts of rice, drain it slowly, and keep it warm for hours. The spores that survive the initial cooking germinate and multiply in the warm, moist environment. Further cooking, such as with fried rice, does not destroy the toxin. Conversely, the heat-labile enterotoxin produced by *B. cereus*, cannot withstand further cooking and thus is a problem in dishes such as cream sauces. Ingestion of heat-labile toxin results in diarrhea within 8 to 24 hours.

Differential characteristics *B. cereus* produces acid from glucose, maltose and salicin fermentation, and is lecithinase- and gelatinase-positive.

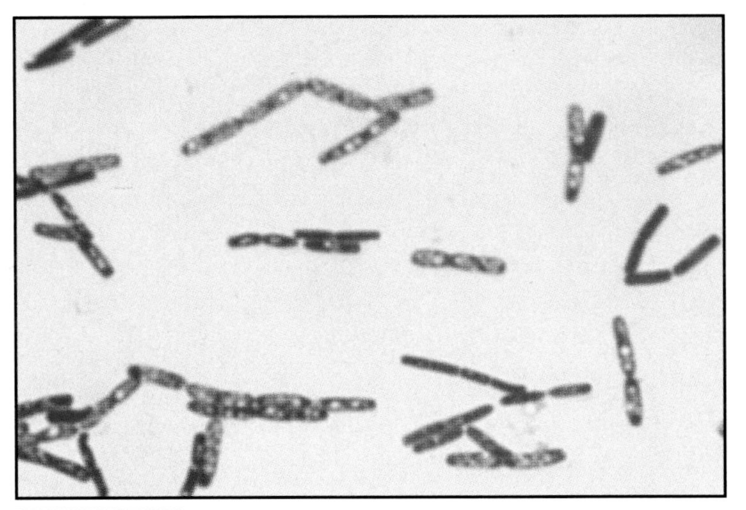

FIGURE 11-3 Gram stain of *Bacillus cereus* (X1600). Note the unstained endospores and parasporal bodies, which appear as clear spots in the cell.

Bacteroides fragilis

Bacteroides fragilis is a nonmotile, anaerobic, encapsulated, nonsporing, pleomorphic, Gram-negative rod. It constitutes part of the human intestinal flora. In fact an entire group of bacteria from the genus *Bacteroides*, indigenous to the lower gut, is called the "*B. fragilis* group." Concentrations in excess of 10^{11} bacterial cells per gram of feces are not uncommon. Not surprisingly, *B. fragilis* is the principal opportunistic pathogen of the group, and the most common anaerobe isolated from human infections. The most significant infection it causes is intra-abdominal abscess (*peritonitis*) due to spillage of intestinal contents into the abdominal cavity from a ruptured appendix or a penetrating wound.

Differential characteristics *B. fragilis* does not ferment arabinose, melezitose or salicin. It grows well in the presence of 20% bile, is indole-negative and catalase-positive.

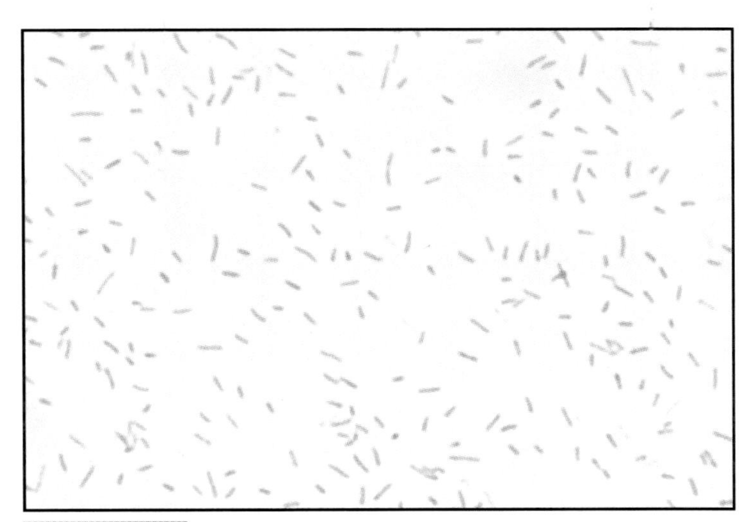

FIGURE 11-4 Gram stain of *Bacteroides fragilis* (X1600). Note the pleomorphism.

Bordetella pertussis

Bordetella pertussis is a small, nonmotile, obligately aerobic, Gram-negative coccobacillus. It is the cause of the disease known as *pertussis* or "whooping cough," characterized by violent coughing, vomiting and gasping for breath. It can infect anyone with no immunity or diminished immunity, but is most severe and communicable among infants less than one year of age. Humans are the exclusive host of this organism, and asymptomatic or unrecognized symptomatic adults are the likely reservoirs. This highly communicable organism infects greater than 90% of unimmunized people exposed. The bacteria enter the mouth or nasopharynx as aerosols then attach themselves to the respiratory cilia. Although the active disease caused by *B. pertussis* is a superficial infection, it secretes an array of virulence factors which destroy the underlying epithelial tissue and act to impair the body's natural defenses.

Differential characteristics *B. pertussis* is oxidase-positive, urease-negative, and nitrate-negative.

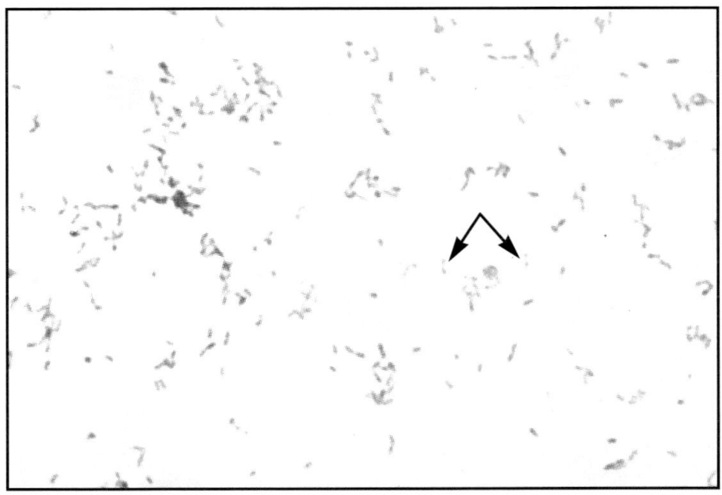

FIGURE 11-5 Gram stain of *Bordetella pertussis* (X1600). Cells are usually arranged singly or in pairs and exibit bipolar staining (arrows).

Borrelia burgdorferi

Borrelia burgdorferi is a motile, microaerophilic, Gram-negative spirochete that causes *Lyme disease.* Transmission is by bites of certain ticks from the genus *Ixodes* which ordinarily live and feed on rodents and deer. The resulting infection is characterized by three distinct stages: Stage 1. Localized skin lesion accompanied by headache, fatigue and malaise; Stage 2. Weeks to months later, the inflammation and pain become generalized with the possible development of meningoencephalitis or myocarditis; Stage 3. After months or even years of latency the infection becomes chronic, producing severe headaches, muscle and joint pain, and secondary skin lesions. Although the mechanism is not completely understood, the organism is capable of surviving in the nervous system and joints for years.

Differential characteristics The most effective tests for clinical identification of *B. burgdorferi* are serologic tests such as indirect immunofluorescence assay (IFA) or ELISA using specific bacterial antigens.

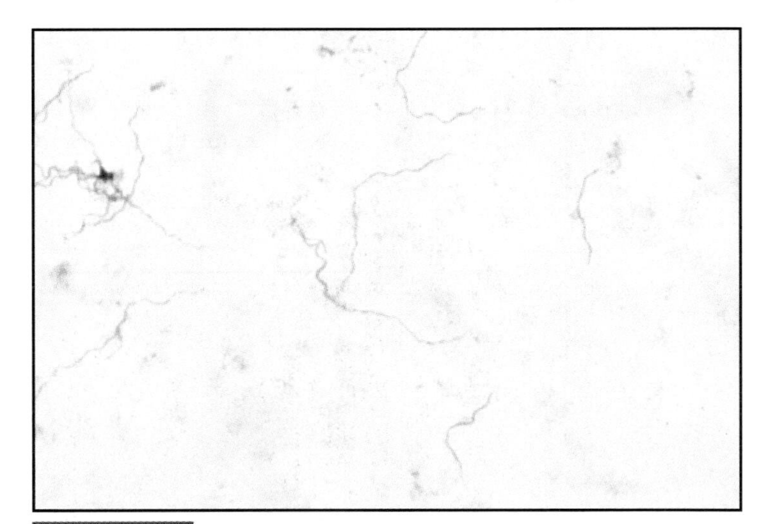

FIGURE 11-6 Giemsa stain of *Borrelia burgdorferi* (X1600).

Brucella melitensis

Brucella melitensis, the most virulent species of the genus *Brucella*, is a small, nonmotile, aerobic, Gram-negative coccobacillus. Although most brucellae cause *brucellosis* (sometimes called undulant fever), *B. melitensis* is responsible for the most severe symptoms. Each of the infective agents of brucellosis has a domestic animal reservoir: *B. melitensis* (goats and sheep), *B. abortus* (cattle), *B. suis* (swine), and *B. canis* (dogs). Each organism has a different carrier but all infect humans three principal ways: ingestion, direct contact through abraded skin, or inhalation. *B. melitensis* is an intracellular parasite which can continue to multiply even within phagocytic cells. It eventually destroys the phagocyte thereby liberating more bacteria which can be transported throughout the body by peripheral circulation (*hematogenic* dissemination). Acute and chronic brucellosis can lead to invasion of virtually all organ systems.

Differential characteristics The most effective test for clinical identification of *B. melitensis* is tube agglutination using a specific bacterial antigen. Other serologic tests include: radioimmunoassay (RIA), complement fixation, and ELISA.

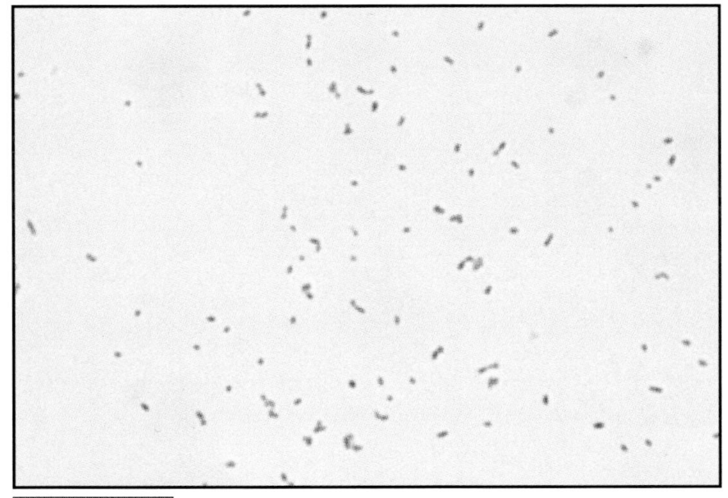

FIGURE 11-7 Gram stain of *Brucella melitensis* (X1600). Usually arranged singly, these poorly staining cells were colored by allowing the counterstain to stand for 5 minutes.

Campylobacter jejuni

Campylobacter jejuni is a motile, micraerophilic, capnophilic (requires increased CO_2), nonsporing, Gram-negative helical rod. It is a common worldwide human pathogen found in house pets, domestic animals, poultry and waterfowl. Transmission is usually by ingestion of raw milk, undercooked poultry or meat, or contaminated water. Once inside the small intestine the organism corkscrews into the mucous layer, multiplies, and secretes a cholera-like enterotoxin which causes watery diarrhea. In most cases the infection is self-limiting and the diarrhea ends after a few days. Some strains, however, secrete a cytotoxin which destroys local cells and causes bloody diarrhea. Dissemination by the bloodstream may result in salmonellosis-like enteritis and extraintestinal infection. In recent years, this organism has been implicated in a variety of infections including septic arthritis, meningitis, proctocolitis, and most recently, Guillain-Barré syndrome, a degenerative nerve disorder.

Differential characteristics *C. jejuni* reduces nitrate to nitrite and hydrolyzes hippurate.

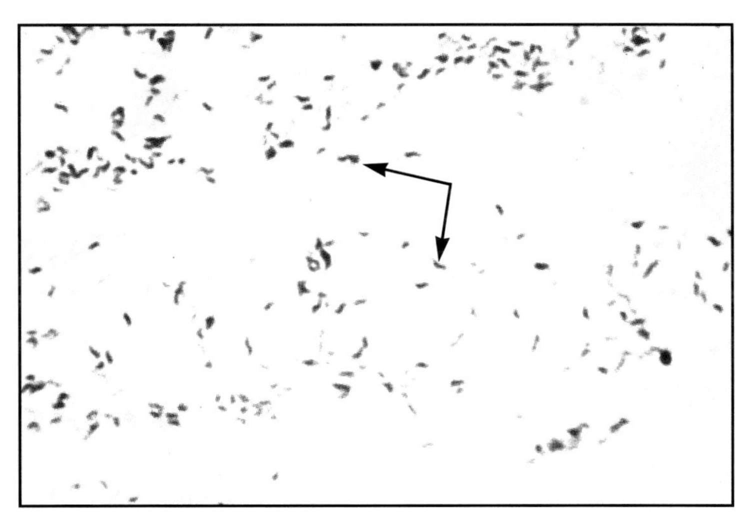

FIGURE 11-8 Simple crystal violet stain of *Campylobacter jejuni* (X1600). Note the characteristic "S" and "gull wing" shapes of the paired organisms (arrows).

Citrobacter diversus

Citrobacter diversus is a motile, facultatively anaerobic, Gram-negative rod. This organism — part of the normal human intestinal flora — is a common opportunistic pathogen, having been isolated from wound infections, soil, sewage, food, water, and urine. Able to form capsules and evade phagocytosis it is a cause of neonatal meningitis and sepsis. Transmission is primarily through ingestion of contaminated food or water.

Differential characteristics *C. diversus* ferments glucose and cellobiose, is oxidase- and lysine decarboxylase-negative, and ornithine decarboxylase- and indole-positive.

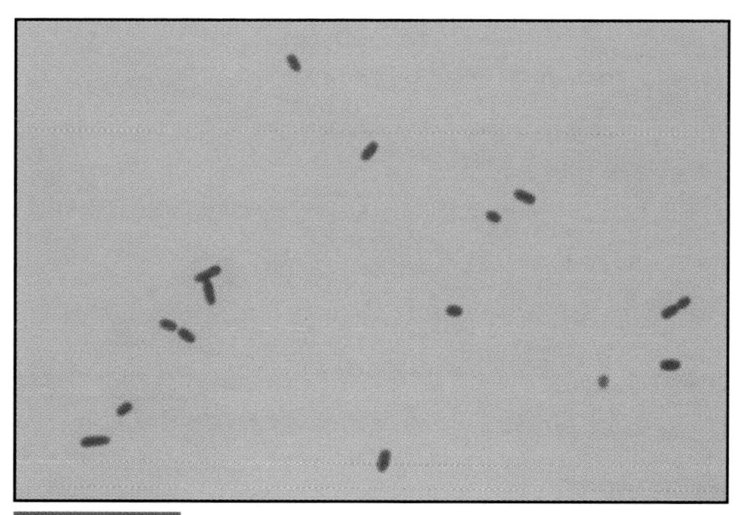

FIGURE 11-9 Simple crystal violet stain of *Citrobacter diversus* (X1600). These straight rods usually appear singly.

Citrobacter freundii

Citrobacter freundii, like *C. diversus* (left), is a motile, encapsulated, facultatively anaerobic, Gram-negative rod. Although it is a normal inhabitant of the human lower intestine it is an important opportunistic pathogen when introduced into other organ systems. Organisms have been isolated from water, food, urine, sewage, and wound infections. It is a well-known causative agent of sepsis and gastrointestinal infection. As with several other members of *Enterobacteriaceae*, *C. freundii* infections are transmitted primarily by fecally contaminated food or water.

Differential characteristics *C. freundii* ferments glucose, does not ferment cellobiose, and is negative for oxidase, ornithine decarboxylase, lysine decarboxylase, indole, and urease.

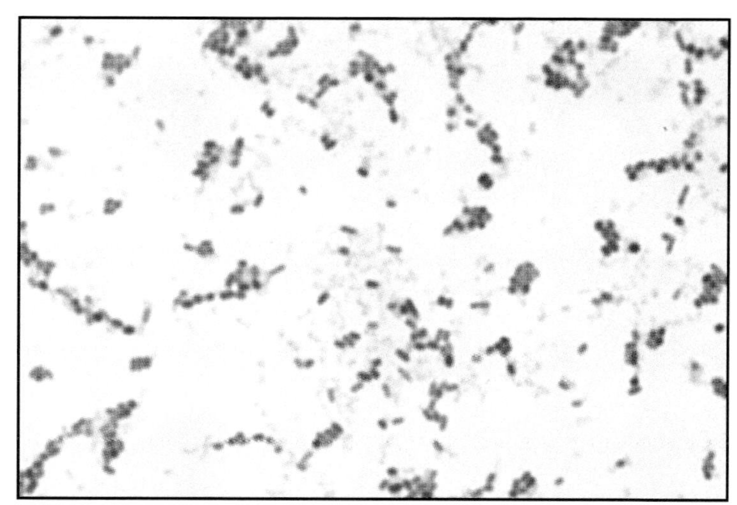

FIGURE 11-10 Simple crystal violet stain of *Citrobacter freundii* (X1600). These straight rods usually appear singly or in pairs.

Clostridium botulinum

Clostridium botulinum is a motile, anaerobic, endospore-forming, Gram-positive rod. It ordinarily lives in animal intestines and can be found in soil worldwide. Under proper growth conditions — usually insufficiently heated, home-canned foods — different strains of this organism produce at least seven distinct and potent neurotoxins. Four of these heat-labile toxins are known to cause *botulism* in humans. When ingested, the toxin prevents release of the neurotransmitter acetylcholine, creating flaccid paralysis, and eventually respiratory failure. Although rare, two other forms of botulism occur — infant botulism and wound botulism. In infant botulism, the organism actually colonizes the intestine where it *then* produces the toxin. In wound botulism, spores contaminating a wound germinate and grow, producing toxins which are then absorbed by the body.

Differential characteristics For final identification of *C. botulinum*, a fermentation profile taken from commercially available prereduced anaerobically sterilized (PRAS) media or the determination of specific metabolic products using gas-liquid chromatography is recommended.

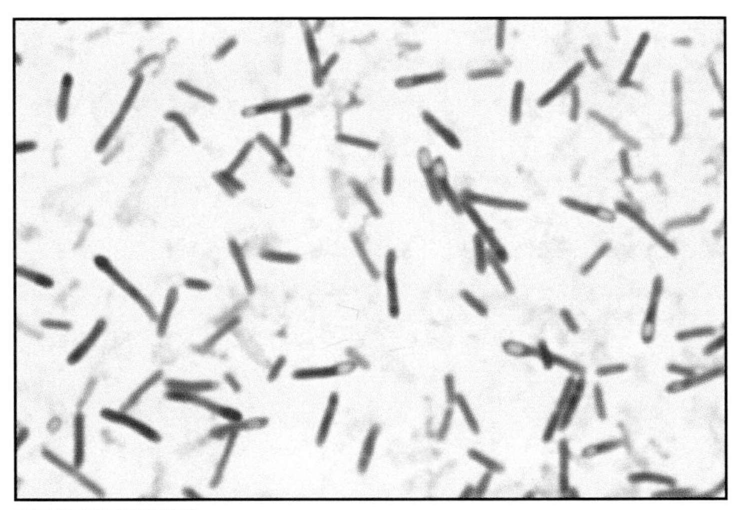

FIGURE 11-11 Gram stain of *Clostridium botulinum* (X1600). Note the oval subterminal endospores which distend the cells.

Clostridium dificile

Clostridium dificile is a motile, anaerobic, endospore-forming, Gram-positive rod. It can be found throughout the world in water, soil, a variety of animal intestines, and the intestinal tracts of nearly half the healthy infants under one year of age. It is the causative agent of antibiotic-associated diarrhea (AAD), pseudomembranous colitis (PMC), and the more severe, antibiotic-associated colitis (AAC). The spores it produces are highly resistant to antibiotics and virtually impossible to eliminate from the environment. Not surprisingly, it is the most common cause of hospital-acquired diarrhea. In humans, when the normal intestinal flora have been suppressed by antibiotic therapy, previously dormant intestinal *C. dificile* will proliferate. It does not enter host cells but produces cytotoxins which damage the intestinal cells and facilitate fluid production.

Differential characteristics For final identification of *C. dificile*, a fermentation profile taken from commercially available prereduced anaerobically sterilized (PRAS) media or the determination of specific metabolic products using gas-liquid chromatography is recommended.

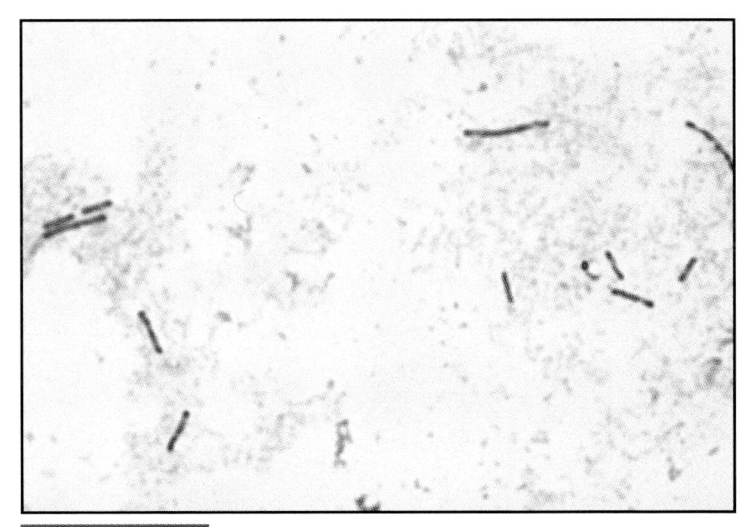

FIGURE 11-12 Gram stain of *Clostridium dificile* (X1600). Note the oval subterminal endospores.

Clostridium perfringens

Clostridium perfringens is a nonmotile, anaerobic, endospore-forming, Gram-positive rod. Although spores are produced by this organism, they are rarely seen in stained preparations. *C. perfringens* is very common in soil but can also be found in a variety of animal intestinal tracts. In humans it has been associated with food poisoning resulting in severe gastroenteritis with necrosis of the small and large intestine, sepsis, and widespread systemic infection. Its best known mode of attack, however, is wound contamination, frequently resulting in myonecrosis (*gas gangrene*). Damaged tissue, especially that encountered during wartime, provides the perfect anaerobic environment for this organism to thrive. When conditions are favorable the organism produces a highly cytolytic toxin which destroys host muscle tissue locally and quickly spreads throughout the body creating a fever, sweating, low blood pressure, and eventually renal failure.

Differential characteristics For final identification of *C. perfringens*, a fermentation profile taken from commercially available prereduced anaerobically sterilized (PRAS) media or the determination of specific metabolic products using gas-liquid chromatography is recommended.

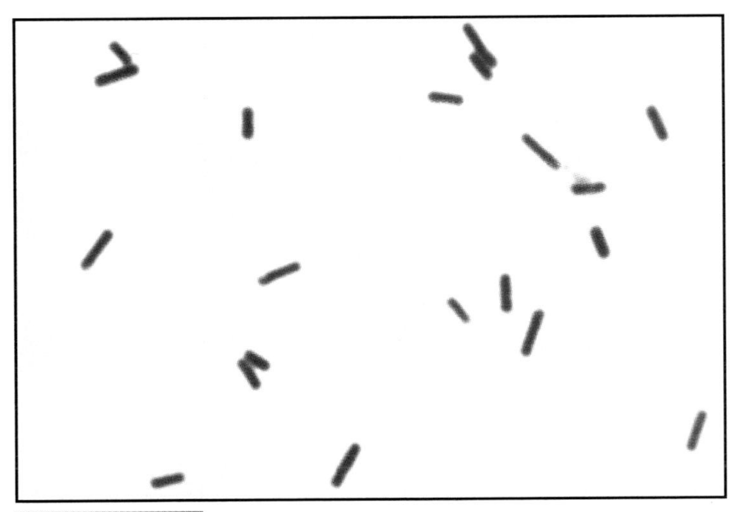

FIGURE 11-13 Gram stain of *Clostridium perfringens* (X1600). Note the blunt ends of the cells and the absence of visible endospores.

Clostridium septicum

Clostridium septicum is a motile, anaerobic, endospore-forming, Gram-positive rod. It is a causative agent of bacteremia, severe necrosis of the appendix and intestine, and less commonly, gas gangrene. Although a high percentage of adults carry it asymptomatically in their appendices, its habitat has not been fully established. Entry to the bloodstream is believed to occur from this area. Although not completely understood, *C. septicum* is frequently associated with certain types of cancer. Patients with *C. septicum* bacteremia are also likely to be suffering from another underlying disease such as leukemia, lymphoma or carcinoma of the large intestine.

Differential characteristics For final identification of *C. septicum*, a fermentation profile taken from commercially available prereduced anaerobically sterilized (PRAS) media or the determination of specific metabolic products using gas-liquid chromatography is recommended.

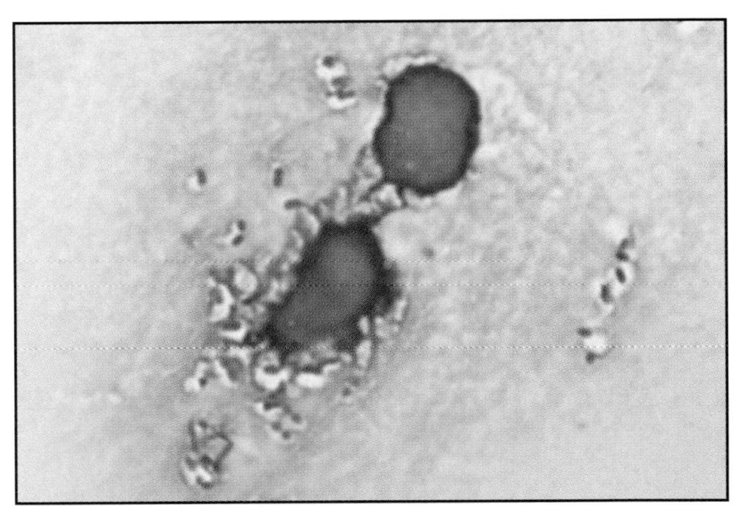

FIGURE 11-14 Simple crystal violet stain of *Clostridium septicum* in a human blood smear (X1600). Although not shown in this preparation, the spores are oval, subterminal and distend the cell.

Clostridium tetani

Clostridium tetani is a motile, anaerobic, endospore-forming, Gram-positive rod. It is very common in the soil and can also be found in a variety of animal intestinal tracts, including humans. Of all the important pathogens in the genus, *C. tetani* is possibly the best known because of the wide-ranging familiarity with the disease it is responsible for — tetanus. Commonly called "lockjaw," tetanus has all but been eradicated in the United States by tetanus toxoid vaccine, although it remains a serious problem in undeveloped countries. Like botulism, tetanus is strictly a toxigenic disease. Usually, *C. tetani* spores enter through a traumatic or puncture wound where they germinate, grow and release the toxin *tetanospasmin*. Tetanospasmin is absorbed and transmitted by motor neurons to the central nervous system where it permanently binds to neurons and blocks the release of inhibitory neurotransmitters. The result is a descending severe muscle spasm which eventually paralyzes the chest muscles resulting in respiratory failure.

Differential characteristics For final identification of *C. tetani*, a fermentation profile taken from commercially available prereduced anaerobically sterilized (PRAS) media or the determination of specific metabolic products using gas-liquid chromatography is recommended.

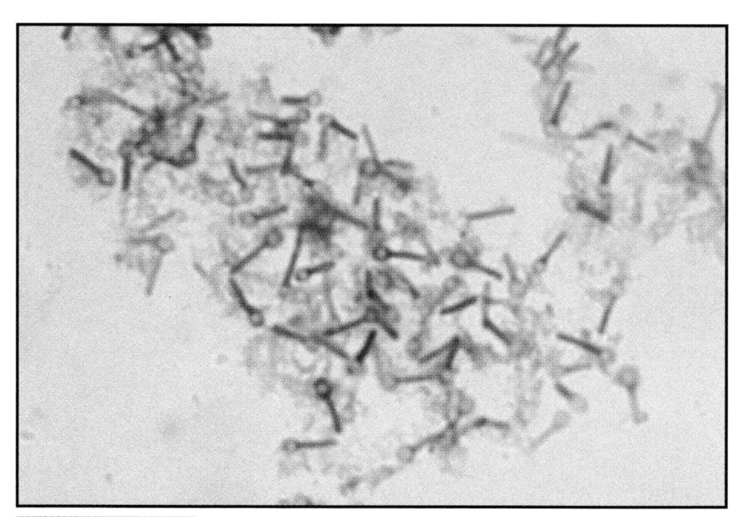

FIGURE 11-15 Gram stain of *Clostridium tetani* (X1600). Note the round, terminal endospores. Occasionally the spores are oval or subterminal.

Corynebacterium diphtheriae

Corynebacterium diphtheriae is a nonmotile, nonsporing, Gram-positive rod. It is the toxigenic agent of *diphtheria* and *cutaneous diphtheria*. Although it can live for months in the environment, *C. diphtheriae* is transmitted most often from person to person in aerosol droplets. Diphtheria is characterized by two distinct syndromes: local respiratory infection and systemic poisoning from absorption of the cytotoxin produced at the local site. In the former, *C. diphtheriae* produces a thick membrane which obstructs the upper respiratory airway. In the latter, cell-killing toxin is dispersed throughout the body, frequently resulting in myocarditis and congestive heart failure. Cutaneous diphtheria is a result of bacterial entry through an open wound causing a local infection with similar systemic effects as the inhalation disease.

Differential characteristics *C. diphtheriae* produces acid from glucose and maltose fermentation, and no acid from sucrose. It is negative for urease, pyrazinamidase, and alkaline phosphatase.

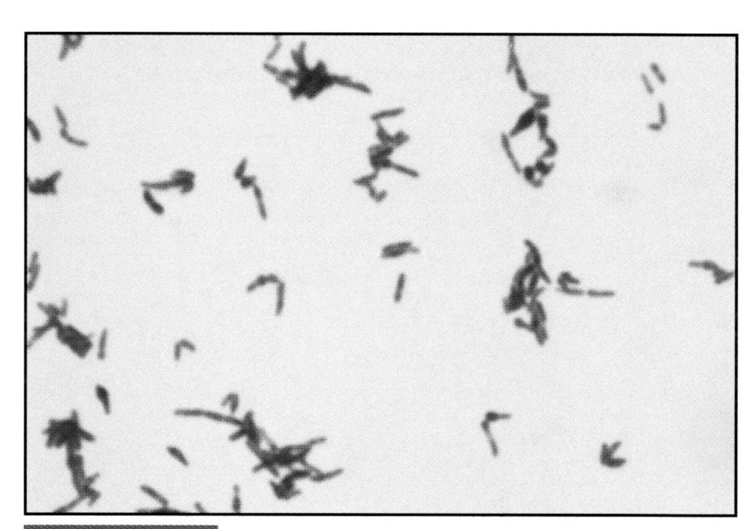

FIGURE 11-16 Gram stain of *Corynebacterium diphtheriae* (X1600). Cells are usually club-shaped and frequently arranged as palisades or "V" configurations.

Enterobacteriaceae

Enterobacteriaceae (the enterics) is a large family of Gram-negative rods, all of which are inhabitants of the human intestinal tract. All enterics are facultative anaerobes and all ferment glucose. With the exception of some strains of *Salmonella*, *Shigella*, *Yersinia*, and *Escherichia coli*, the enterics rarely cause gastrointestinal problems. However, many strains can cause primary or secondary infections outside of the intestinal tract. Important nosocomial pathogens from *Enterobacteriaceae* include species of *Enterobacter*, *Citrobacter*, *Klebsiella*, *Proteus*, *Serratia*, *Providencia, Morganella*, and *Hafnia*. These organisms, ordinarily not a significant threat to healthy individuals, can cause severe infections when introduced into open wounds or otherwise sterile regions of the body. Understandably, with the increasing numbers of antibiotic resistant strains over the last few decades, these organisms have risen in clinical significance. Certain species have been isolated from the CSF of meningitis patients and others are known to cause pneumonia. It is estimated that 80% of all Gram-negative isolates in clinical laboratories are members of *Enterobacteriaceae* which accounts for nearly three-fourths of urinary tract infections and nearly half of septicemia cases.

Differential characteristics Several commercial multiple test systems are available for the identification of the individual species of *Enterobacteriaceae*, including Enterotube® II, API 20 E®, and BBL Crystal™.

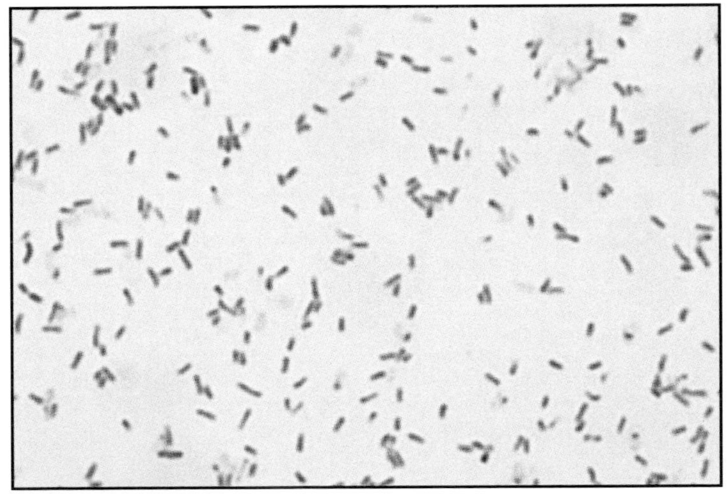

FIGURE 11-17 Gram stain of *Enterobacter aerogenes* from the family *Enterobacteriaceae* (X1600). All members of *Enterobacteriaceae* are straight Gram-negative rods.

Escherichia coli

Escherichia coli is a motile, facultatively anaerobic, Gram-negative rod. Although it is a normal inhabitant of the human intestine, it also may be pathogenic, causing urinary tract infections, sepsis, meningitis, and various diarrheal diseases. There are several different strains of diarrhea-causing *E. coli* that are differentiated by their virulence properties. These are: enteropathogenic *E. coli* (EPEC) that causes diarrhea in infants, enterotoxigenic *E. coli* (ETEC) that is responsible for infant diarrhea and traveler's diarrhea, enterohemorrhagic *E. coli* (EHEC) that is associated with hemorrhagic colitis and hemolytic uremic syndromes, enteroinvasive *E. coli* (EIEC) that produces a shigellosis-like disease, and enteroaggregative *E. coli* (EAEC) that causes acute and chronic diarrhea. These strains mostly cause disease in developing countries, but EHEC O157:H7 has been responsible for outbreaks of hemorrhagic colitis in the United States due to undercooked beef.

Differential characteristics *E. coli* is positive for indole and methyl red tests, and negative for Voges-Proskauer, citrate, hydrogen sulfide, urease, and phenylalanine deaminase.

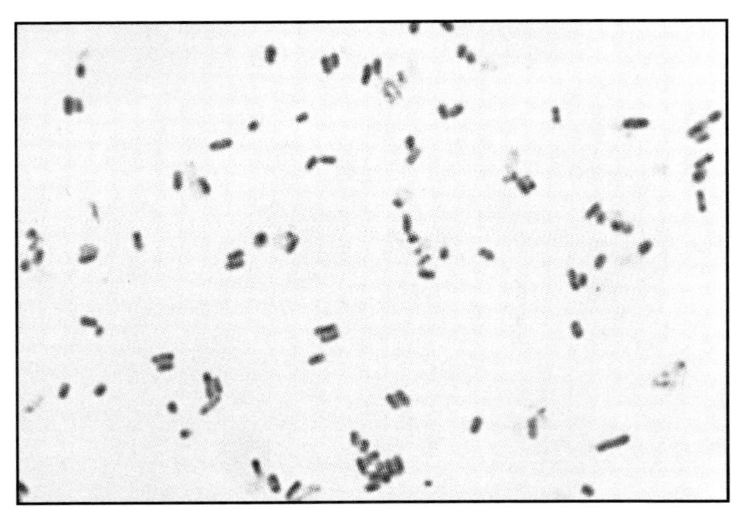

FIGURE 11-18 Gram stain of *Escherichia coli* (X1600). Cells are usually arranged singly or in pairs. Note the characteristic bipolar staining.

Fusobacterium nucleatum

Fusobacterium nucleatum is a nonmotile, anaerobic, nonsporing, spindle-shaped, Gram-negative rod. Normally found in the human oral cavity, it is a commonly encountered opportunistic pathogen. It causes dental infections, upper respiratory infections (including chronic sinusitis, aspiration pneumonia, and lung abscesses), and infections of the central nervous system (including brain abscesses). Most recently it has been discovered to be the causative agent of severe systemic infections in cancer patients following chemotherapy. The virulence of fusobacteria is largely due to endotoxins which evoke a vigorous immune reaction that can lead to toxic shock with widespread systemic collapse.

Differential characteristics *F. nucleatum* is indole-positive. It does not grow on bile agar and does not ferment mannitol, lactose or rhamnose. It is negative for esculin hydrolysis, catalase, lecithinase, lipase, starch hydrolysis, milk proteolysis, DNase and gelatinase.

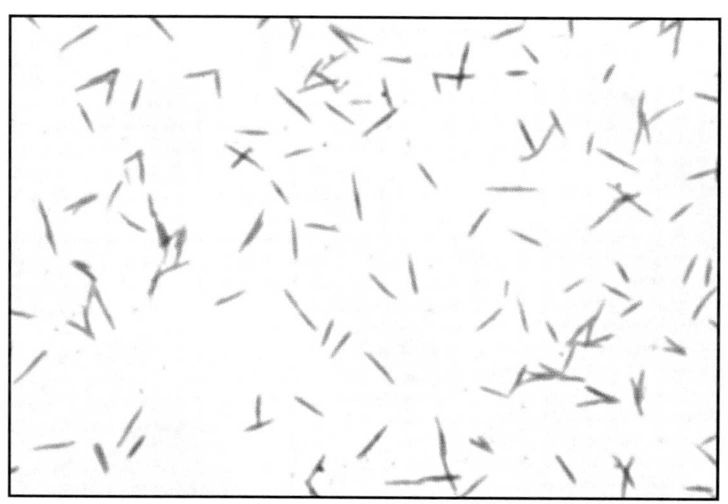

FIGURE 11-19 Simple crystal violet stain of *Fusobacterium nucleatum* (X1600). Note the long thin spindle-shaped cells.

Haemophilus influenzae

Haemophilus influenzae is a small, nonmotile, facultatively anaerobic, pleomorphic, Gram-negative rod. It can be found in the nasopharynx of virtually everyone over the age of 3 months and is transmitted by aerosol droplets. Given the name "*Haemophilus*" (blood loving) because it requires specific blood factors to live, it is an obligate human pathogen. In humans of any age, *H. influenzae* is responsible for an array of serious infections including: pneumonia, bronchitis, endocarditis, conjunctivitis, acute sinusitis, and urethritis possibly leading to postpartum bacteremia. Due largely to an immature system, it is also the most common cause of bacterial meningitis in children under age five. (In contrast, a healthy adult immune system would likely destroy the pathogen.) There seem to be two distinct types in the species — one that produces capsules and one that does not. The strains that live commensally in the nasopharynx are not encapsulated. Those found in the CNS are encapsulated and can survive phagocytosis, eventually disseminating through the bloodstream to all organ systems, including the brain.

Differential characteristics A battery of three tests — indole, urease, and ornithine decarboxylase — is used to differentiate the eight biotypes of *H. influenzae*. Each strain demonstrates a unique combination of positives and negatives.

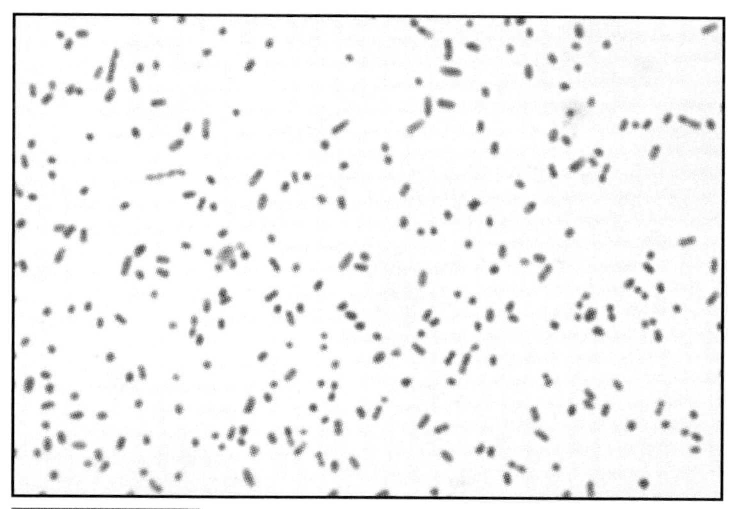

FIGURE 11-20 Simple crystal violet stain of *Haemophilus influenzae* (X1600). Note the extensive pleomorphism.

Helicobacter pylori

Helicobacter pylori is a motile, curved, spiral, or straight, slightly plump, Gram-negative rod. Although it has been isolated from the gastrointestinal tract of humans and primates, how it is transmitted is still unknown. It is believed to be a major cause of gastritis and peptic ulcers although it resides solely on the gastric mucosa and does not enter cells. The host immune response is believed responsible in part for the severity of the symptoms. Possibly the most notable feature of this organism is its ability to hydrolyze urea very rapidly. It has been hypothesized that the neutralizing effect of the ammonia released from this hydrolysis produces a more favorable environment for colonization in the stomach. *H. pylori*, although not new, has been under serious investigation for only 15 years. Much research is still needed.

Differential characteristics This organism is rapidly urease-positive. It is also catalase- and oxidase-positive.

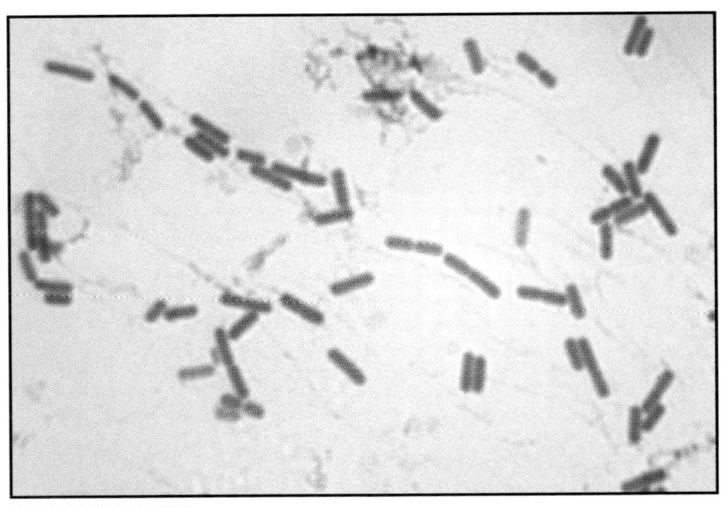

FIGURE 11-21 Gram stain of stock culture of *Helicobacter pylori* (X1600). *H. pylori* is Gram-negative, however, young cultures grown *in vitro* frequently stain Gram-positive.

Klebsiella pneumoniae

Klebsiella pneumoniae is a nonmotile, encapsulated, facultatively anaerobic, Gram-negative rod. It is found in soil, water, grain, fruits and vegetables, and the intestinal tracts of humans and a variety of animals. Plasmid mediated antibiotic resistance in this species has increased dramatically over the last ten years, making it a pathogen of increasing importance. Harbored in the nasopharynx and oropharynx of humans, it is frequently transmitted nosocomially as aerosol droplets from person to person. It is responsible for a serious primary pneumonia characterized by bleeding and widespread necrosis. Other infections caused by *K. pneumoniae* include septicemia, enteritis and meningitis in infants, and urinary tract infections.

Differential characteristics *K. pneumoniae* is negative for indole, arginine and ornithine decarboxylase, and positive for lysine decarboxylase.

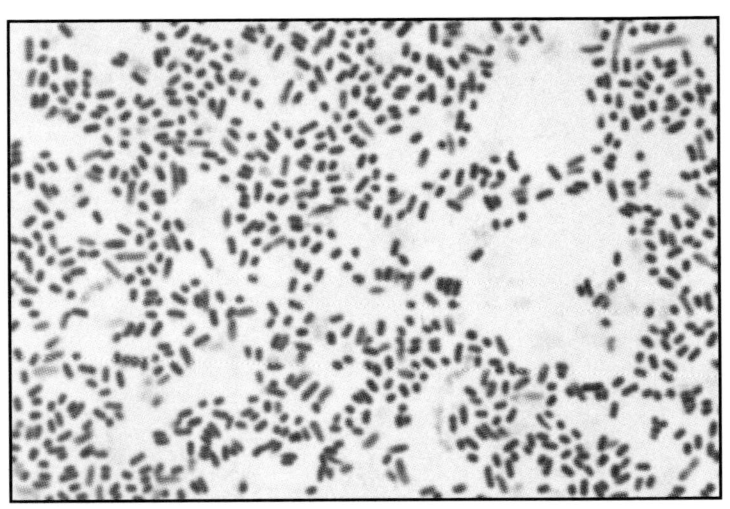

FIGURE 11-22 Gram stain of *Klebsiella pneumoniae* (X1600).

Legionella pneumophila

Legionella pneumophila is a motile, nonsporing, aerobic, Gram-negative rod. It is a comparatively "new" organism, having only been discovered following the outbreak of Legionnaires' disease in 1976. In the hotels and hospitals where outbreaks have occurred, the organism has been isolated from fresh water sources and various plumbing fixtures. Since then it has been discovered in abundance in aquatic habitats around the world. *L. pneumophila* is not passed directly from person to person but is inhaled as aerosols from environmental sources. *Legionellosis* is the general heading used to include all of the diseases caused by this organism (*i.e.* Legionnaires' disease — a pneumonia-like disease, Pontiac fever — a milder, short term flu-like illness, and a variety of other systemic infections). Although capable of surviving extracellularly, it is classified as an intracellular pathogen because of its ability to survive and multiply inside the pulmonary macrophages which attempt to destroy it. The multiplying bacteria eventually kill the macrophages, spread, and repeat the process. Research on this relatively new genus is far from complete.

Differential characteristics Final identification of *L. pneumophila* is typically made with direct immunofluorescence assay or DNA probe.

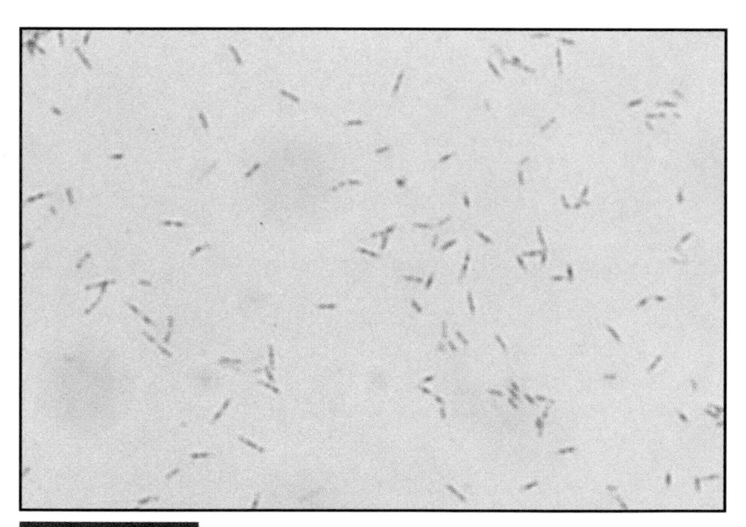

FIGURE 11-23 Gram stain of *Legionella pneumophila* (X1600).

Listeria monocytogenes

Listeria monocytogenes is a small, motile, β-hemolytic, nonsporing, Gram-positive rod. It is a very common soil saprophyte and vegetable matter decomposer, but can also be found in red meat, poultry, fish, and the intestinal tract of many animals including humans. Surprisingly *L. monocytogenes* is a relatively infrequent agent of human disease, albeit a very serious one. Studies have recorded *listeriosis* case-fatality rates variously from 20% to 70%. It primarily affects pregnant women and their fetuses, immunocompromised individuals, and the elderly. *L. monocytogenes* gains entry to the body as a food-borne contaminant or through openings in the skin. It spreads by parasitizing host phagocytes which eventually die and release the bacteria to infect adjacent cells. *Listeriosis,* the heading under which all human diseases caused by this organism are placed, is composed mainly of septicemia, meningitis, and encephalitis.

Differential characteristics *L. monocytogenes* is catalase-positive. It ferments glucose, trehalose, and salicin, and hydrolyzes esculin.

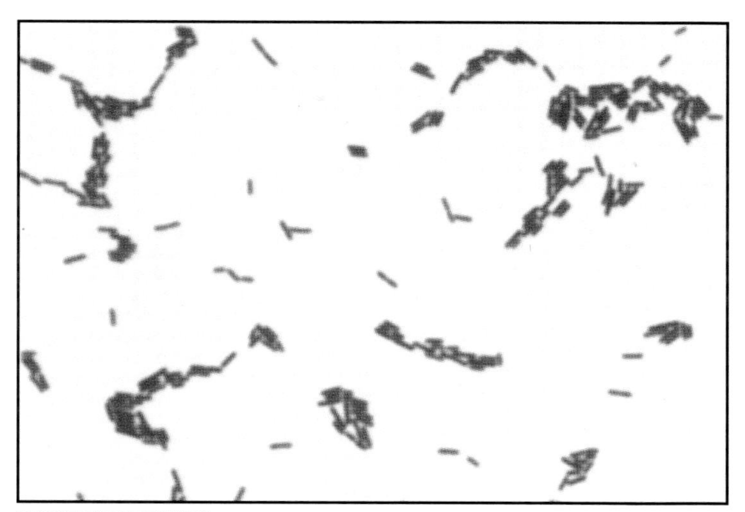

FIGURE 11-24 Gram stain of *Listeria monocytogenes* (X1600). This organism typically appears as single cells, diplobacilli, chains, or clusters of parallel cells.

Mycobacterium leprae

Mycobacterium leprae is a nonmotile, acid-fast, nonsporing, weakly Gram-positive rod. It is the causative agent of *leprosy* — now rare in the United States, but still a serious problem in Asia and Africa. Although it can be cultivated in a few laboratory animals, it is believed to occur naturally only in humans and in the nine-banded armadillo of Texas and Louisiana. Like *Legionella* and *Listeria*, *M. leprae* is an intracellular parasite. Once ingested by a macrophage it survives by chemically suppressing the cell's defensive activity. Two distinctive forms of leprosy are known to occur — *lepromatous leprosy* and *tuberculoid leprosy*. Lepromatous leprosy, the more contagious, is characterized by a suppressed immune response, rapid proliferation of the organism, severe disfigurement, and loss of nerve function. In tuberculoid leprosy, a vigorous host immune response results in the formation of granulomas on the face, trunk, and extremities. Several intermediate forms of the disease also exist. The mode of transmission is not well understood, however, a victim with the lepromatous form can shed billions of bacterial cells from the nose in a single day.

Differential characteristics *M. leprae* is uncultivable *in vitro*, therefore, standard biochemical tests are not useful in identifying it. Clinical diagnoses are based on characteristics of the disease, staining, and biopsies from skin lesions or nasal secretions. A skin test is also available using *lepromin*, an antigen extraction from lepromatous lesions.

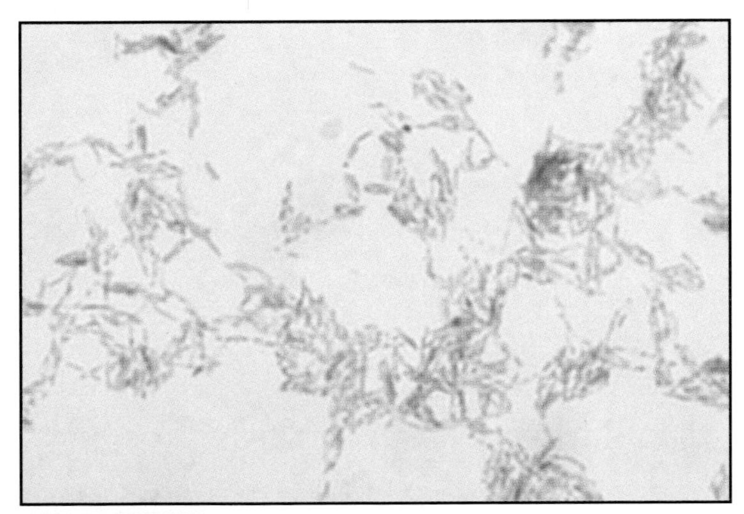

FIGURE 11-25 Acid-fast stain of *Mycobacterium leprae* (X1600). These rods can be straight or curved and are typically found in clusters.

Mycobacterium tuberculosis

Mycobacterium tuberculosis is a nonmotile, acid-fast, nonsporing, weakly Gram-positive rod. It is the pathogen responsible for *tuberculosis*. Humans are its principal host and reservoir although it has been isolated from other primates. It can be passed directly from person to person or inhaled as droplet nuclei (bacteria carried on airborne particles). Two manifestations of the disease exist: primary tuberculosis and secondary tuberculosis. Primary tuberculosis, the condition produced upon initial exposure to the bacillus, is for most healthy individuals no more than a mild flu-like illness. In this initial stage, the bacteria enter the alveoli and are ingested by resident macrophages. They multiply intracellularly and spread to other areas of the lung, killing the macrophages in the process. Eventually, the host immune response kills most of the bacteria, but some remain alive inside small granulomas called *tubercles*. In otherwise healthy individuals, these tubercles usually remain intact for a lifetime, holding the bacteria in check. In immunocompromised individuals, however, the organism soon disseminates hematogenically to all organ systems of the body. Secondary tuberculosis is the condition that occurs as the aging immune system weakens or is compromised by other factors. This condition, characterized by progressive, necrotic lung inflammation, is the form of tuberculosis most people associate with the disease.

Differential characteristics An acid-fast stain is an important initial step in identifying *M. tuberculosis*. Following that, a variety of specialized commercial media and biochemical tests are available for final determination.

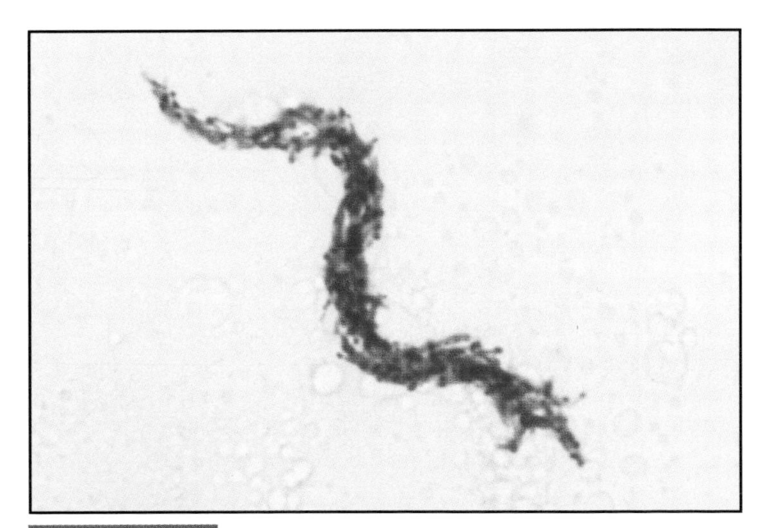

FIGURE 11-26 Acid-fast stain of *Mycobacterium tuberculosis* demonstrating characteristic cordlike orientation (X1600).

Neisseria gonorrhoeae

Neisseria gonorrhoeae is an aerobic, Gram-negative diplococcus that sometimes demonstrates "twitching motility." It is a strictly human pathogen and is the causative agent of the sexually transmitted disease (STD) *gonorrhea*. It attaches to urethral or vaginal columnar epithelial cells by pili and other surface proteins. The composition of these surface components is controlled genetically and can therefore be changed. Because of this, the organism is able to evade host antibodies which might otherwise attack it. Most infections caused by *N. gonorrhoeae* are confined to the lower reproductive area. However, if allowed to reach the bloodstream, serious consequences can arise. Certain strains of this organism contain a surface antigen similar to that of red blood cells and, thereby, evade host serum antibodies. When this occurs, disseminated gonococcal infection (DGI) is the result, frequently including dermatitis-arthritis-tenosynovitis syndrome and occasionally, endocarditis or meningitis. Other infections of this organism include endometritis, epididymitis, pelvic inflammatory disease (PID), proctitis, pharyngitis, conjunctivitis, peritonitis, and perihepatitis.

Differential characteristics *N. gonorrhoeae* produces acid from glucose utilization and produces no acid from maltose, sucrose, lactose, or fructose.

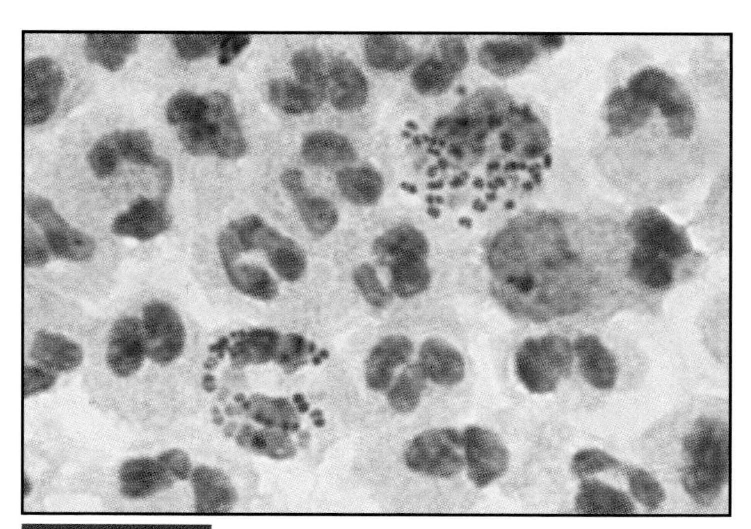

FIGURE 11-27 Gram stain of *Neisseria gonorrhoeae* inside a polymorphonuclear leukocyte (X1600). This organism is typically seen in pairs (diplococci) with adjacent sides flattened.

Neisseria meningitidis

Neisseria meningitidis is one of two human pathogens in the genus. Like its close relative *N. gonorrhoeae*, it is an aerobic, Gram-negative diplococcus which sometimes demonstrates "twitching motility." It resides in mucous membranes of the nasopharynx, oropharynx, and the anogenital region. Since it does not remain viable for long outside the human body, it must be transferred sexually or by direct contact with infected respiratory secretions. It is the infective agent of *meningococcemia* and the accompanying *meningococcal meningitis*, a devastating disease primarily of children and young adults. The organism's virulence can be attributed to at least four factors: surface pili which help it attach to host mucous membranes, a heavy capsule which helps it survive and multiply inside phagocytes, a hemolysin which facilitates the destruction of red blood cells (RBCs), and in some strains, surface components (similar to those of RBCs) that do not stimulate a serum antibody response. In most healthy individuals the organism produces a localized infection or no symptoms at all. In the absence of an early antibody response, however, hematogenic dissemination may occur resulting in *fulminant sepsis* and/or meningitis.

Differential characteristics *Neisseria meningitidis* produces acid from glucose and maltose, and produces no acid from sucrose, lactose, or fructose.

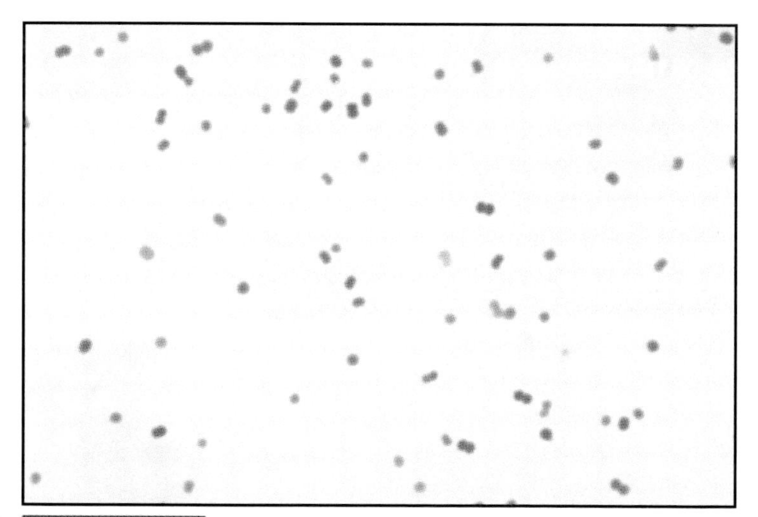

FIGURE 11-28 Gram stain of *Neisseria meningitidis* (X1600). This organism is typically seen in pairs (diplococci) with adjacent sides flattened.

Nocardia asteroides

Nocardia asteroides is a nonmotile, Gram-positive to Gram-variable, partially acid-fast rod once believed to be a fungus because of its ability to produce branching vegetative hyphae. However, in contrast to the fungi, its mycelia fragment into rod and coccus-like elements which contain no membrane bound organelles. Found on vegetation and in the soil, *N. asteroides* is the causitive agent of *nocardiosis*. Although *N. asteroides* occasionally causes primary nocardiosis in otherwise healthy individuals, it has developed into a significant opportunistic pathogen among immunocompromised patients. In the majority of cases transmission is by inhalation of aerosol droplets leading to pulmonary nocardiosis (chronic pneumonia). Dissemination of the organism by the bloodstream typically leads to central nervous system nocardiosis (brain abscesses) and infection of virtually all organ systems.

Differential characteristics *Norcardia asteroides* is lysozyme-resistant and urease-positive. It reduces nitrate, does not hydrolyze casein, tyrosine, xanthine, hypoxanthine, gelatin, or starch, and does not produce acid from lactose, xylose, or arabinose.

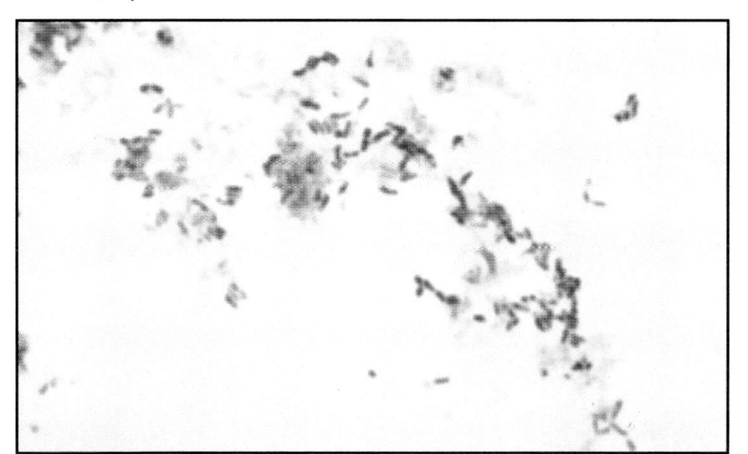

FIGURE 11-29 Acid-fast stain of *Nocardia asteroides* (X1600). Note the uneven staining resulting in a mixture of red and blue cells.

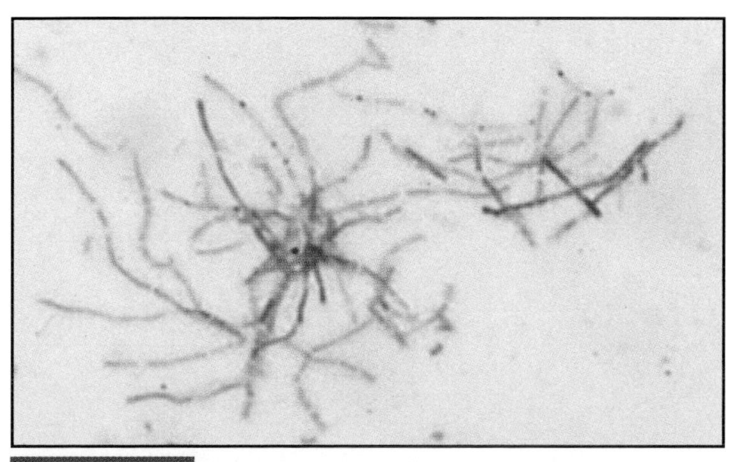

FIGURE 11-30 Gram stain of an older stock culture of *Nocardia asteroides* (X1600). Note the extensive branching filaments.

Proteus mirabilis

Proteus mirabilis is a facultatively anaerobic, highly motile, straight, Gram-negative rod. A normal inhabitant of the human intestinal tract, it can also be found in soil, polluted water and the intestines and feces of a variety of other animals. *P. mirabilis*, like all *Proteus* species, has the ability to periodically migrate on the surface of solid media. This cycle of alternating migration and consolidation is due to a characteristic called "swarming motility" and produces a series of concentric rings on agar plates (Fig. 1-12). *P. mirabilis* is a common nosocomial pathogen isolated from septic wounds and urinary tract infections. Transmission is by direct contact with a carrier or other contaminated source. It is of particular importance as a urinary tract pathogen because of its ability to produce urease. Urease splits urea, thus creating an alkaline environment and promoting the formation of kidney stones. Complicated urinary tract infections caused by *P. mirabilis* may lead to fatal bacteremia.

Differential characteristics *P. mirabilis* is negative for indole, esculin hydrolysis and salicin fermentation. It is positive for lipase and ornithine decarboxylase, and strongly positive for urease.

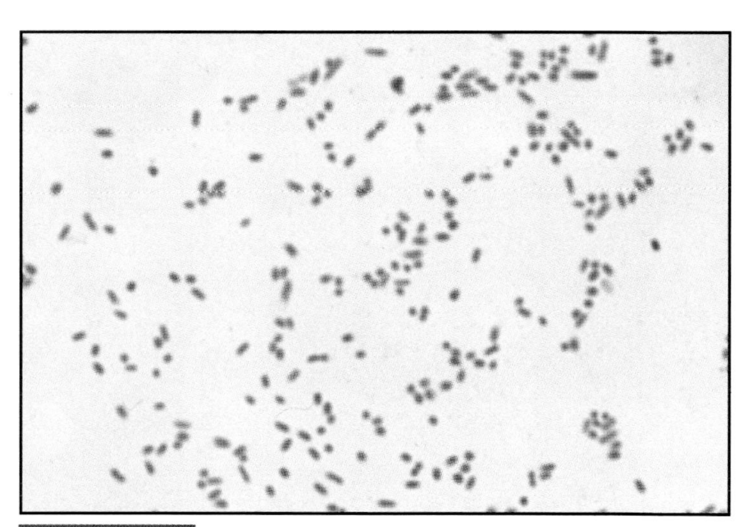

FIGURE 11-31 Gram stain of *Proteus mirabilis* (X1600).

Pseudomonas aeruginosa

Pseudomonas aeruginosa is an aerobic, highly motile, straight or slightly curved, Gram-negative rod. Common in water, soil, and on plants, it also has been isolated from hospital sinks and tubs, dialysis equipment, contact lens solution, aerators, irrigation fluids, hot tubs, ointments, insoles of shoes, and even soaps and cleaning solutions. It is the most significant opportunistic pathogen of its genus and is an especially troublesome nosocomial agent because of its ability to survive eradication attempts, even in marginal environments. Entry into the host is by ingestion, inhalation, or through openings in the skin. *P. aeruginosa* employs surface pili to attach to host cells and secretes an array of tissue damaging enzymes. Although healthy individuals are rarely affected by the organism, immunosuppressed patients are particularly susceptible. Among the nosocomial infections caused by this organism are pneumonia, wound sepsis, bacteremia, and urinary tract infections. Other infections include: corneal ulcers, swimmers ear, folliculitis (from contaminated swimming pools or hot tubs), and osteomyelitis of the calcaneus in children due to puncture wounds through the shoe.

Differential characteristics *P. aeruginosa* is a carbohydrate nonfermenter. It utilizes glucose oxidatively and does not utilize lactose or esculin. It reduces nitrate, is positive for oxidase and arginine decarboxylase, and negative for ONPG, urease, and lysine decarboxylase.

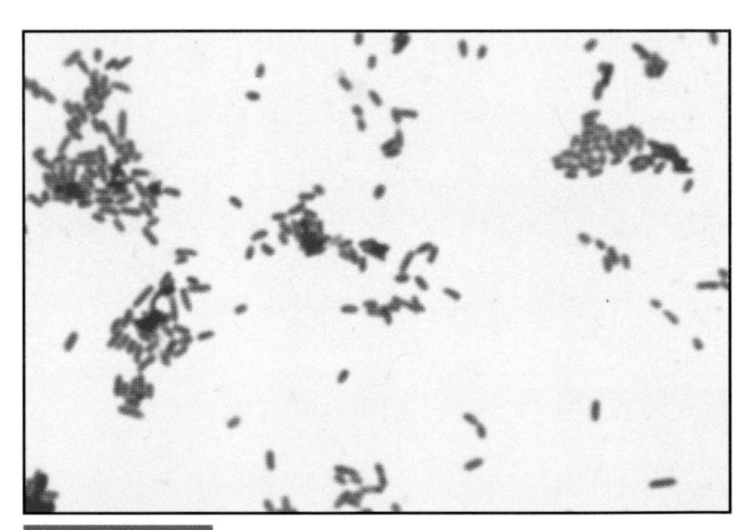

FIGURE 11-32 Simple crystal violet stain of *Pseudomonas aeruginosa* (X1600).

Salmonella Enteritidis

Salmonella Enteritidis (abbreviated, by current convention, from *Salmonella enterica* subspecies *enterica* serotype Enteritidis) is a motile, facultatively anaerobic, nonsporing, straight, Gram-negative rod. The gastrointestinal tract of many animals including poultry, rodents and wild birds harbor this organism. It is one of approximately 2200 nontyphoidal *Salmonellae*, all of which cause *gastroenteritis* in humans. The mechanisms described here for *Salmonella* Enteritidis apply to those serotypes as well. Entry into the body is by ingestion of food or water contaminated with feces. *Salmonellae* are susceptible to acidic conditions and thus require a large number of organisms ($>10^6$) to infect an individual with normal stomach acidity. They clearly favor individuals with low gastric acidity as hosts. Once through the hostile environment of the stomach, they are taken up by intestinal epithelial cells and are released into the underlying connective tissue where they begin to multiply. The mechanism by which diarrhea is induced is not fully understood, however evidence suggests that it is due to the production of a cholera-like or Shiga-like toxin. *Salmonella* Enteritidis is not well suited for intracellular conditions, therefore, except in cases where the host immune system has been compromised, gastrointestinal infections are generally short lived. Lacking an appropriate host immune response, however, dissemination of the organism may occur resulting in widespread systemic infection.

Differential characteristics Because the vast number of strains differ primarily in antigenic structure, serogrouping by a reference laboratory is necessary for final identification.

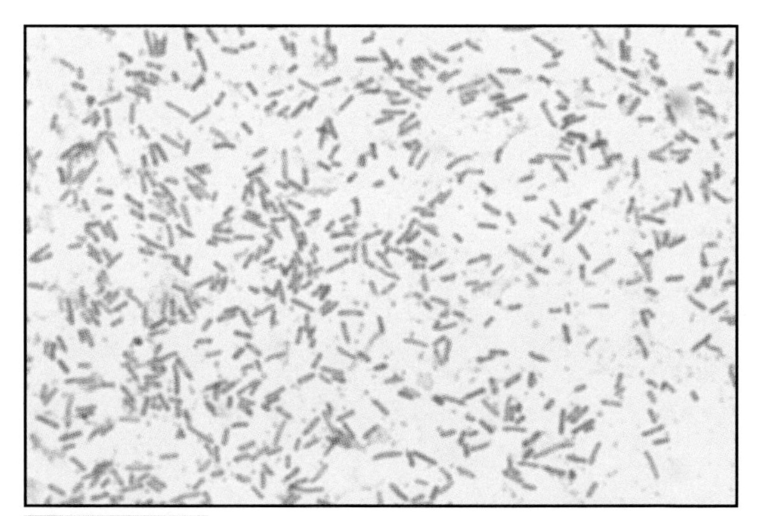

FIGURE 11-33 Gram stain of *Salmonella* Enteritidis (X1600).

Salmonella typhi

Salmonella typhi is a motile, encapsulated, facultatively anaerobic, nonsporing, straight, Gram-negative rod. It is the causative agent of *typhoid fever* in humans. The organism is typically transmitted by fecally contaminated food or water. As described in *Salmonella* Enteritidis above, it initially attacks cells of the small intestine, is ushered into the underlying connective tissue and regional lymph nodes, and begins to multiply. It then enters the bloodstream where it produces acute bacteremia and infects the liver, spleen, bone marrow, and eventually the kidneys and gall bladder. This phase, accompanied by high fever and sometimes diarrhea, is long lasting and continuous (up to 8 weeks in untreated cases). In a small percentage of patients the organism is harbored asymptomatically in the gall bladder ("carriers") and sloughed in the feces for a year or more.

Differential characteristics Serogrouping by a reference laboratory is necessary for final identification of *Salmonella spp*.

FIGURE 11-34 Gram stain of *Salmonella typhi* (X1600).

Shigella dysenteriae

Shigella dysenteriae is a nonmotile, facultatively anaerobic, straight, Gram-negative rod. It is one of four *Shigella* species (*S. dysenteriae*, *S. flexneri*, *S. boydii*, and *S. sonnei*), all of which are responsible for *bacillary dysentery* (*shigellosis*) in humans and a few other primates. *S. dysenteriae* is endemic in Africa, Asia and Latin America; *S. flexneri* and *S. sonnei* are found primarily in developed areas including the United States; and *S. boydii* is mostly restricted to India. Transmission is by direct person-to-person contact or ingestion of food or water contaminated by human feces. It is highly communicable and can cause illness with as few as 200 organisms. Although all species of *Shigella* cause the disease, *S. dysenteriae* alone produces the cell-killing *Shiga toxin* and is, therefore, responsible for the most severe symptoms. Unlike *Salmonellae*, *Shigella spp.* are resistant to the stomach's acidic environment which accounts, in part, for the low infectious dose. Once in the intestine, they are taken up by host epithelial cells where they multiply and then spread, in a process which kills the cells and forms mucosal ulcerations. This process combined with an acute immune response is responsible for the purulent bloody diarrhea characteristic of the disease.

Differential characteristics *S. dysenteriae* does not ferment lactose, mannitol, raffinose, sucrose, or xylose and is negative for ONPG and ornithine decarboxylase.

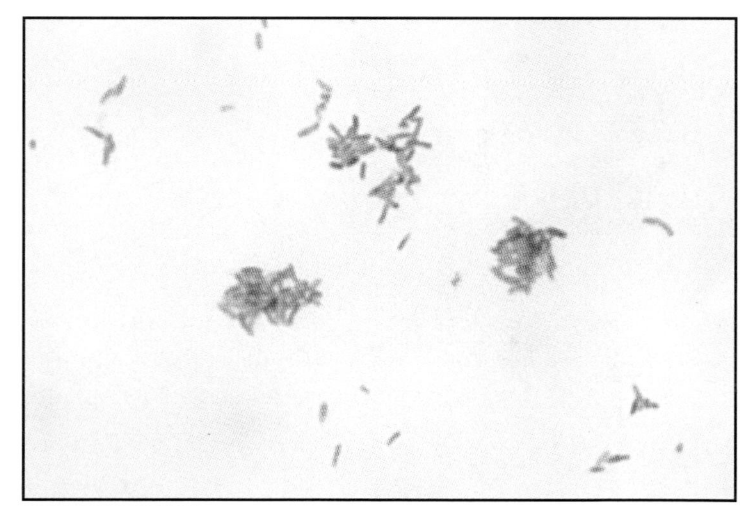

FIGURE 11-35 Gram stain of *Shigella dysenteriae* (X1600).

Staphylococcus aureus

Staphylococcus aureus is a nonmotile, facultatively anaerobic, Gram-positive coccus. Ubiquitous in nature, it constitutes part of the normal human flora. Most commonly found in the anterior nares of human adults, it also is known to inhabit the skin and vagina. It is a common nosocomial pathogen that causes toxic shock syndrome, food poisoning, scalded skin syndrome, and abscesses virtually everywhere in the body. Factors which increase its virulence include: antiphagocytic proteins, lipase production (which aids entry to the skin), coagulase (which enhances the formation of abscesses), enterotoxins (which induce vomiting and diarrhea), and exotoxins (which destroy polymorphonuclear leukocytes, aid necrosis, and produce fever, chills, shock and rash). *S. aureus* is transmitted by direct human-to-human contact, aerosols, or environmental factors. It is a robust organism that resists cleaning solutions, antimicrobial agents, and can survive for weeks in the environment.

Differential characteristics *S. aureus* is positive for both the slide coagulase test (bound coagulase) and the tube coagulase test (free coagulase).

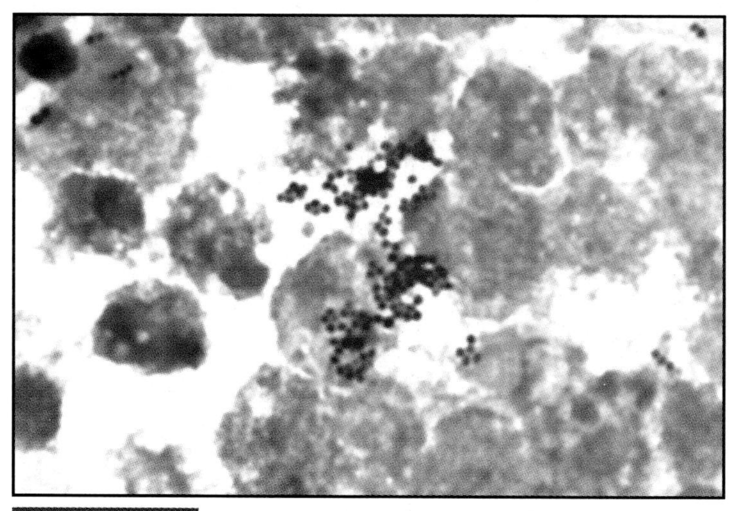

FIGURE 11-36 Gram stain of *Staphylococcus aureus* in synovial fluid (X1600). Note the characteristic grape-like clusters.

Staphylococcus epidermidis

Staphylococcus epidermidis is a nonmotile, facultatively anaerobic, Gram-positive coccus. A commensal inhabitant of human skin, it has become a significant nosocomial pathogen. In fact, it is the most common coagulase-negative *Staphylococcus* encountered clinically. Most strains produce a slime layer which may enable them to attach to certain hospital apparati used in invasive procedures, thereby gaining entry to the body. Infections originating at the site of prosthetic implantation are frequently caused by *S. epidermidis*. Due to multiple antibiotic resistance and the generally weakened condition of a convalescing patient, disseminated *S. epidermidis* infection can be quite severe and is frequently fatal.

Differential characteristics *S. epidermidis* ferments maltose but does not ferment sucrose, xylose, or trehalose. It is positive for alkaline phosphatase production and negative for coagulase.

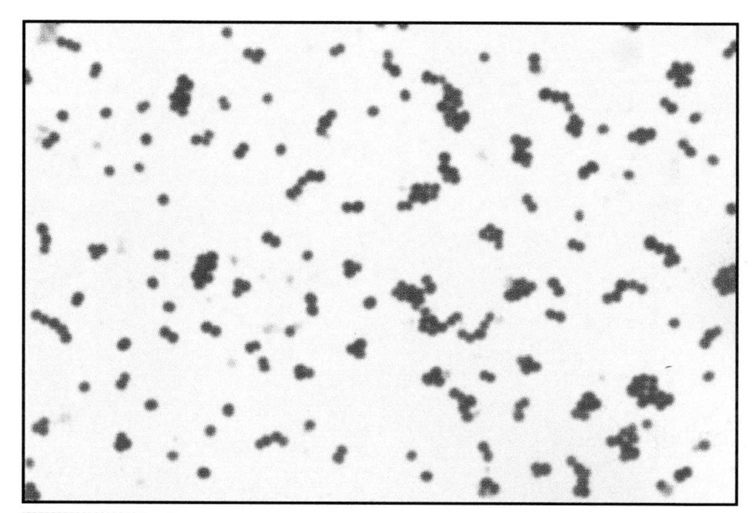

FIGURE 11-37 Gram stain of *Staphylococcus epidermidis* grown in broth (X1600). Cells are usually seen in pairs, tetrads or grape-like clusters.

Streptococcus agalactiae

Streptococcus agalactiae is a β-hemolytic or nonhemolytic, nonmotile, encapsulated, facultatively anaerobic, Gram-positive coccus. Known also as group B Streptococci, strains of *S. agalactiae* are the major cause of neonatal meningitis in the United States. The reservoir for this organism is believed to be the intestinal tracts of humans and animals, but is often found in the vagina of pregnant women. The organism is typically acquired by the infant *in utero* through a damaged membrane, from the birth canal during childbirth, or from contact with contaminants after birth. Its virulence is attributable to a polysaccharide capsule which allows it to survive phagocytosis, multiply and eventually spread by way of the bloodstream. Disseminated disease also causes pneumonia and septic shock.

Differential characteristics *S. agalactiae* is positive for the CAMP test, negative for Voges-Proskauer and PYR, and is bacitracin and SXT resistant.

FIGURE 11-38 Gram stain of *Streptococcus agalactiae* (X1600). Cells are usually seen as pairs or in chains.

Streptococcus mutans

Streptococcus mutans is one member of the group known as the "oral streptococci." The oral streptococci are further divided into four small groups: the "*S. mutans* group," "*S. sanguis* group," "*S. mitis* group," and "*S. salivarius* group." Each group includes several organisms, all of which are α-hemolytic or nonhemolytic, nonmotile, facultatively anaerobic, Gram-positive cocci. The oral streptococci are found in the mouth, upper respiratory tract and urogenital tract of humans where they constitute part of the normal flora. Members of this group are the most common cause of subacute endocarditis in patients with existing heart valve problems or prosthetic heart valves. They are also responsible for bacteremia following dental or urogenital invasive procedures, and in immunosuppressed patients undergoing chemotherapy and bone marrow transplantation. Clinically, the most common encounter with oral streptococci is in the dentist's chair. Several of these organsims are capable of hydrolyzing sucrose and forming dental plaque, which in turn provides the anaerobic environment ideal for fermentation. Acid produced by this fermentation and that of certain *Lactobacilli* erodes the tooth enamel and is responsible for the formation of dental caries (see "Snyder Test" in Section 7).

Differential characteristics Species identification is usually not clinically necessary for the α-hemolytic and nonhemolytic streptococci. The four groups can be differentiated from other groups based on their reactions in six biochemical tests: arginine hydrolysis, esculin hydrolysis, urease, Voges-Proskauer, and acid production from mannitol and sorbitol fermentation.

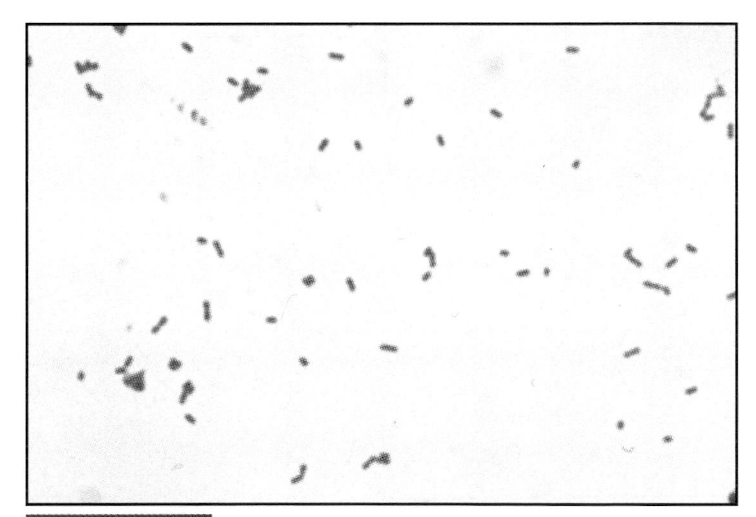

FIGURE 11-39 Gram stain of *Streptococcus mutans* (X1600). Cells usually appear in short chains of cocci but may also form short rods.

Streptococcus pneumoniae

Streptococcus pneumoniae is an α-hemolytic, nonmotile, encapsulated, facultatively anaerobic, Gram-positive coccus. In children under two years of age and elderly adults it is the principal cause of community-acquired bacterial pneumonia. In adults it is also the major cause of bacterial meningitis. The organism typically colonizes the nasopharynx where it either is eliminated from the body, spreads to the lungs and develops into pneumonia, or is harbored asymptomatically for up to several months (the carrier state). Transmission is usually by direct contact with a carrier or contaminated aerosols. At least 80 different serotypes of *S. pneumoniae* exist and are defined antigenically by their capsules. Some serotypes are more virulent than others due to their ability to avoid phagocytosis by host cells and the degree to which they stimulate antibody production. In the lungs it stimulates a vigorous immune response marked by copious fluid production. The invading organisms are cleared with no long term effects in the majority of infections. However, if complicated by bacteremia, meningitis and other secondary infections are the likely result.

Differential characteristics *S. pneumoniae* is negative for arginine hydrolysis, esculin hydrolysis, and acid production from mannitol and sorbitol. It is also urease and Voges-Proskauer negative.

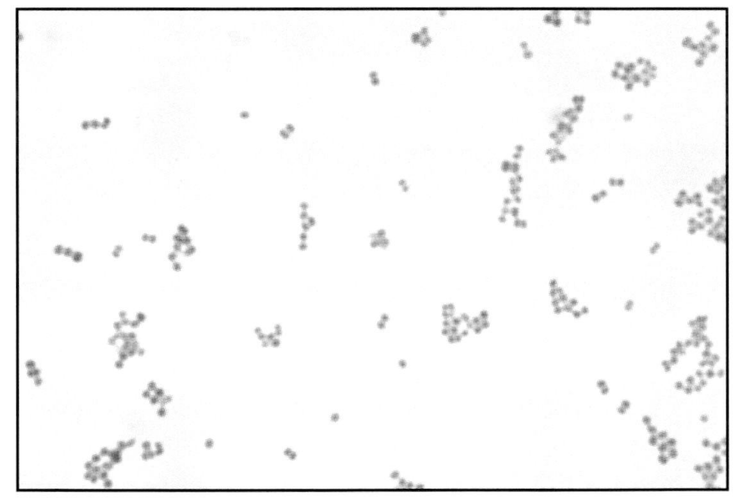

FIGURE 11-40 Gram stain of a stock culture of *Streptococcus pneumoniae* (X1600). Specimens from sputum samples are typically seen as singles or in pairs.

Streptococcus pyogenes

Streptococcus pyogenes (a group A streptococcus) is a β-hemolytic nonmotile, encapsulated, facultatively anaerobic, Gram-positive coccus. It is responsible for streptococcal pharyngitis ("strep throat"), impetigo, middle ear infections, mastoiditis, and an array of infections resulting from hematogenic dissemination of the organism, including glomerulonephritis and acute rheumatic fever (ARF). The reservoir for *S. pyogenes* is the human nose, throat and skin. It is transmitted by direct person-to-person contact or by contaminated aerosols. A variety of virulence factors allow it to attach to epithelial cells and also help it avoid phagocytosis. Once attached to the host cells it releases several toxins which elicit a vigorous inflammatory response, resulting in severe local inflammation sometimes accompanied by tissue necrosis. Once thought to be waning after the discovery of antibiotics, group A streptococci have made a comeback over the last 13 years. Since outbreaks of ARF in 1985 there have been reports of other infections including postpartum endomyometritis (puerpural fever), necrotizing fasciitis ("flesh eating"), and toxic shock-like syndrome (TSLS), which is very similar to staphylococcal toxic shock syndrome.

Differential characteristics *S. pyogenes* ferments lactose, salicin and trehalose but not inulin, mannitol, raffinose, fibose or sorbitol. It hydrolyzes arginine and PYR but not hippurate. Alkaline phosphatase is produced, but acetoin and α-galactosidase are not.

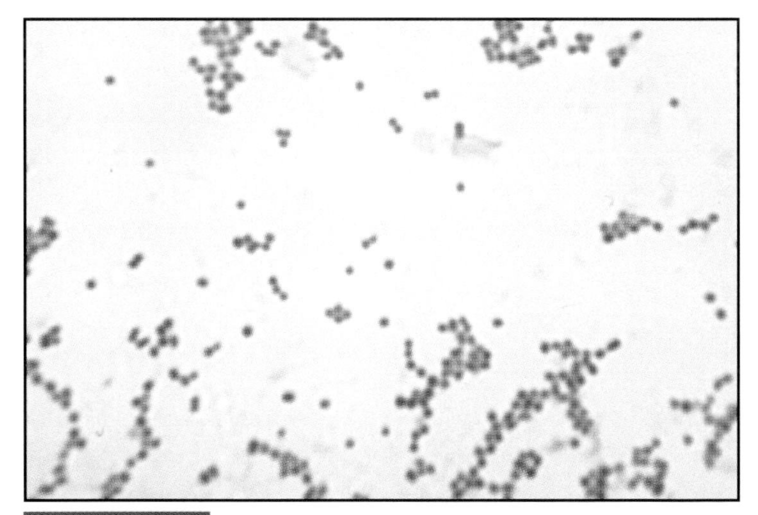

FIGURE 11-41 Gram stain of a stock culture of *Streptococcus pyogenes* (X1600). Specimens isolated from patients are usually seen in pairs and short chains.

Treponema pallidum

Treponema pallidum, a motile, microaerophilic, Gram-negative helical rod, is the causative agent of *syphilis*. For reasons largely socioeconomic and behavioral, the organsim has recently shown a resurgence in activity and is now responsible for approximately 100,000 cases annually. It is primarily a sexually transmitted disease (STD), however, intravenous drug use (especially crack cocaine) has broadened the epidemic. Although, easily treated with penicillin, the number of congenital syphilis cases (when the organism crosses the placenta and infects the fetus) has also increased dramatically. The organism enters the body through mucous membranes, abrasions, or fissures in the epithelia. Some of the organisms attach to host epithelial cells and multiply, while others are carried to lymph nodes where they enter the bloodstream and disseminate throughout the body. The disease develops in three distinct stages: primary, secondary and tertiary syphilis. Primary syphilis lasts about two to six weeks and is characterized by the formation of a local lesion called a "chancre." Hematogenic dissemination to all body regions (including the CNS) takes place in this stage. An asymptomatic period lasting up to six months follows. Secondary syphilis is characterized by formation of lesions in the liver, lymph nodes, muscles, and skin. A latent period lasting anywhere from five years to several decades follows. Tertiary syphilis is characterized by the destruction of neural and cardiovascular tissue and the formation of tumors throughout the body.

Differential characteristics Serological tests have proven to be the most effective method of determining the presence of *T. pallidum*. See Figures 9-10 and 9-11.

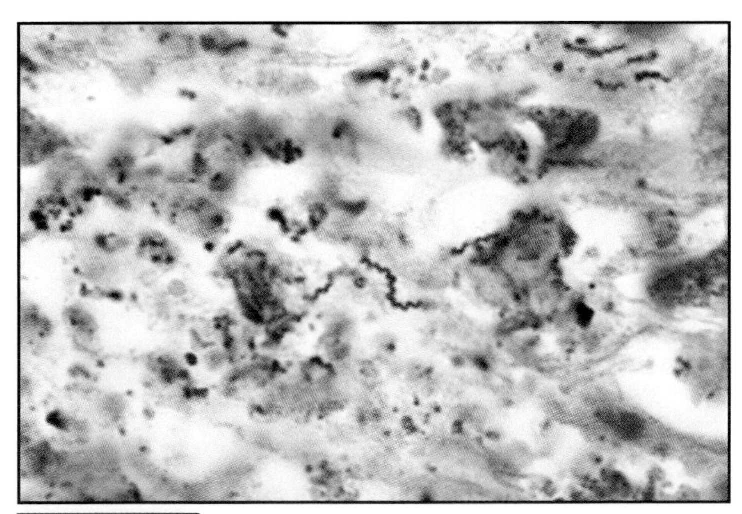

FIGURE 11-42 Silver stain of *Treponema pallidum* (center) in animal tissue (X1600). This spirochete demonstrates "corkscrew motility" by means of periplasmic flagella.

Vibrio cholerae

Vibrio cholerae is a motile, facultatively anaerobic, Gram-negative straight or curved rod. *V. cholerae* strains are typed according to their cell wall composition (O antigen). Although there are 139 serotypes of this species, only two (the O1 serogroup) have been responsible for the seven cholera pandemics since 1817, including the current one. The strains have been somewhat arbitrarily divided into the O1 *V. cholerae* and the non-O1 *V. cholerae* to distinguish the cholera causing organisms from the rest. In 1992, however, a new strain appeared in Madras, India and quickly spread throughout India, Bangladesh and southeast Asia. This new strain, called serogroup O139, is believed to be the etiologic agent of the eighth cholera pandemic. Although immunity to O1 strains does not confer immunity to O139, the diseases caused are indistinguishable. The organism enters the body by the fecal-oral route, frequently by way of undercooked contaminated seafood. A large inoculum (10^{10} cells) is required to infect healthy individuals because sufficient numbers of bacteria must survive the stomach's acidity and reach the small intestine to cause disease. Once in the small intestine the organism attaches to the mucosal layer and secretes *cholera toxin*. The toxin begins a cascade of reactions which ultimately alter electrolyte levels and stimulate a vigorous outpouring of fluids into the intestinal lumen. The result is the characteristic "watery" or "secretory diarrhea," frequently fatal within a few hours. Because the infection is self-limiting, fluid and electrolyte replacement is often sufficient treatment.

Differential characteristics *V. cholerae* ferments sucrose and can also be grown in nutrient broth without the addition of 1% NaCl. It can be further identified using various ELISA techniques to identify cholera toxin.

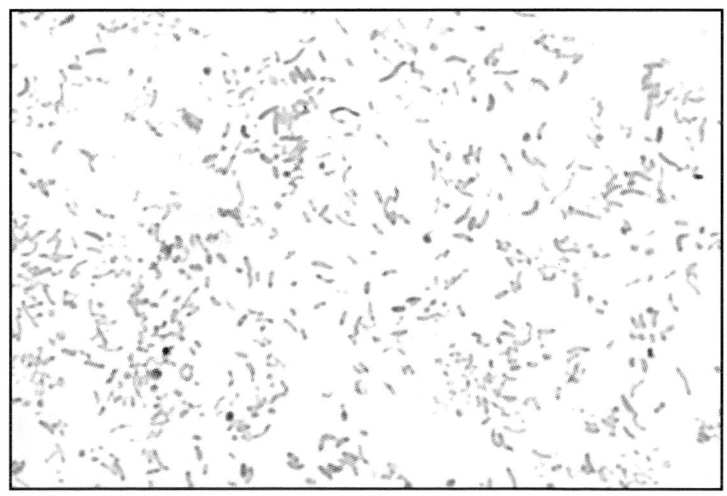

FIGURE 11-43 Gram stain of *Vibrio cholerae* (X1600). Note the curved rods.

Yersinia pestis

Yersinia pestis is a nonmotile, facultatively anaerobic, Gram-negative rod or coccobacillus. It forms an envelope (not a capsule) when grown at 37°C. It has been responsible for dozens of plague epidemics and pandemics over the last several hundred years, one of which took the lives of 25 million Europeans in the fourteenth century. Existing on every continent except Australia, its habitat is any of a variety of small animals including rats, ground squirrels, rabbits, mice, and prairie dogs. In urban outbreaks, rats are the principal carriers. *Y. pestis* produces several antiphagocytic factors which enable it to survive and multiply both intracellularly and extracellularly. It also produces exotoxins and endotoxins which are believed to be responsible for acute inflammation and necrosis. Although the organism can be inhaled as droplet nuclei (which causes *pneumonic plague*), it is most commonly transmitted by fleas. The fleas ingest it when feeding on infected animal blood and deposit it in another animal or human during a subsequent blood meal. *Y. pestis* produces a coagulase which clots the blood inside the flea's stomach. When the flea attempts to feed again, it regurgitates the coagulated material and contaminated blood back into the bite wound. If the bacteria are deposited directly into the bloodstream by the flea (usually in children), *septicemic plague* is the likely outcome. The more common *bubonic plague* is characterized by fever and chills, with severe inflammation and hemorrhagic necrosis of the inguinal or axillary lymph nodes called "bubos." It is the mildest form of the disease, yet the fatality rate in untreated cases is approximately 75%.

Differential characteristics *Y. pestis* is negative for indole production and ornithine decarboxylase. It does not ferment sucrose, rhamnose, or cellobiose.

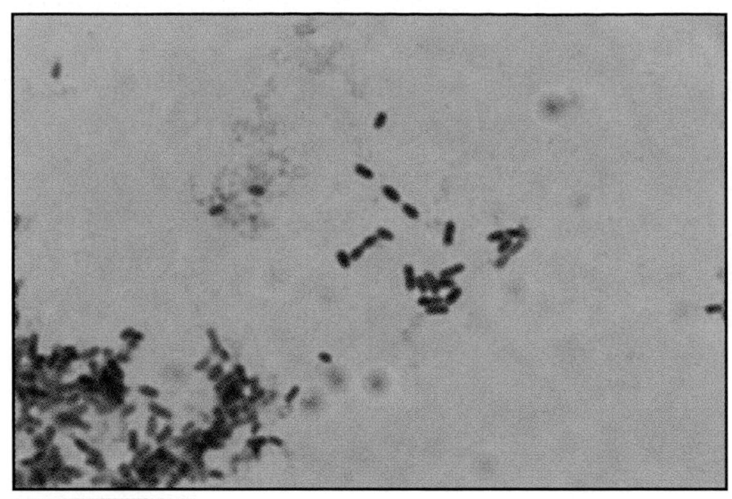

FIGURE 11-44 Gram stain of *Yersinia pestis* (X1600). This organism is typically seen as straight rods or coccobacilli.

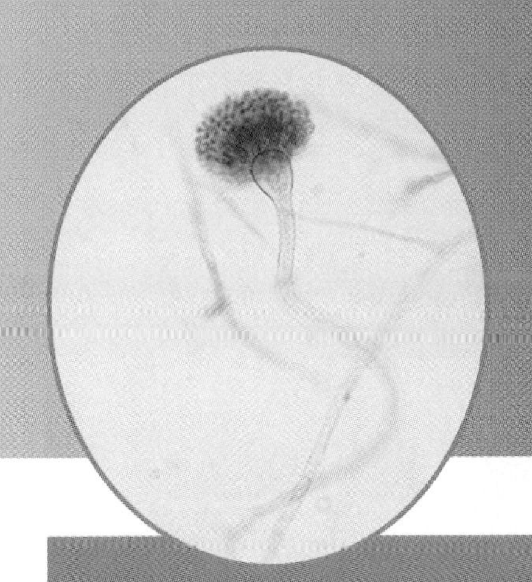

Survey of Fungi 12

Fungi are eukaryotic absorptive heterotrophs. They produce exoenzymes that digest nutrients in the environment, then absorb the products rather than ingest their food like animals. They also have a chitinous cell wall. Most are *saprophytes* that decompose dead organic matter, but some are *parasites* of plants, animals, or humans.

The fungi are unique enough to be designated to their own kingdom. Fungi are somewhat informally divided into unicellular *yeasts* and filamentous *molds* based on their overall appearance. *Dimorphic fungi* have both mold *and* yeast life cycle stages. Filamentous fungi that produce fleshy reproductive structures — mushrooms, puffballs, and shelf fungi — are referred to as *macrofungi* (even though the majority of the fungus is filamentous and hidden underground or within decaying matter).

Fungal filaments are called *hyphae*. Collectively, the hyphae of a fungus form a *mycelium*. Gametes are produced by *gametangia*. Spores are produced by a variety of *sporangia*. Typically, the only diploid cell in the fungal life cycle is the zygote, which then undergoes meiosis to produce haploid spores characteristic of the fungal group. Various asexual spores may also be produced during the life cycle of many fungi. If they form at the ends of hyphae, they are called *conidia*. Other asexual spores are *blastospores*, which are produced by budding and *arthrospores*, which are produced when a hypha fragments. *Chlamydospores* (*chlamydoconidia*) are formed at the end of some hyphae and are a resting stage.

More formal taxonomic categories based primarily on the pattern of sexual spore production are also used. Zygomycetes are terrestrial and produce nonmotile sporangiospores and zygospores. Ascomycetes produce a sac (an *ascus*) in which the zygote undergoes meiosis to produce haploid *ascospores*. Basidiomycetes that undergo sexual reproduction produce a *basidium* which undergoes meiosis to produce four *basidiospores* attached to its surface. Deuteromycetes are an unnatural assemblage of fungi in which sexual stages are either unknown or are not used in classification. The majority of deuteromycetes resemble ascomycetes.

Following is a survey of fungi likely to be encountered in an introductory microbiology class and selected medically important fungi. Within each group, species are listed alphabetically.

YEASTS OF MEDICAL OR ECONOMIC IMPORTANCE

Candida albicans

Candida albicans (Fig. 12-1) is part of the normal respiratory, gastrointestinal and female urogenital tract floras. Under the proper circumstances, it may flourish and produce pathological conditions, such as *thrush* in the oral cavity, *vulvovaginitis* of the female genitals, and *cutaneous candidiasis* of the skin. *Systemic candidiasis* may follow infection of the lungs, bronchi or kidneys. Entry into the blood may result in *endocarditis*. Individuals most susceptible to *Candida* infections are diabetics, those with immunodeficiency (*e.g.*, AIDS), catheterized patients, and individuals taking antimicrobial medications. Asexual reproduction produces blastoconidia which may remain attached and form a hypha with septa or a *pseudohypha* if septa are absent.

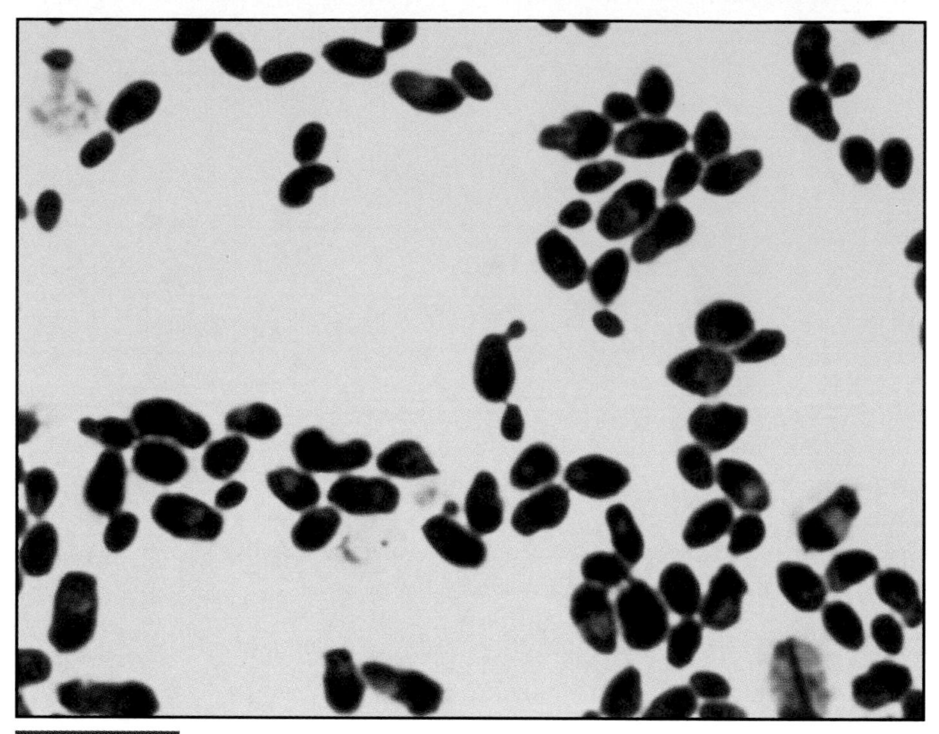

FIGURE 12-1 *Candida albicans* vegetative cells (X2640).

Cryptococcus neoformans

Cryptococcus neoformans causes *cryptococcosis* and is the only known serious human pathogen of the genus. It is an encapsulated, nonfermenting, aerobic yeast (Fig. 12-2) often found in soil mixed with accumulated bird droppings that enrich it with nitrogen. Urban sites where birds roost may also harbor the organism in the dried feces. Infection by *C. neoformans* from inhalation leads to pneumonia and then meningitis. Cryptococcosis, at one time a disease of poultry workers, is increasing in incidence and is among the more common opportunistic infections of AIDS patients. Colonies of *C. neoformans* are shiny, cream-colored and mucoid (due to the capsule).

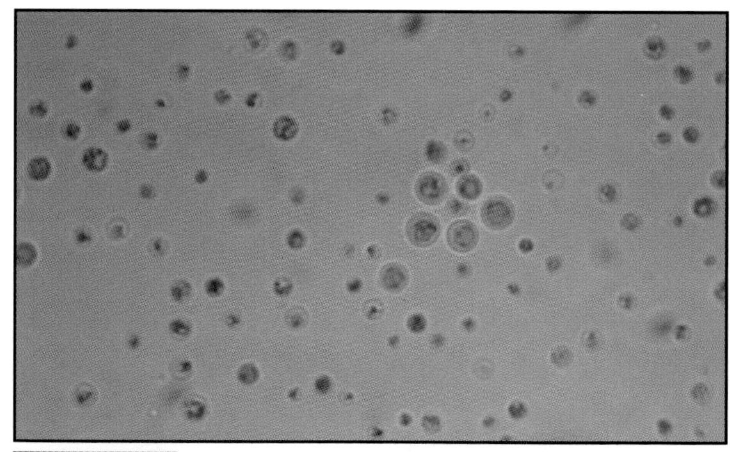

FIGURE 12-2 *Cryptococcus neoformans* (Lactophenol blue stain, X1350). *Cryptococcus neoformans* is characterized by spherical, encapsulated cells that range in size from 2 to 15 µm. The capsule is faintly visible in this preparation.

Pneumocystis carinii

Pneumocystis carinii is an opportunistic pathogen whose transmission is thought to occur through the air, with the primary infection occurring in the lungs. It produces AIDS-related pneumocystis pneumonia (PCP) and was a leading cause of death in AIDS patients prior to development of prophylactic medications. Before the AIDS epidemic, *P. carinii* was only known to produce interstitial plasma cell pneumonitis in malnourished infants and immunosuppressed individuals.

Comparison of ribosomal RNA and other biochemical traits has led to the conclusion that *Pneumocystis carinii*, long classified as a protozoan, is more closely related to the fungi. Its protozoan "roots" are still evident, however, in the terminology associated with it. *P. carinii* exists as a trophozoite and a multinucleate cyst (Fig. 12-3).

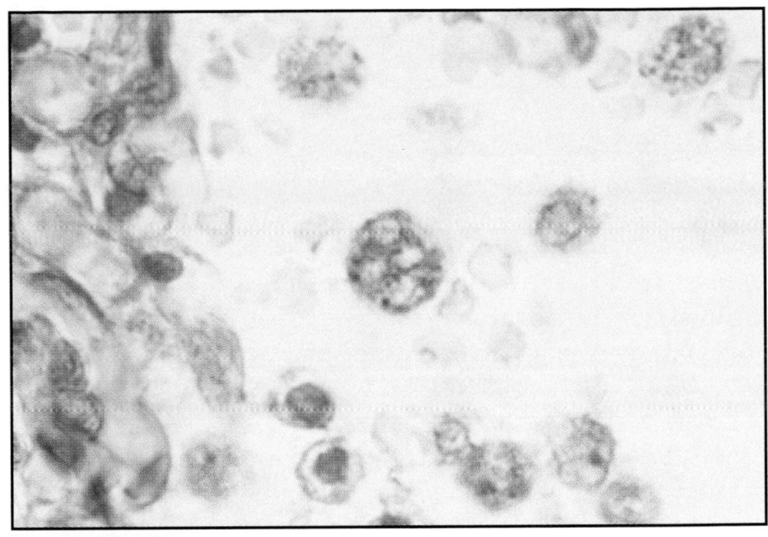

FIGURE 12-3 Section of *Pneumocystis carinii* cyst in an infected lung (X1264).

Saccharomyces cerevisiae

Saccharomyces cerevisiae is an ascomycete used in production of bread, wine and beer, but is not an important human pathogen. It does not form a mycelium, but rather produces a colony similar to bacteria (Fig. 12-4). Asexual reproduction occurs by budding which produces *blastoconidia* (Fig. 12-5). Meiosis produces four ascospores within the vegetative cell which acts as the ascus (Fig. 12-6). Ascospores may fuse to form another generation of diploid vegetative cells or they may be released to produce a population of haploid cells that are indistinguishable from diploid cells. Haploid cells of opposite mating types may also combine to create a diploid cell.

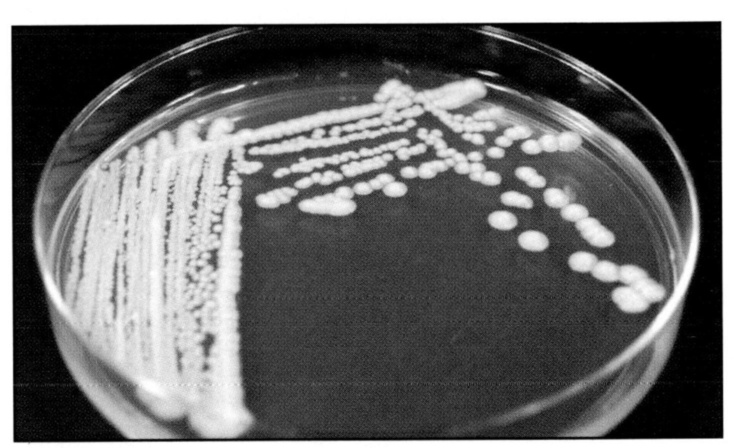

FIGURE 12-4 *Saccharomyces cerevisiae* colony. Note the appearance is similar to a typical bacterial colony, not "fuzzy" like mold colonies.

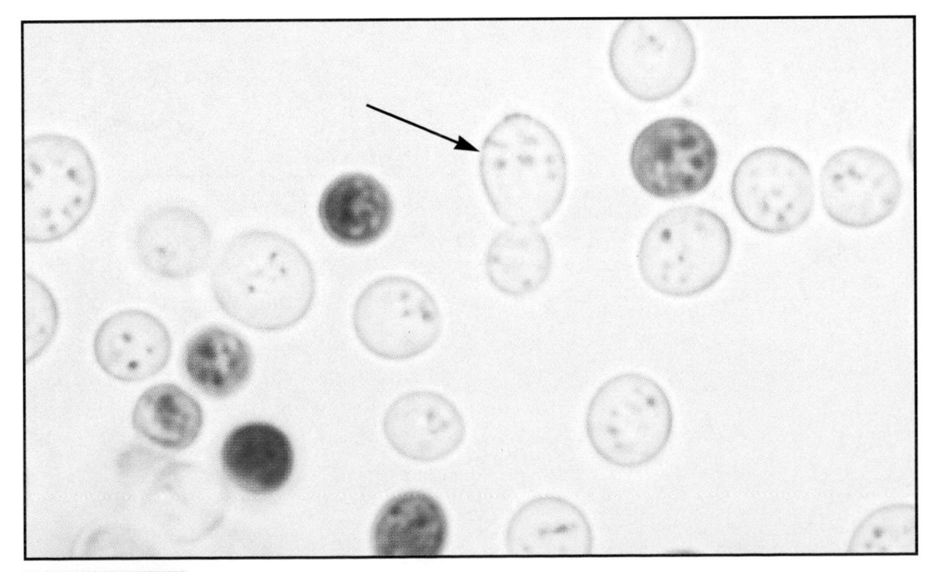

FIGURE 12-5 Wet mount of *Saccharomyces cerevisiae* vegetative cells (X2640). Note the budding cell (blastoconidium) in the center of field (arrow).

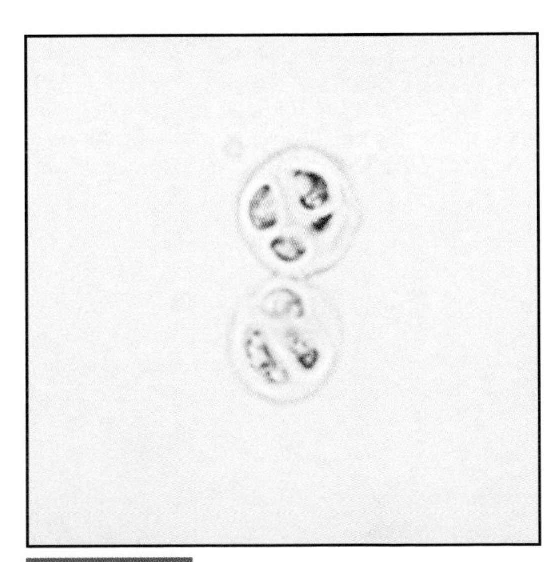

FIGURE 12-6 *Saccharomyces cerevisiae* ascospores (X1650). Four ascospores are produced in the original cell (acting as an ascus) by meiosis.

MOLDS OF MEDICAL OR ECONOMIC IMPORTANCE

Aspergillus

Species of *Aspergillus* are considered deuteromycetes by some and ascomycetes by others. The genus is characterized by green to yellow or brown granular colonies with a white edge. One species, *A. niger*, produces distinctive black colonies (Fig. 12-7). The *Aspergillus* fruiting body is distinctive, with chains of conidia arising from *phialides* attached to a swollen vesicle at the end of a conidiophore (Fig. 12-8). Fruiting body structure and size, and conidia color are useful in species identification.

A. fumigatus and other species are opportunistic pathogens that cause *aspergillosis*, an umbrella term covering many diseases. One form of pulmonary aspergillosis (referred to as *fungus ball*) involves colonization of the bronchial tree or tissues damaged by tuberculosis. *Allergic aspergillosis* may occur in individuals who are in frequent contact with the spores and become sensitized to them. Subsequent contact produces symptoms similar to asthma. *Invasive aspergillosis* is the most severe form. It results in necrotizing pneumonia and may spread to other organs.

Some species of *Aspergillus* are of commercial importance. Fermentation of soybeans by *A. oryzae* produces soy paste. Soy sauce is produced by fermenting soybeans with a mixture of *A. oryzae* and *A. soyae*. *Aspergillus* is also used in commercial production of citric acid.

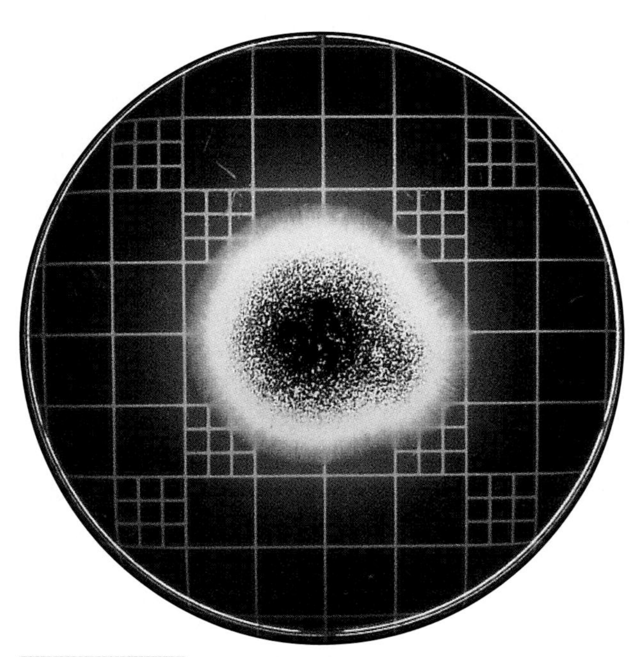

FIGURE 12-7 *Aspergillus niger* colony from above. Notice the white apron at the periphery and the peppered appearance of the central portion due to the black conidia. Most *Aspergillus* colonies are greenish.

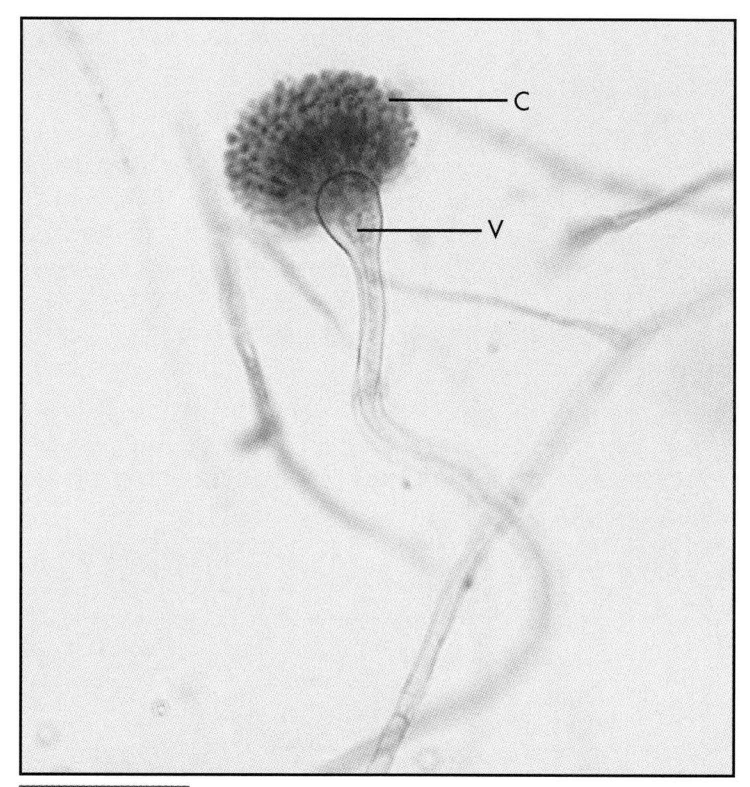

FIGURE 12-8 *Aspergillus* conidiophore with chains of conidia (C) at the end of a swollen vesicle (V) (X634).

Penicillium

Members of the genus *Penicillium* produce distinctive green, radially furrowed colonies with a white apron (Fig. 12-9). While they are classified by some as deuteromycetes because their sexual reproductive structures are not known or are not used in classification, it is clear that they are actually ascomycetes. Sexual reproduction results in the formation of ascospores within an ascus (Figs. 12-10 and 12-11). Asexual reproduction occurs with the typical *Penicillium* fruiting body which consists of spherical conidia in rows at the ends of elongated, branching cells (Fig. 12-12).

Penicillium is best known for its production of the antibiotic penicillin. Other species of *Penicillium* are of commercial importance for fermentations used in cheese production. Examples include *P. roquefortii* (Roquefort cheese) and *P. camembertii* (Camembert and Brie cheeses).

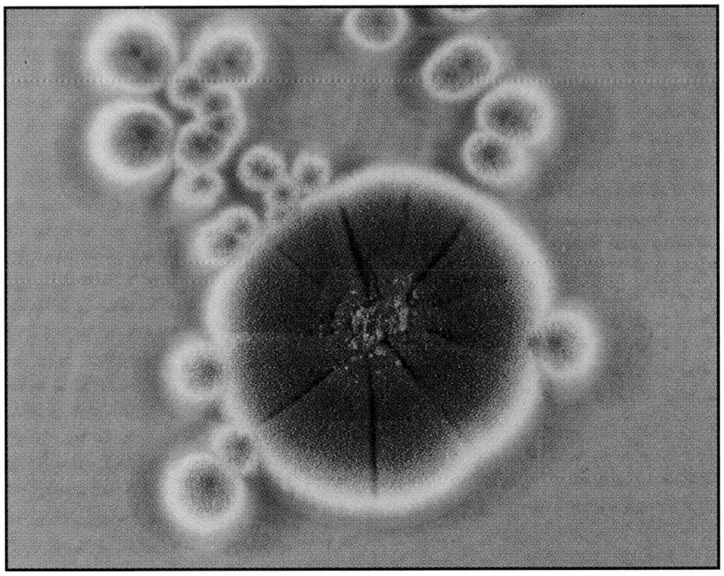

FIGURE 12-9 *Penicillium notatum* colony from above. The green, granular surface with radial furrows and a white apron are typical of the genus.

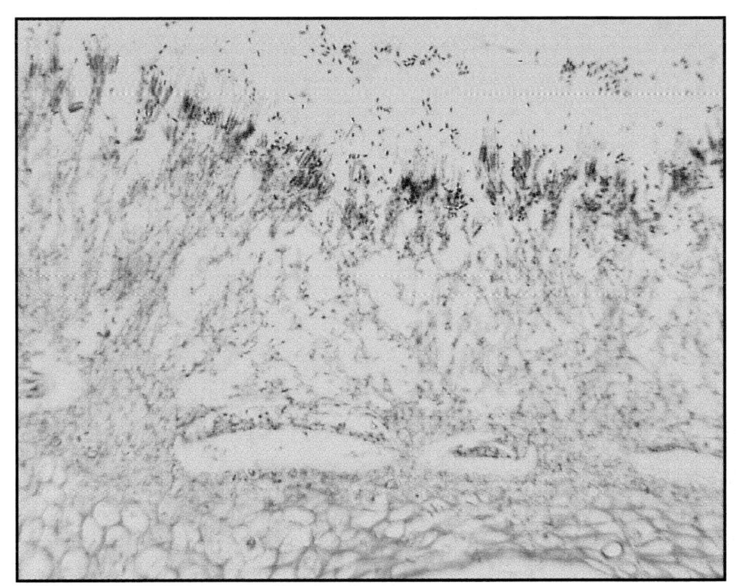

FIGURE 12-10 Section through *Penicillium* (X132). Red ascospores are visible at the upper surface.

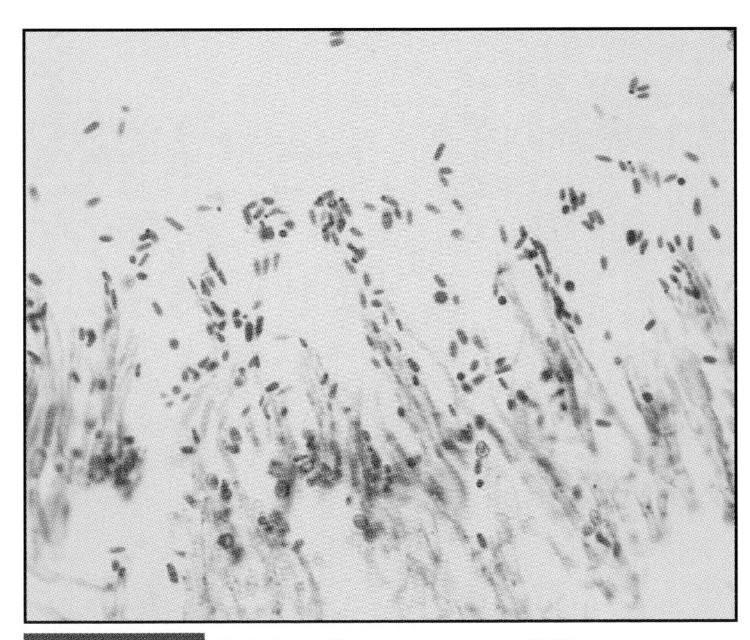

FIGURE 12-11 Red *Penicillium* ascospores (X528).

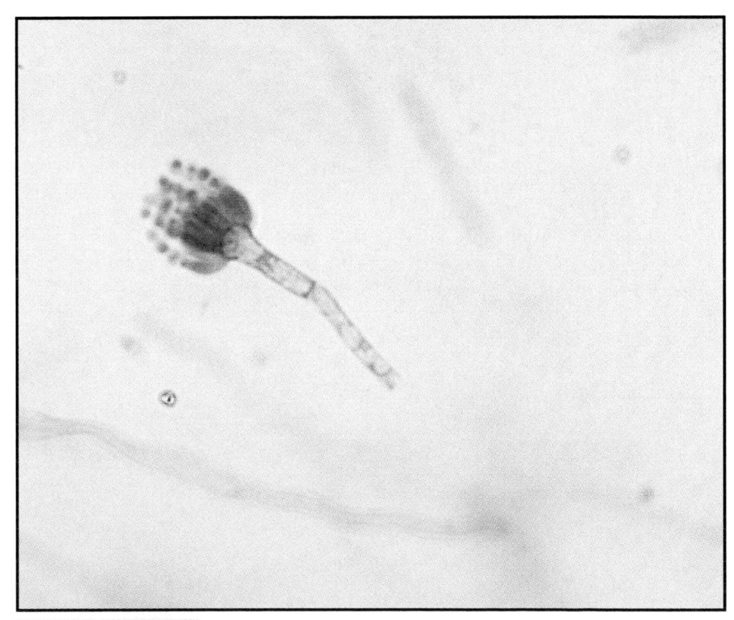

FIGURE 12-12 *Penicillium* conidiophore with chains of asexual spores (conidia) at the end (Lactophenol blue stain, X792).

Rhizopus

Rhizopus stolonifer (Fig. 12-13) is the common bread mold and is a member of the Zygomycetes. Its hyphae are haploid and nonseptate (*coenocytic*), and cytoplasmic streaming within them is common. Surface hyphae (*stolons*) are anchored by *rhizoids* where the hyphae contact the substrate.

The life cycle is illustrated in Figure 12-14. Asexual *sporangiospores* are produced by sporangia (Fig. 12-15) at the ends of elevated *sporangiophores*. These spores develop into hyphae identical to those that produced them. On occasion, sexual reproduction occurs when hyphae of different mating types (+ and − strains) make contact. Initially, *progametangia* (Fig. 12-16) extend from each hypha. Upon contact, a septum separates the end of each progametangium into a gamete (Fig. 12-17). The walls between the two gametangia dissolve and a thick-walled *zygospore* develops (Figs. 12-18 and 12-19). Fusion of nuclei occurs within the zygospore and produces one or more diploid nuclei, or *zygotes*. After a dormant period, meiosis of the zygotes occurs. The zygospore then germinates and produces a sporangium similar to the asexual sporangia. Haploid spores are released which develop into new hyphae, and the life cycle is completed.

Some species of *Rhizopus* are responsible for producing *zygomycosis*. Inhalation of spores leads to invasion of the blood and eventually necrosis of tissues, especially in diabetics and immunocompromised patients.

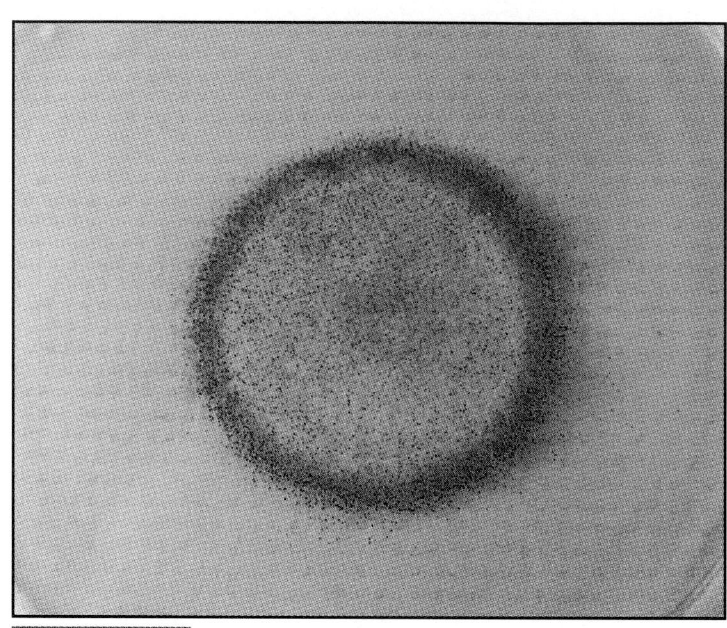

FIGURE 12-13 *Rhizopus stolonifer* colony. The black asexual sporangia of this bread mold are visible.

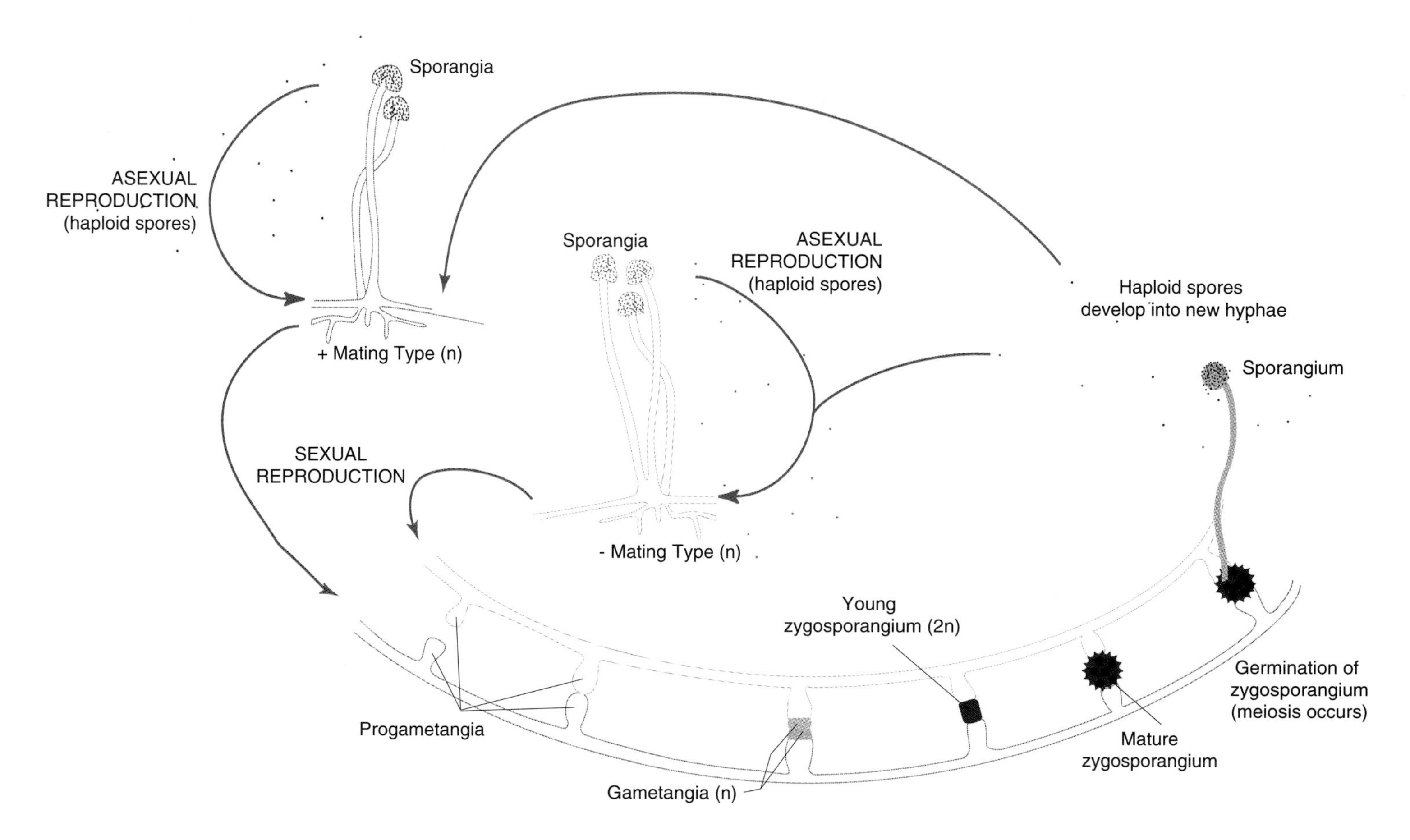

FIGURE 12-14 *Rhizopus* life cycle. (See text for details.)

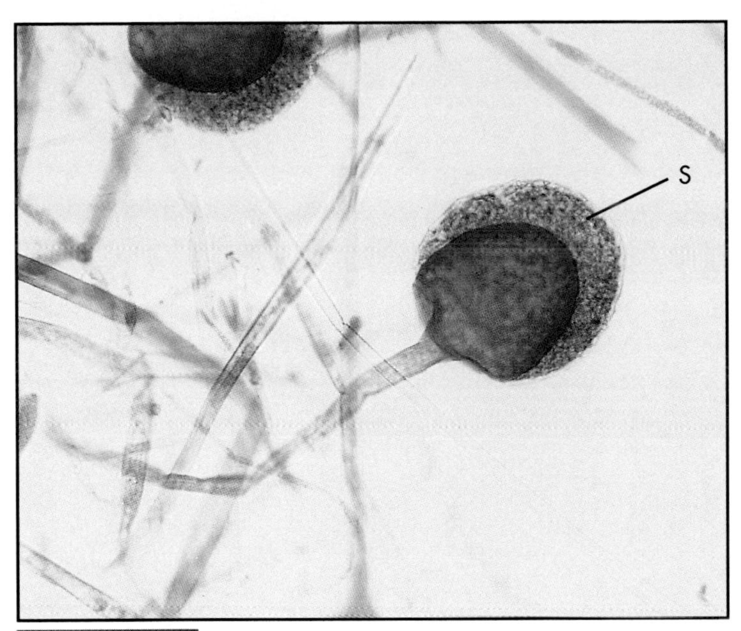

FIGURE 12-15 *Rhizopus* sporangiophore (X264). The sporangium produces haploid asexual sporangiospores (S) at its surface.

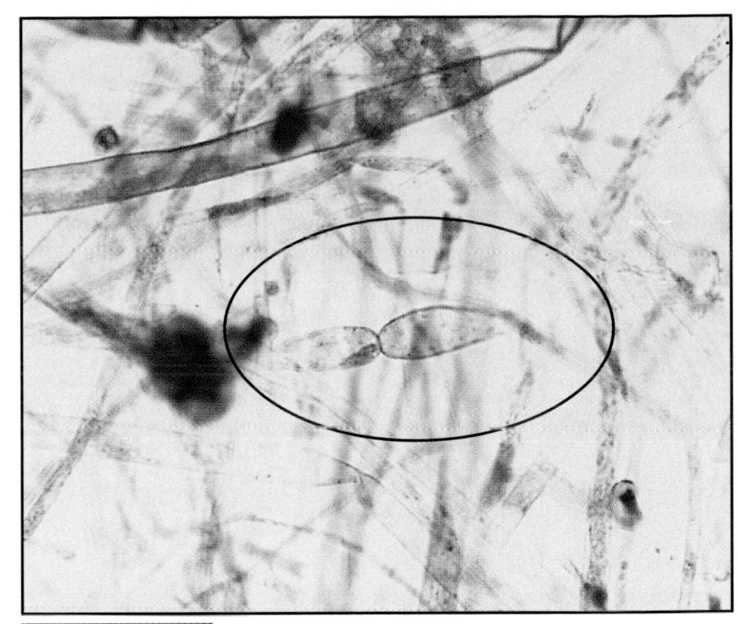

FIGURE 12-16 *Rhizopus* progametangia from different hyphae shown in the center of the field (X264).

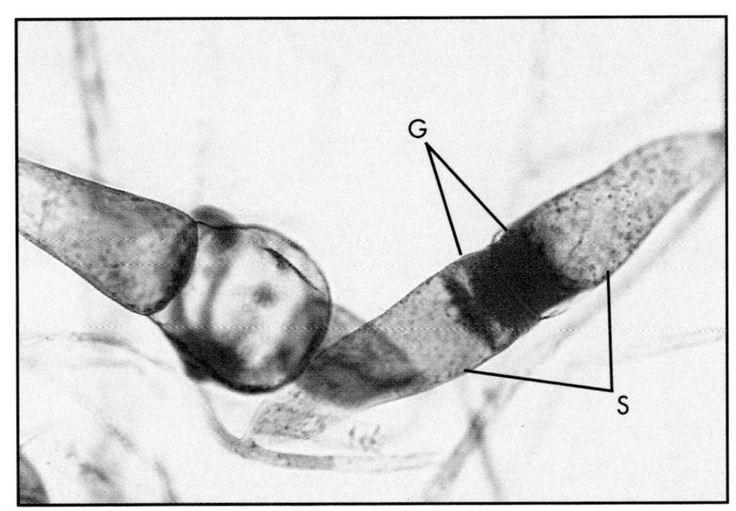

FIGURE 12-17 *Rhizopus* gametangia (G) and suspensors (S) in the center of the field (X264). Gametangia contain haploid nuclei from each mating type.

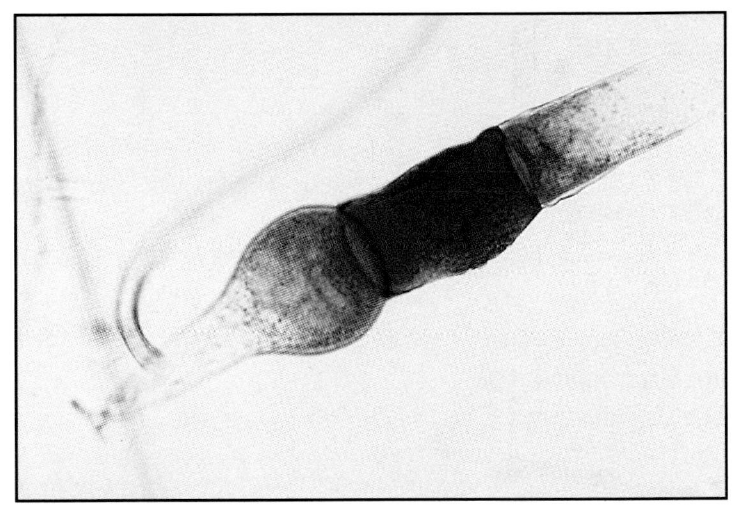

FIGURE 12-18 Young *Rhizopus* zygospore (X264). The zygospore forms when the cytoplasm from the two mating strains fuse (*plasmogamy*).

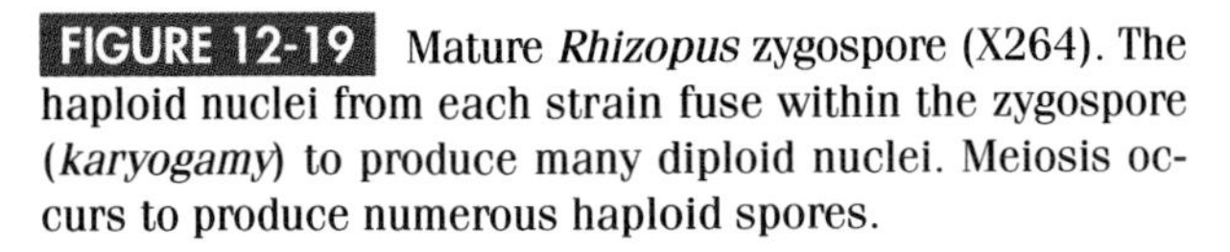

FIGURE 12-19 Mature *Rhizopus* zygospore (X264). The haploid nuclei from each strain fuse within the zygospore (*karyogamy*) to produce many diploid nuclei. Meiosis occurs to produce numerous haploid spores.

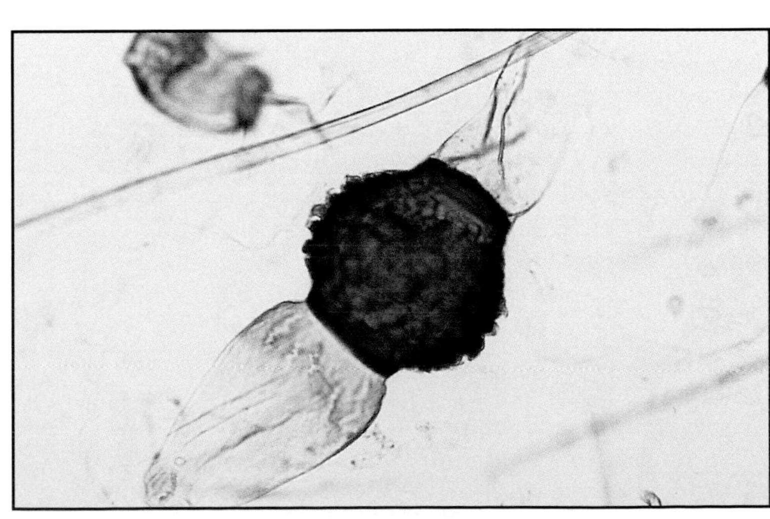

OTHER FUNGI OF MEDICAL IMPORTANCE

Coccidioides immitis

Coccidioides immitis causes *coccidioidomycosis* ("valley fever"), a lung disease associated with desert regions of the southwestern United States, northern Mexico, and parts of Central and South America. The mold form consists of hyphae with alternating cells developing into barrel shaped *arthroconidia* which are released when the neighboring cells die (Fig. 12-20). Infection occurs when these become airborne and are inhaled. Once in the host, thick walled *spherules* containing endospores develop. Most infections are asymptomatic and self-limiting, but in some individuals, symptoms are influenza-like and may include hypersensitivity reactions. Rarely, the disease becomes disseminated and then may be lethal.

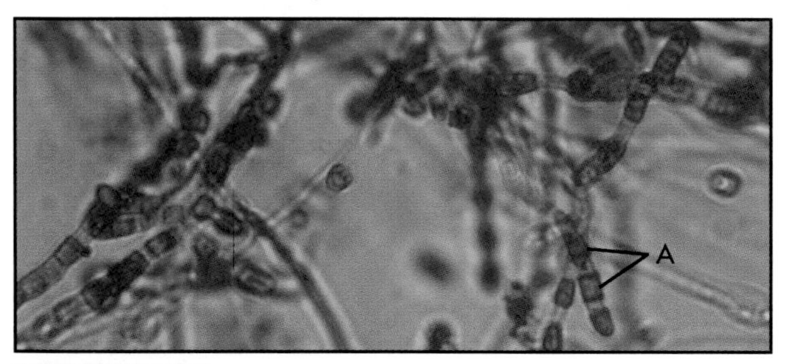

FIGURE 12-20 Barrel shaped arthroconidia (A) in hyphae of *Coccidioides immitis* (X1125). Typically, the arthroconidia develop in alternating cells.

Epidermophyton, Microsporum, and Trichophyton

The dermatophytes include *Epidermophyton, Microsporum,* and *Trichophyton,* related genera of fungi that infect keratinized tissues of mammals — epidermis, nails and hair. They produce a group of conditions known as *tinea* or *ringworm.* The latter name comes from their growth pattern in which the infection spreads outwards from the center of initial infection. The actively growing cells are at the edge of the growth, whereas healing tissue is towards the middle, thus giving the impression of a worm at the lesion's periphery. Various forms of tinea are recognized based on body location: *tinea capitis* occurs on the head, *tinea corporis* occurs on the face and trunk, *tinea cruris* occurs on the groin, *tinea unguium* occurs on the nails, and *tinea pedis* occurs on the feet (producing a condition known as athlete's foot).

These dermatophytes may infect more than one body region and often produce similar symptoms, so they are considered together here. Inflammation of infected tissue due to fungal antigens is the primary symptom and is often manifested as scaling, discoloration and itching. Transmission is through contact with infected skin or contaminated items, such as towels, clothing, combs or bedding.

Microsporum spp.

Terminal, septate macroconidia with thick, rough walls (Fig. 12-21) characterize the genus *Microsporum.* Microconidia may be present, but are not common. Chlamydoconidia (Fig. 12-22), a resting stage, are produced by some species. *M. gypseum*, *M. audouinii*, and *M. canis* are the most common species affecting humans.

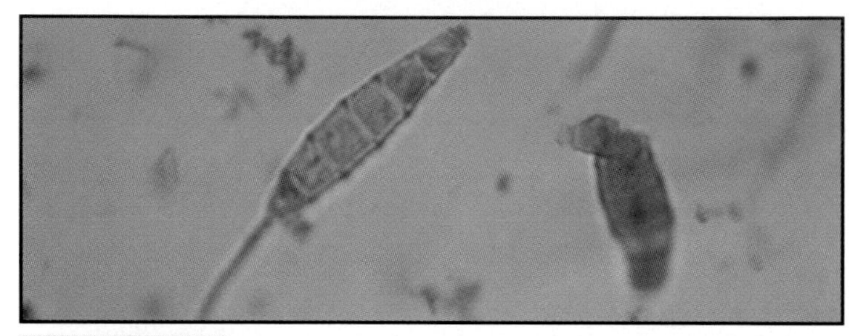

FIGURE 12-21 Macroconidia of *Microsporum gypseum* (X1125). *M. gypseum* typically has fewer than six cells in each macroconidium.

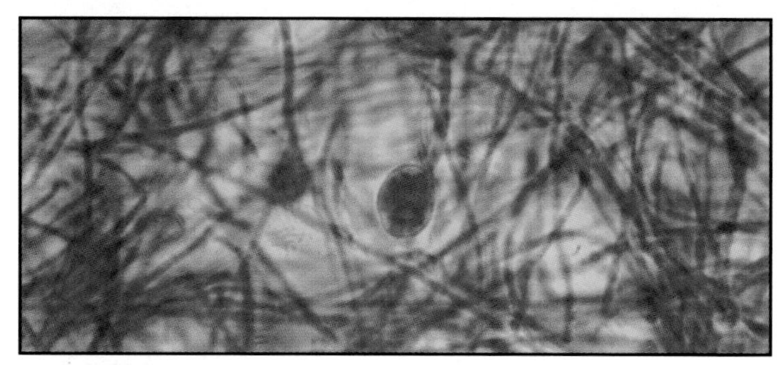

FIGURE 12-22 When grown in culture, *M. audouinii* (X1125) often produces chlamydospores which are resistant resting cells.

Trichophyton spp.

Trichophyton (Fig. 12-23) species are the most important dermatophytes causing infection in adults. They are responsible for most cases of tinea pedis and tinea unguium, and occasionally tinea corporis and tinea capitis. Macroconidia are rare, but are located at the ends of hyphae and are club shaped, smooth, thin walled and septate with up to ten cells. Spherical *microconidia* are more common.

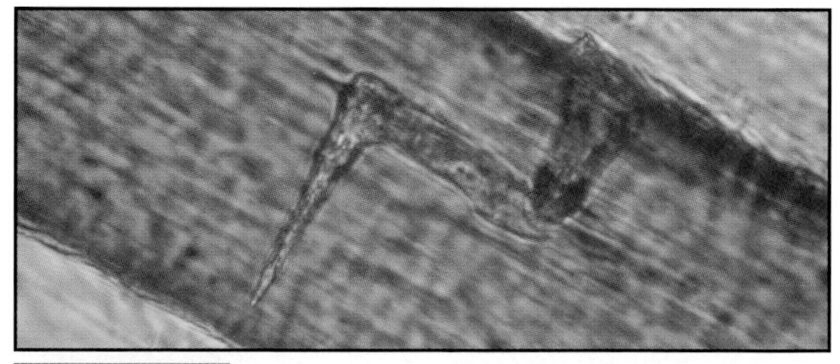

FIGURE 12-23 *Trichophyton mentagrophytes* infection may be diagnosed by a hair perforation test. Note the indentation in the hair (Lactophenol blue stain, X1125).

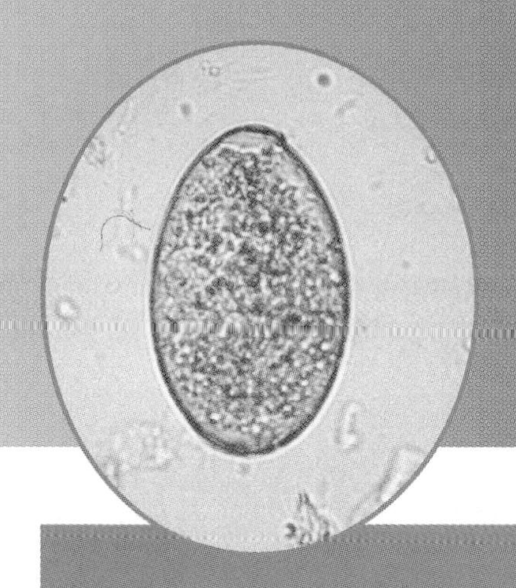

Parasitic Protozoans and Helminths 13

PROTOZOAN SURVEY

Protozoans are unicellular, eukaryotic, heterotrophic microorganisms. A typical life cycle includes a vegetative *trophozoite* and a resting *cyst* stage. Some have additional stages, making their life cycles more complex.

Currently, protozoans are classified as follows: Phylum Sarcomastigophora (including Subphylum Mastigophora [the flagellates] and Subphylum Sarcodina [the amebas]), Phylum Ciliophora (the ciliates), and Phylum Apicomplexa (sporozoans and others). Sarcodines move by sending out extensions of cytoplasm called *pseudopods*. Division is by binary fission. Ciliates owe their motility to the numerous cilia covering the cell. Reproduction is by transverse fission. Members of Mastigophora are characterized by one or more flagella and division by longitudinal fission. Sporozoans are typically nonmotile and usually have complex life cycles involving asexual reproduction in one host and sexual reproduction in another. Figures 13-1 through 13-4 show nonpathogenic protozoan representatives of Mastigophora, Sarcodina, and Ciliophora.

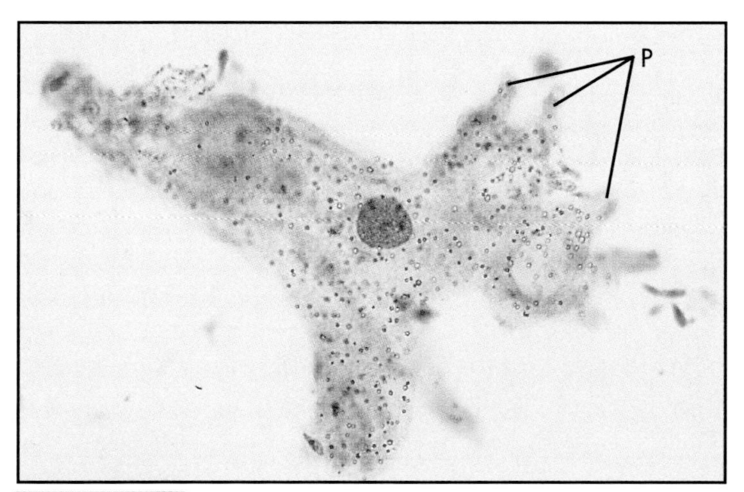

FIGURE 13-1 *Amoeba*, a sarcodine (X212). Note the numerous pseudopods (P).

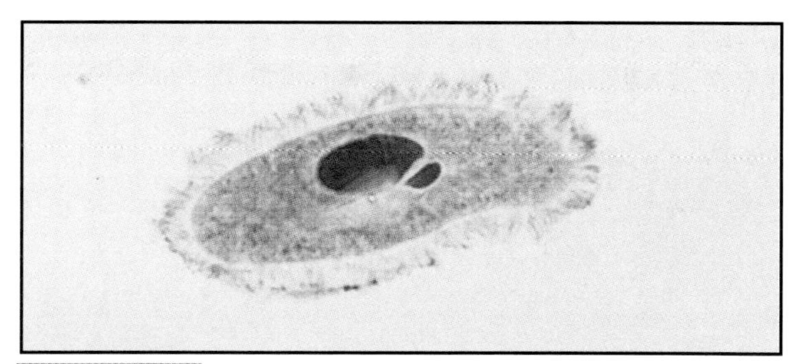

FIGURE 13-2 *Paramecium bursaria*, a ciliate (X528). Note the cilia around the edge of the cell. The macronucleus and micronucleus are also visible.

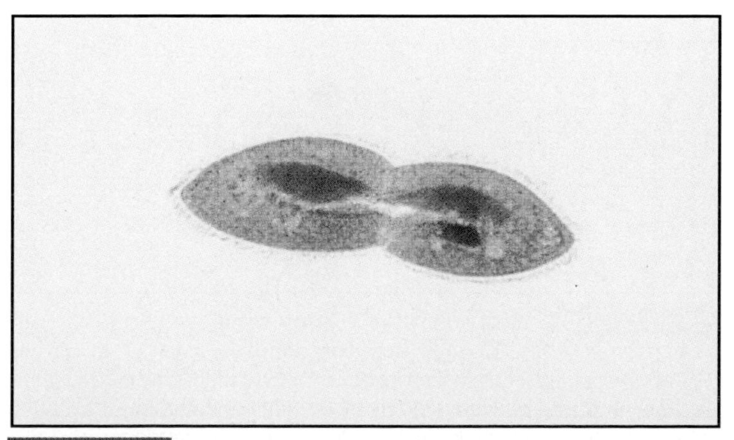

FIGURE 13-3 *Paramecium* undergoing transverse fission (X264).

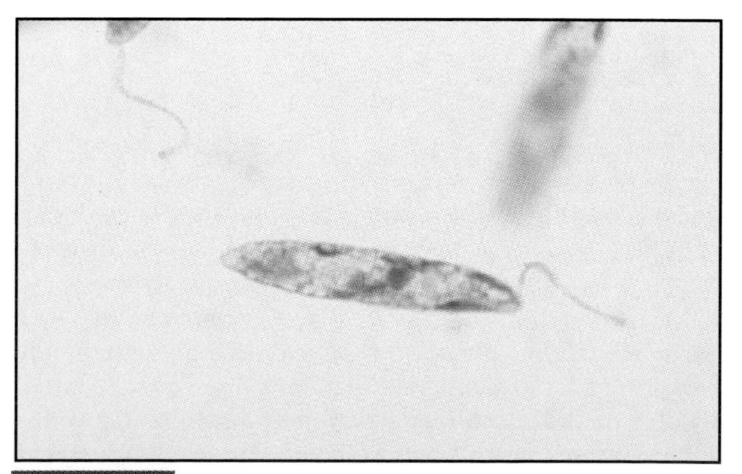

FIGURE 13-4 *Euglena*, a flagellate (X1584).

PROTOZOANS OF CLINICAL IMPORTANCE

Amoeboid Protozoans Found in Clinical Specimens

Blastocystis hominis

Blastocystis hominis is a commensal ameba that occupies the large intestine of up to 20% of the population. In most cases, patients with *B. hominis* show no symptoms, but in some situations where *B. hominis* has been abundant in stool samples of patients exhibiting mild diarrhea, and when all other possible parasites have been ruled out, it has been considered to be the causative agent. Identification is made by finding the central body form (Fig. 13-5) in a stool sample.

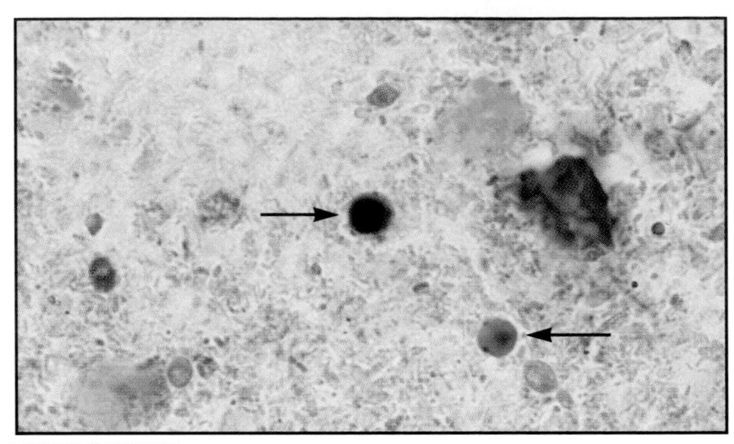

FIGURE 13-5 *Blastocystis hominis* trophozoites (X1000, trichrome stain,). Trophozoites vary greatly in size over the range of 6 to 40 µm. A large central body surrounded by several small nuclei are distinctive features of the trophozoite. Staining properties also may vary, as shown in these two specimens (arrows).

Entamoeba histolytica

Entamoeba histolytica is the causative agent of *amoebic dysentery* (*amebiasis*), a disease most common in areas with poor sanitation. Identification may be made by finding either trophozoites (Fig. 13-6) or cysts (Fig. 13-7) in a stool sample. The diagnostic features of each are described in the captions.

Infection occurs when cysts are ingested by a human host, either through fecal-oral contact or more typically, contaminated food or water. Cysts (but not trophozoites) are able to withstand the acidic environment of the stomach. Upon entering the less acidic small intestine, the cysts undergo *excystation*. Mitosis produces eight small trophozoites from each cyst.

The trophozoites parasitize the mucosa and submucosa of the colon causing ulcerations. They feed on red blood cells and bacteria. The extent of damage determines whether the disease is acute, chronic, or asymptomatic. In the most severe cases, infection may extend to other organs, especially the liver, lungs or brain. Abdominal pain, diarrhea, blood and mucus in feces, nausea, vomiting, and hepatitis are among the symptoms of amebic dysentery.

Cysts develop when the fecal material becomes too solid to be favorable for the trophozoites. Initially uninucleate, mitosis produces the mature quadranucleate cyst. Cysts are shed in the feces and may infect new hosts. They may also persist in the original host resulting in an *asymptomatic carrier* — a major source of contamination and infection.

Other members of the genus *Entamoeba* deserve mention here. *Entamoeba hartmanni* resembles *E. histolytica,* but is nonpathogenic and has smaller trophozoites and cysts. *Entamoeba coli* is a fairly common, nonpathogenic intestinal commensal that must be differentiated from *E. histolytica* in stool samples. Its characteristic features are given in the captions to Figures 13-8 and 13-9. *Entamoeba dispar* (not shown) is a newly formed species comprising nonpathogenic strains of *E. histolytica* identifiable by electrophoresis of certain proteins.

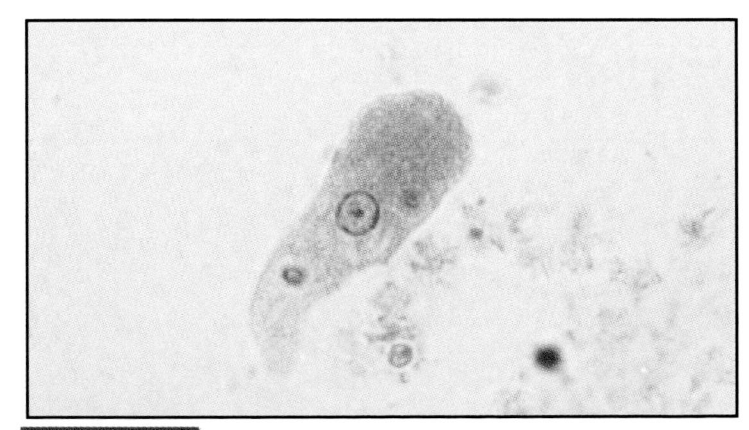

FIGURE 13-6 *Entamoeba histolytica* trophozoite (X800, iron hematoxylin stain). Trophozoites range in size from 12 to 60 µm. Notice the small, central karyosome, the beaded chromatin at the nucleus' margin, the ingested red blood cells and the finely granular cytoplasm. Compare with an *Entamoeba coli* trophozoite in Figure 13-8.

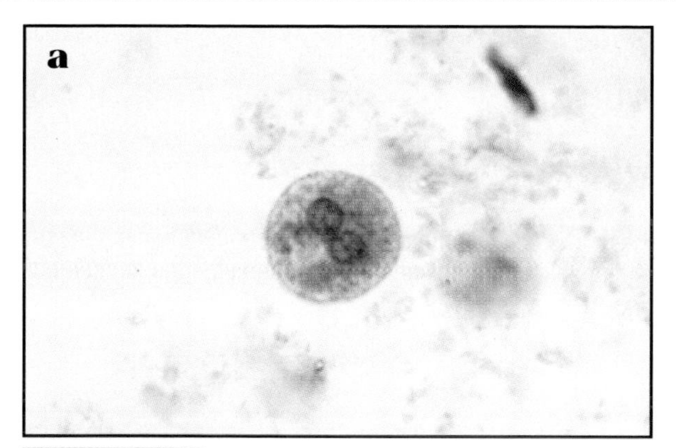

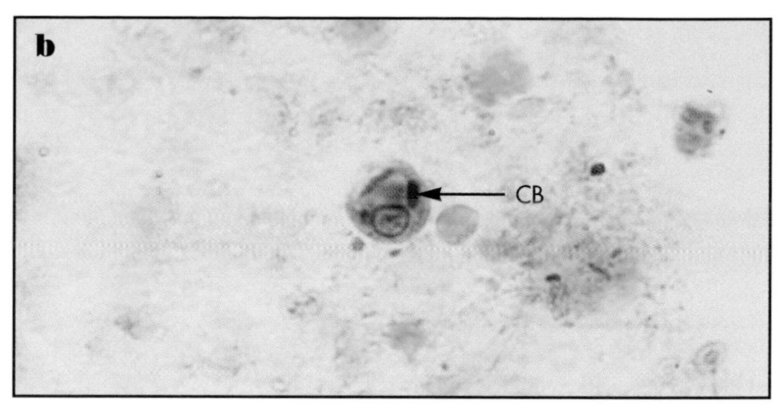

FIGURE 13-7 *Entamoeba histolytica* cysts. *(a)* Cysts are spherical with diameter of 10 to 20 µm. Two of the four nuclei are visible; other nuclear characteristics are as in the trophozoite. Compare with an *Entamoeba coli* cyst in Figure 13-9 (X1320, iron hematoxylin stain). *(b) E. histolytica* cyst (X1200, trichrome stain) with cytoplasmic chromatoidal bars (CB). These are found in approximately 10% of the cysts, have blunt ends and are composed of ribonucleoprotein.

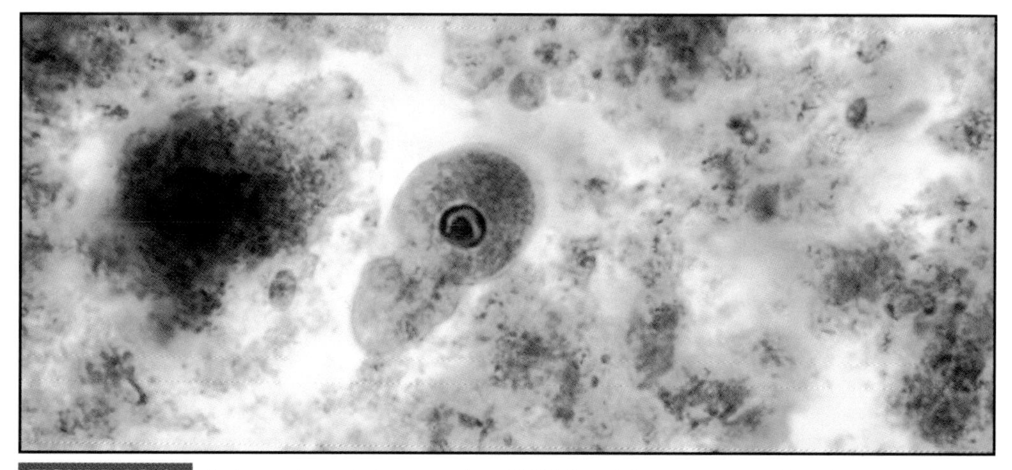

FIGURE 13-8 *Entamoeba coli* trophozoite (X1000, trichrome stain). Trophozoites have a size range of 15 to 50 µm. Notice the relatively large and eccentrically positioned karyosome, the unclumped chromatin at the nucleus' periphery, and the vacuolated cytoplasm lacking ingested RBCs. The usually nonpathogenic *E. coli* must be distinguished from the potentially pathogenic *E. histolytica*, so compare with Figure 13-6.

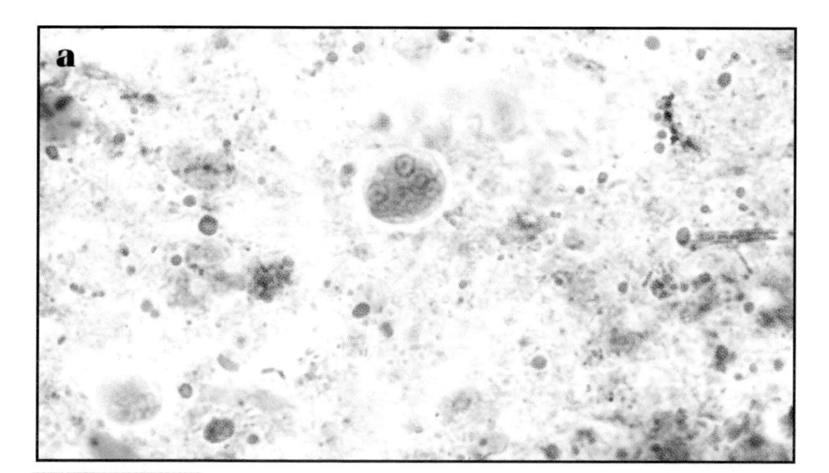

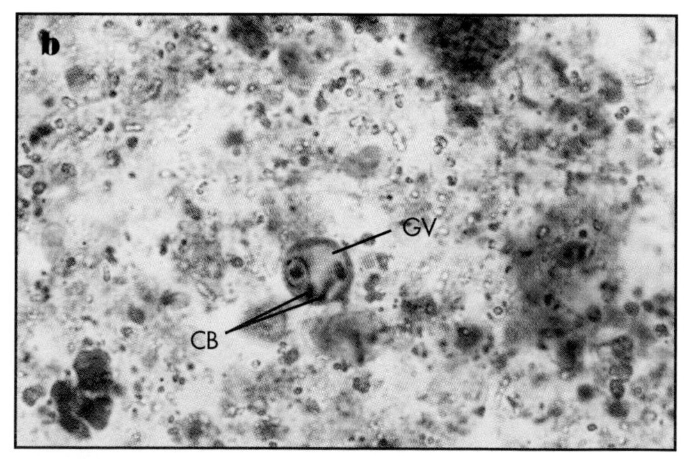

FIGURE 13-9 *Entamoeba coli* cysts. Cysts are typically spherical and are between 10 to 35 µm in diameter. They contain 8, or sometimes 16, nuclei. This makes differentiation from *E. histolytica* cysts simpler, since they never have more than 4 nuclei. *(a)* Five nuclei are visible in this specimen (X1000, trichrome stain). *(b)* Chromatoidal bars (CB) and a large glycogen vacuole (GV) characteristic of immature cysts are visible in this specimen (X1000, trichrome stain).

Endolimax nana

Endolimax nana is a fairly common commensal ameba that resides mainly in the cecum of humans. It exists as a trophozoite and cyst (Figs. 13-10 and 13-11) and its life cycle resembles that of *Entamoeba histolytica*. Infection occurs by ingestion of cysts in fecally contaminated food or water. Identification is made by finding the trophozoites and/or cysts in stool samples.

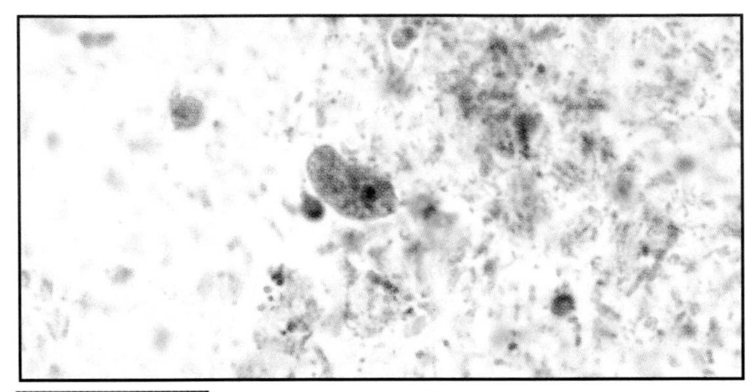

FIGURE 13-10 *Endolimax nana* trophozoite (X1000, trichrome stain). Trophozoites are small, with a size range of 6 to 15 μm. Notice the large karyosome filling the majority of the nuclear space and the absence of peripheral chromatin.

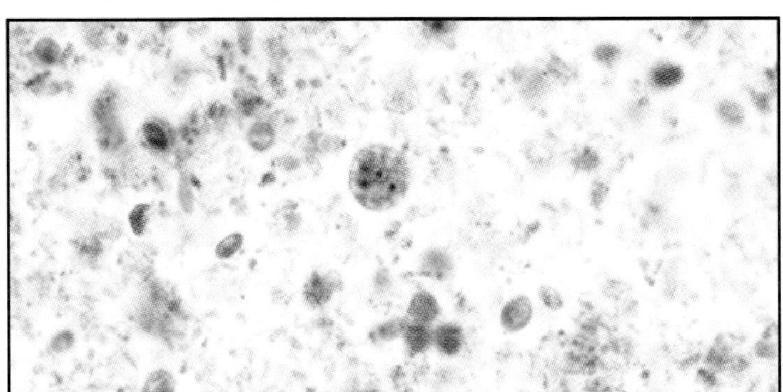

FIGURE 13-11 *Endolimax nana* cyst (X1000, trichrome stain). *E. nana* cysts range in size from 5 to 14 μm. There are typically four nuclei, each with a distinctive large karyosome.

Iodamoeba bütschlii

Iodamoeba bütschlii is less commonly found in humans than *E. coli* or *E. nana,* but when present, it lives in the cecum and feeds on other resident organisms. Transmission is by fecal contamination, but it is nonpathogenic. Laboratory diagnosis is done by identifying trophozoites and/or cysts (Figures 13-12 and 13-13) in stool specimens. Distinguishing characteristics are given in the captions.

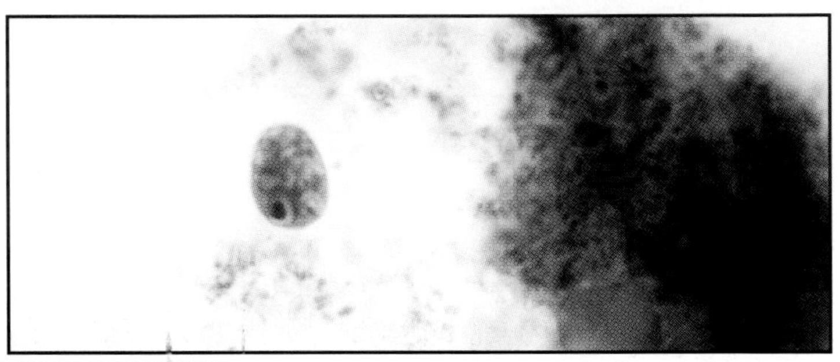

FIGURE 13-12 *Iodamoeba bütschlii* trophozoite (X1000, iron hematoxylin stain). *Iodamoeba bütschlii* trophozoites are 6 to 15 μm in size. The nucleus has a large karyosome, but lacks peripheral chromatin. On occasion, fine karyosome strands may be observed radiating outward from the karyosome to the nuclear membrane.

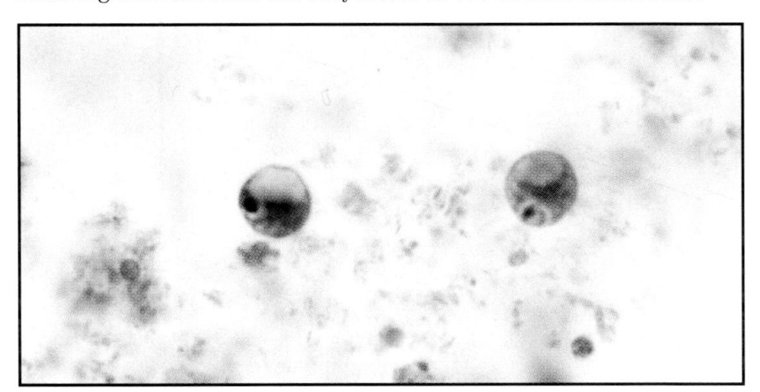

FIGURE 13-13 *Iodamoeba bütschlii* cysts (X1000, iron hematoxylin stain). Cysts of *I. bütschlii* range in size from 6 to 15 μm. The single nucleus has a large karyosome and no peripheral chromatin. A glycogen vacuole is typically found in these cysts.

Naegleria fowleri

Naegleria fowleri is a free-living soil and water ameba with an amoeboid stage (Fig. 13-14), a cyst stage and a flagellated stage. Under the proper conditions, it also is a facultative parasite that causes *primary amebic meningoencephalitis (PAM)*. Infection probably occurs when a human (especially children and young adults) forces contaminated water containing the protozoan up the nasal passages (as in diving). The organisms travel up the olfactory nerves into the cranial vault where they multiply and digest the olfactory bulbs and cerebral cortex. Symptoms occur about a week after infection and include fever, severe headache and coma. Death occurs within about a week from the onset of symptoms.

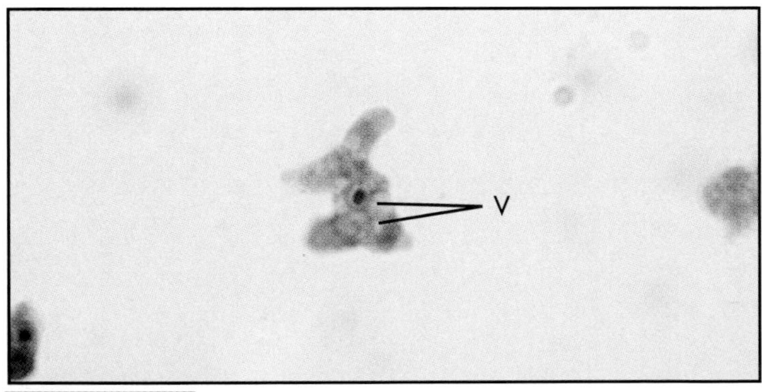

FIGURE 13-14 *Naegleria fowleri* trophozoite from culture (X1000, iron hematoxylin stain). Trophozoites are between 10 and 35 μm in size. Notice the large karyosome within the nucleus and the lobed pseudopods. Vacuoles (V) are also visible in the cytoplasm.

Ciliate Protozoans Found in Clinical Specimens

Balantidium coli

Balantidium coli (Figs. 13-15 and 13-16) is the causative agent of *balantidiasis* and exists in two forms: a vegetative trophozoite and a cyst. Laboratory diagnosis may be made by identification of either the cyst or the trophozoite, with the latter being more commonly found.

The trophozoite is highly motile due to its cilia and has a macro- and a micronucleus. Cysts in sewage-contaminated water are the infective form. Trophozoites may cause ulcerations of the colon mucosa, but not to the extent produced by *Entamoeba histolytica*. Symptoms of acute infection are bloody and mucoid feces. Diarrhea alternating with constipation may occur in chronic infections. Most infections are probably asymptomatic.

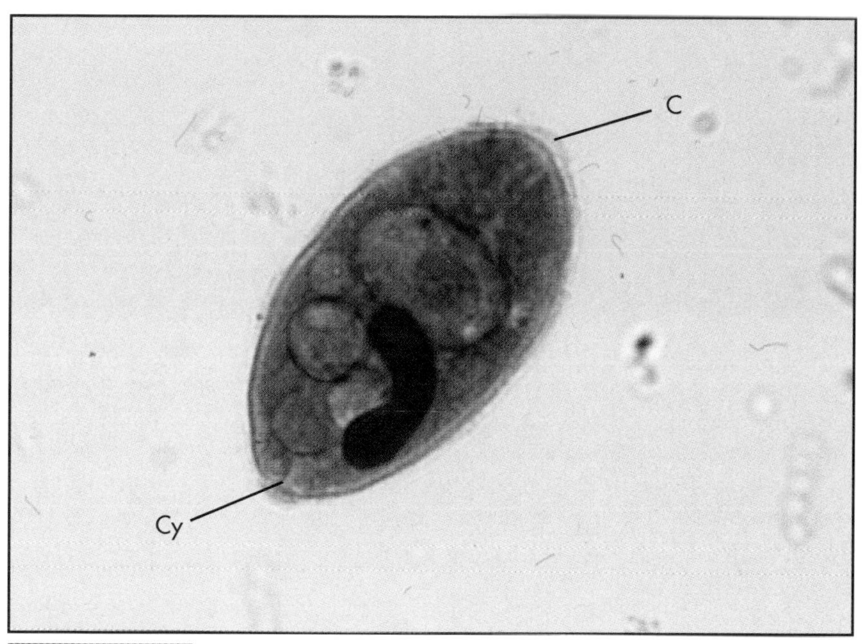

FIGURE 13-15 *Balantidium coli* trophozoite (X800). Trophozoites are oval in shape and have dimensions of 50 to 100 µm long by 40 to 70 µm wide. Cilia (C) cover the cell surface. Internally, the macronucleus is prominent; the adjacent micronucleus is not. An anterior cytostome (Cy) is usually visible.

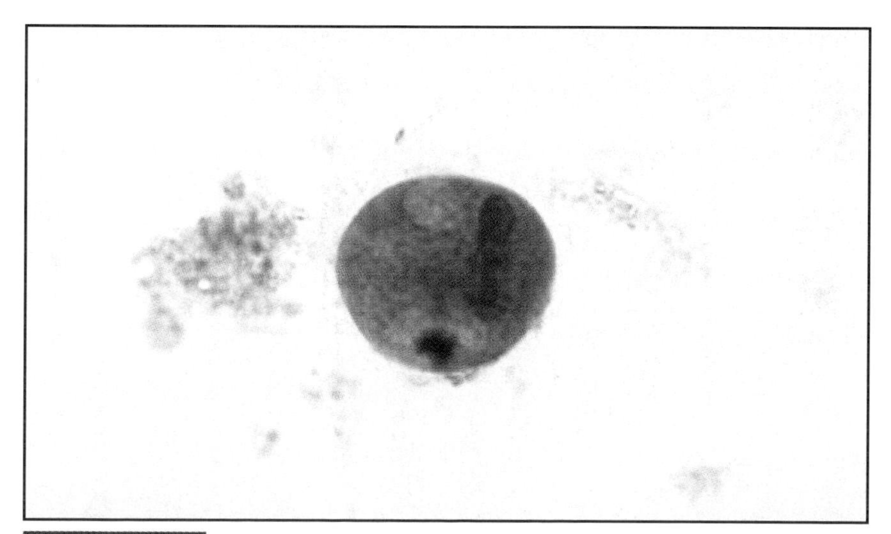

FIGURE 13-16 *Balantidium coli* cyst (X1000). Cysts are usually spherical and have a diameter in the range of 50 to 75 µm. There is a cyst wall and the cilia are absent. As in the trophozoite, the macronucleus is prominent, but the micronucleus may not be.

Flagellate Protozoans Found in Clinical Specimens

Chilomastix mesnili

Chilomastix mesnili exists as a trophozoite and a cyst (Figs. 13-17 and 13-18). Both may be found in stool samples and are used in identifying infection by this nonpathogen. It typically lives in the cecum and large intestine as a commensal.

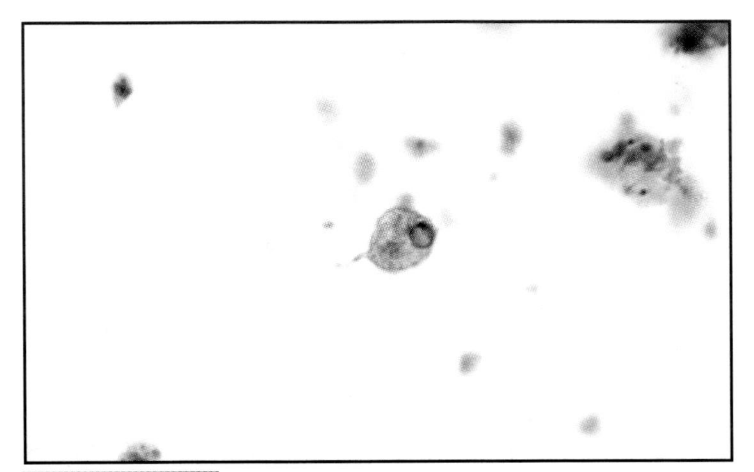

FIGURE 13-17 *Chilomastix mesnili* trophozoite (X1000, iron hematoxylin stain). Trophozozoites are elongated with a tapering posterior and a blunt anterior end that holds the nucleus. The dimensions are 6 to 20 µm long by 5 to 7 µm wide. There are four flagella: three at the anterior end (which may be difficult to see) and one associated with the prominent cytostome.

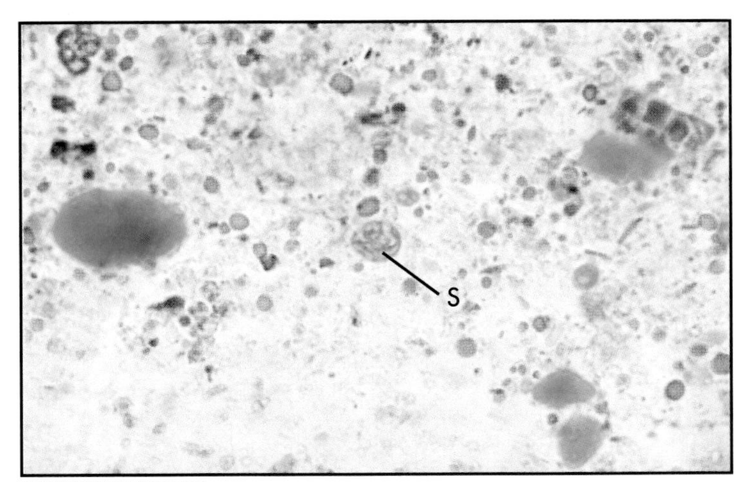

FIGURE 13-18 *Chilomastix mesnili* cyst (X1000, trichrome stain). Cysts are lemon-shaped, often with an anterior knob, and are 6 to 10 µm long. The nucleus may be difficult to see. A distinctive "shepherd's crook" (S) associated with the cytostome is also visible.

Dientamoeba fragilis

At one time, *Dientamoeba fragilis* was considered to be an ameba, but cytological evidence suggests it is better classified as a flagellate. It exists only as a trophozoite (Fig. 13-19) and lives primarily in the cecum where it feeds on the bacterial and yeast flora as well as cellular debris. It is found in approximately 4% of humans and is being identified in stool samples more frequently. Typical symptoms include diarrhea, abdominal pain and anal pruritus. The mode of transmission is unclear, but there is evidence supporting the idea that *D. fragilis* is transmitted in the eggs of the pinworm, *Enterobius vermicularis*.

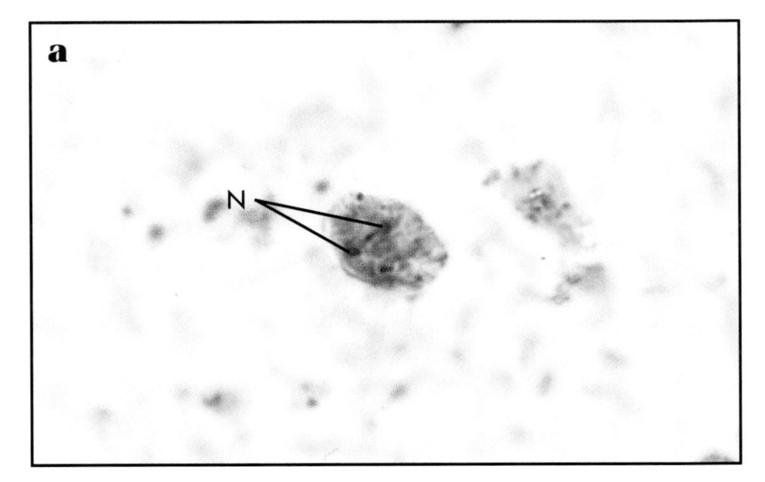

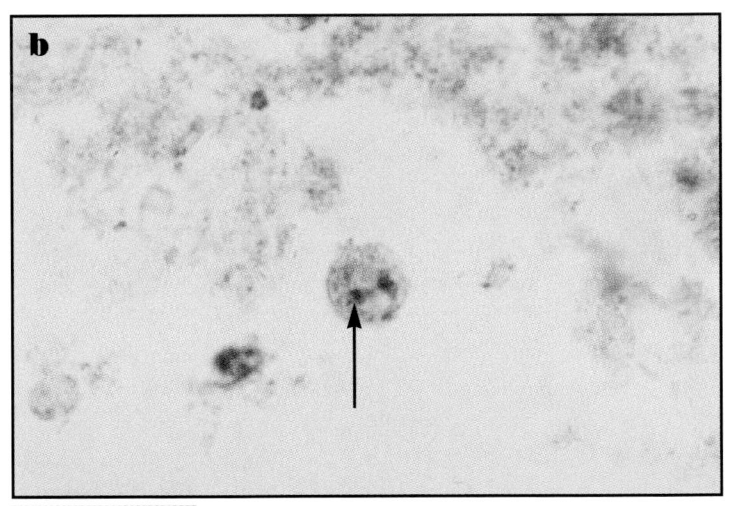

FIGURE 13-19 *Dientamoeba fragilis* trophozoites (X1000, iron hematoxylin stain). Trophozoites are 5 to 12 µm in size. Most cells have two nuclei (N), but many have only one. Each nucleus contains a single karyosome and the nuclear membrane is indistinct. *(a)* A typical trophozoite. *(b)* A trophozoite with a fragmented karyosome (arrow).

Giardia lamblia

Giardiasis is caused *Giardia lamblia* (also known as *Giardia intestinalis)*, a flagellate protozoan. It exists in the duodenum as a heart-shaped vegetative trophozoite (Fig. 13-20) with four pairs of flagella and a sucking disc that allows it to resist gut peristalsis. Multinucleate cysts (Fig. 13-21) are formed when the organism enters the colon. Cysts are shed in the feces and may produce infection of a new host upon ingestion. Transmission typically involves fecally contaminated water or food, but direct fecal-oral contact transmission is also possible.

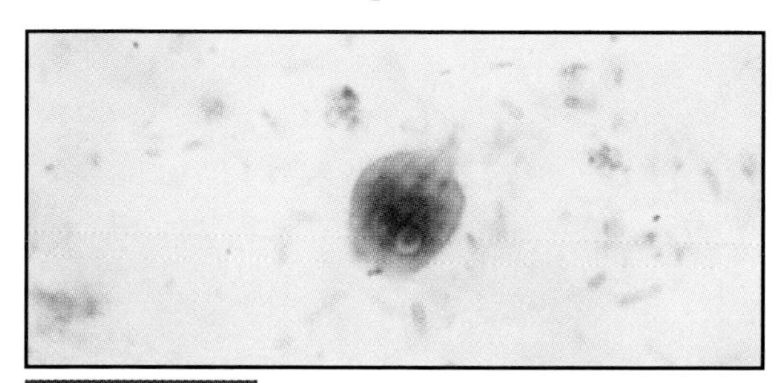

FIGURE 13-20 *Giardia lamblia* trophozoite (X1320, iron hematoxylin stain). Trophozoites have a long, tapering posterior end and range in size from 9 to 21 µm by 5 to 15 µm. There are two nuclei with small karyosomes. The two median bodies and the four pairs of flagella are not visible in this specimen.

The organism attaches to epithelial cells, but does not penetrate to deeper tissues. Most infections are asymptomatic. Chronic diarrhea, dehydration, abdominal pain and other symptoms may occur if the infection produces a large enough population to involve a significant surface area of the small intestine. Diagnosis is made by identifying trophozoites or cysts in stools specimens.

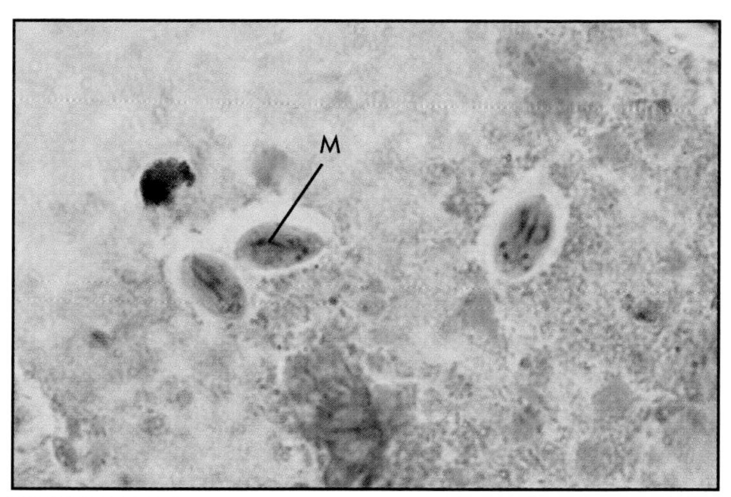

FIGURE 13-21 *Giardia lamblia* cyst (X1000, trichrome stain). *Giardia* cysts are smaller than trophozoites (8 to 12 µm by 7 to 10 µm), but the four nuclei with eccentric karyosomes and the median bodies (M) are still visible.

Leishmania donovani

Leishmania donovani actually represents a number of geographically separate species and subspecies that are difficult to distinguish morphologically. All produce *visceral leishmaniasis* or *kala-azar*, a disease found in tropical and subtropical regions. The organism exists as a nonflagellated *amastigote* (Fig. 13-22) in the mammalian host (humans, dogs, and rodents) and as an infective, motile *promastigote* in the sandfly vector (Fig. 13-23). They are introduced into the mammalian host by sandfly bites. Distribution of the disease is associated with distribution of the appropriate sandfly vector.

Upon introduction into the host by the sandfly, the organism is phagocytized by macrophages and converts to the amastigote stage. Mitotic divisions result in filling of the macrophage, which bursts and releases the parasites. Phagocytosis by other macrophages follows and the process repeats. In this way, the organism spreads through much of the reticuloendothelial system, including lymph nodes, liver, spleen and bone marrow. Kala-azar is a progressive disease and is fatal if untreated.

Amastigotes in an infected host may be ingested by a sandfly during a blood meal. Once inside the sandfly, they develop into the promastigote form and multiply. They eventually occupy the fly's buccal cavitiy where they can be transmitted to a new mammalian host during a

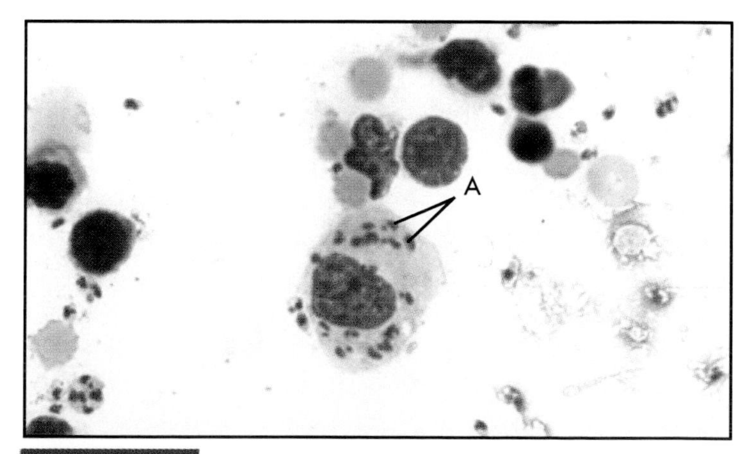

FIGURE 13-22 *Leishmania donovani* amastigotes in spleen tissue (X1000). Amastigotes (A) are 3 to 5 µm in size and multiply within phagocytic cells by binary fission. Amastigotes are also known as Leishman-Donovan (L-D) bodies.

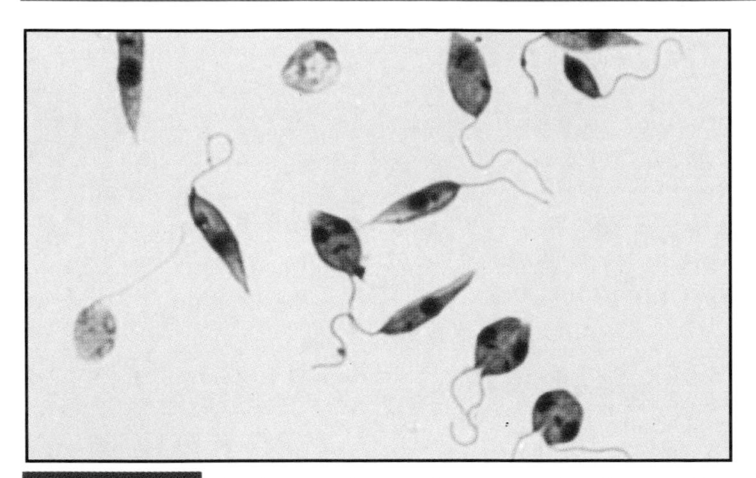

FIGURE 13-23 *Leishmania donovani* promastigotes, the infective stage, obtained from a culture (X1320). Notice the anterior flagellum, the kinetosome at its base, and the nucleus. Two of the cells at the lower right are dividing.

subsequent blood meal. Transmission requires the vector and does not occur by direct contact.

The related *Leishmania tropica* and *L. mexicana* cause *Oriental sore*, a cutaneous infection. *L. braziliensis* causes *New World cutaneous leishmaniasis*, an infection of skin and oral, nasal and pharyngeal mucous membranes. In all cases, infection involves macrophages in the affected region. Unlike *L. donovani*, all three may be transmitted by direct contact or by sandfly bites. Leishmaniasis was a common infection among troops in the Gulf War.

Trichomonas vaginalis

Trichomonas vaginalis is the causative agent of *trichomoniasis* (*vulvovaginitis*) in humans (Fig. 13-24). It has four anterior flagella and an undulating membrane.

Trichomoniasis may affect both sexes, but is more common in females. *T. vaginalis* causes inflammation of genitourinary mucosal surfaces — typically the vagina, vulva and cervix in females and the urethra, prostate and seminal vesicles in males. Most infections are asymptomatic or mild. There may be some erosion of surface tissues and a discharge associated with infection. The degree of infection is affected by host factors, especially the bacterial flora present and the pH of the mucosal surfaces. Transmission typically is by sexual intercourse.

The morphologically similar nonpathogenic *Trichomonas tenax* and *T. hominis* are residents of the oral cavity and intestines, respectively.

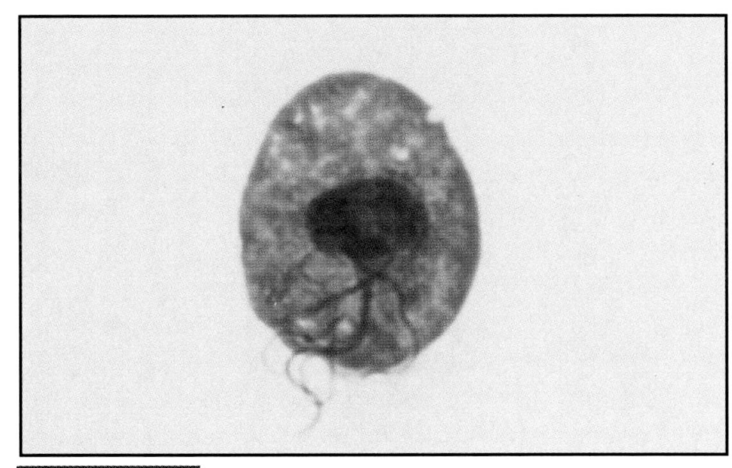

FIGURE 13-24 *Trichomonas vaginalis* (X2027). The trophozoite is the only stage of the *Trichomonas* life cycle. Several flagella are visible.

Trypanosoma brucei

Trypanosoma brucei (Fig. 13-25) is a species of flagellated protozoan divided into two subspecies: *T. brucei gambiense* and *T. brucei rhodesiense*. Both produce *African trypanosomiasis*, also known as *African sleeping sickness*. They are very similar morphologically, but differ in geographic range.

Trypanosomes have a complex life cycle. One stage of the life cycle, the *epimastigote*, multiplies in an intermediate host, the tsetse fly (genus *Glossina*). The infective *trypomastigote* stage is then transmitted to the human host through tsetse fly bites. Once introduced, trypomastigotes multiply and produce a chancre at the site of the bite. They enter the lymphatic system and spread through the blood, ultimately to the heart and brain. Immune response to the pathogen is hampered by the trypanosome's ability to change surface antigens faster than the immune system can produce appropriate antibodies. This antigenic variation also makes development of a vaccine unlikely.

Progressive symptoms include headache, fever and anemia, followed by symptoms characteristic of the infected sites. The sleeping sickness symptoms — sleepiness, emaciation, and unconsciousness — begin when the central nervous system becomes infected. The disease may last for years, but mortality rate is high. Death results from heart failure, meningitis, or severe debility of some other organ(s).

The infective cycle is complete when an infected individual (humans, cattle, and some wild animals are reservoirs) is bitten by a tsetse fly which ingests the organism during its blood meal. It becomes infective for its lifespan.

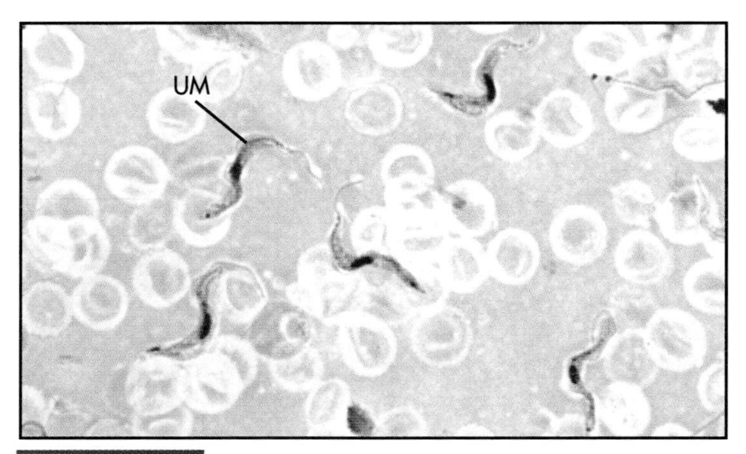

FIGURE 13-25 *Trypanosoma brucei* trypomastigotes in a blood smear (X1000). The nucleus and undulating membrane (UM) are visible.

Trypanosoma cruzi

Trypanosoma cruzi (Fig. 13-26) causes *American trypanosomiasis* (*Chagas' disease*). Cone-nosed ("kissing") bugs are the insect vector. They transmit the infective trypanosome during a blood meal through their feces. Scratching introduces the organism into the bite wound or conjunctiva. A local lesion (*chagoma*) forms at the entry site and is accompanied by fever. Spreading occurs via lymphatics (producing *lymphadenitis*) and trypomastigotes may be found in the blood within a couple of weeks. Trypanosomes then become localized in reticuloendothelial cells of the spleen, liver and bone marrow where they multiply intracellularly. Infected individuals may infect the cone-nosed bugs during a subsequent blood meal.

American trypanosomiasis occurs in South and Central America. It may be fatal, mild, or asymptomatic in adults. It is especially severe in children who often introduce the trypanosome through the conjunctiva, leading to edema of the eyelids and face of the affected side. The disease may spread to the central nervous system or to the heart, causing severe myocarditis.

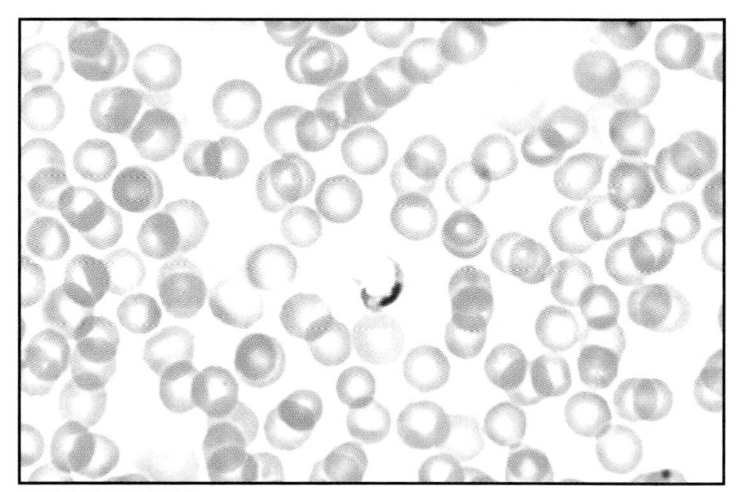

FIGURE 13-26 *Trypanosoma cruzi* in a blood smear (X1000). The nucleus and undulating membrane are visible.

Sporozoan Protozoans Found in Clinical Specimens

Plasmodium spp.

Plasmodia are sporozoan parasites with a complex life cycle, part of which is in various vertebrate tissues, while the other part involves an insect. In humans, the tissues are the liver and red blood cells, while the insect vector is the female *Anopheles* mosquito. A generalized life cycle is shown in Figure 13-27. Diagnostic life cycle stages for the various species are shown in Figures 13-28 to 13-35.

There are four species of *Plasmodium* that cause malaria in humans. These are *P. vivax* (benign tertian malaria), *P. malariae* (quartan malaria), *P. falciparum* (malignant tertian malaria), and *P. ovale* (ovale malaria). The life cycles are similar for each species as is the progress of the disease, so *P. falciparum* will be discussed as an example, with unique aspects compared to the others.

The *sporozoite* stage of the pathogen is introduced into a human host during a bite from an infected female *Anopheles* mosquito. Sporozoites then infect liver cells and produce the asexual *merozoite* stage. Merozoites are released from lysed liver cells, enter the blood, and infect erythrocytes. (Reinfection of the liver occurs at this stage in all but *P. falciparum* infections.) Once in RBCs, merozoites enter a cyclic pattern of reproduction in which more merozoites are released from the red cells synchronously every 48 hours (hence *tertian* — every third day — malaria). These events are tied to the symptoms of malaria: chills, nausea, vomiting and headache correspond to rupture of the erythrocytes. A spiking fever ensues and is followed by a period of sweating, after which the exhausted patient falls

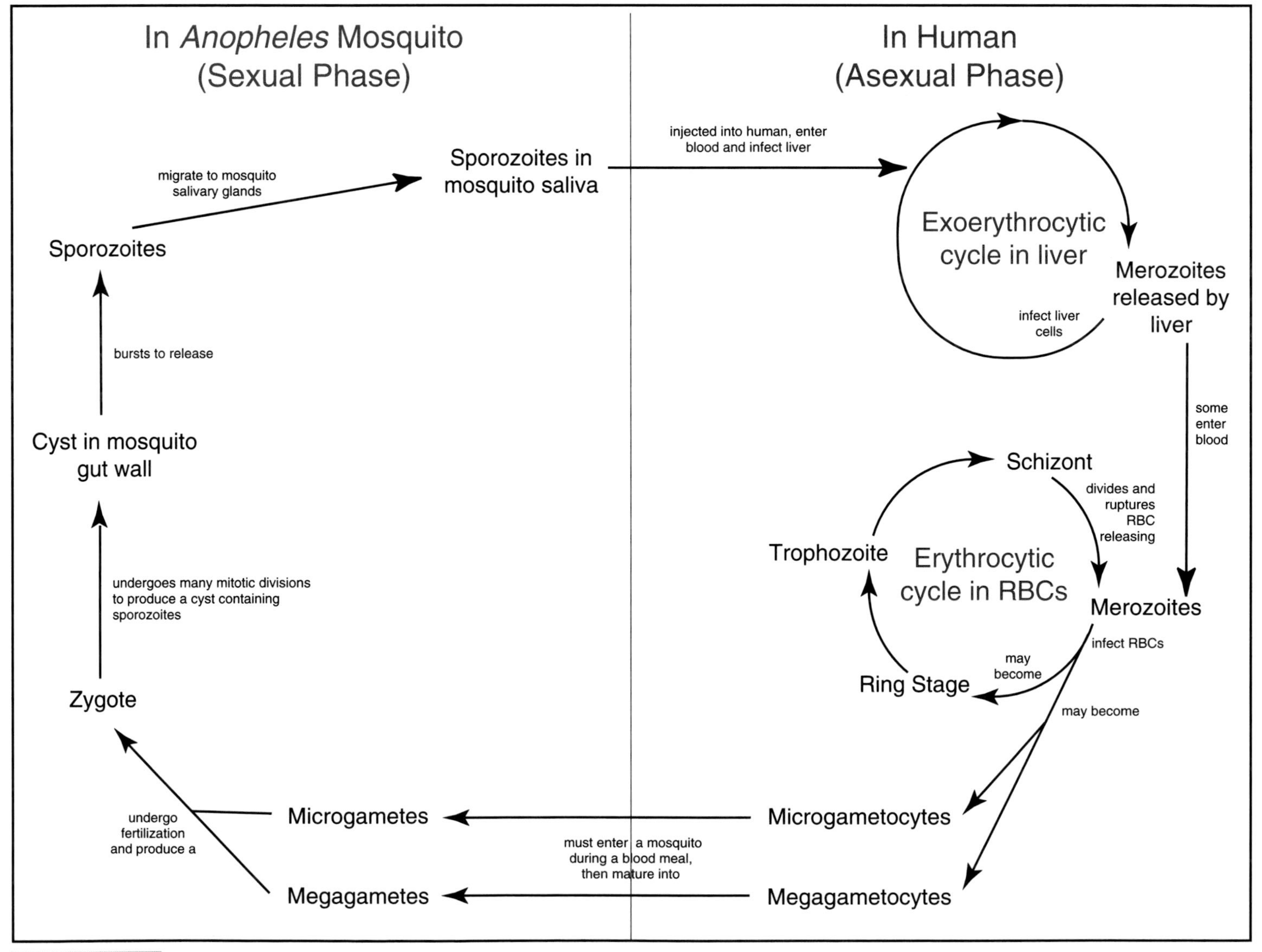

FIGURE 13-27 *Plasmodium* life cycle.

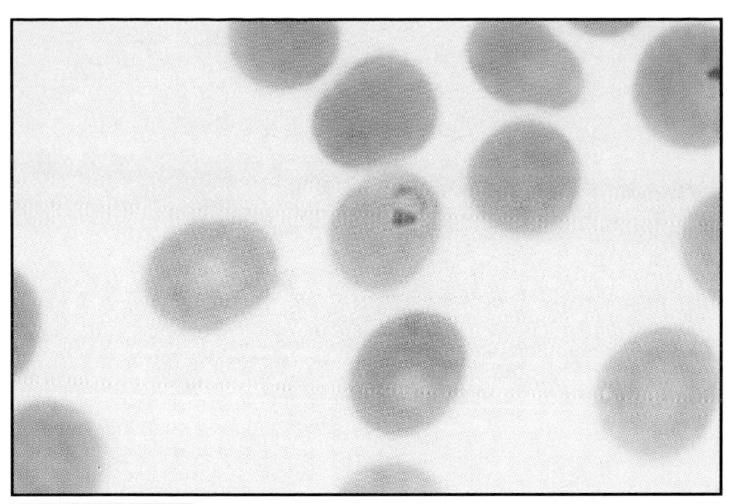

FIGURE 13-28 *Plasmodium falciparum* ring stage in a red blood cell (X2640). The ring is the trophozoite. Note the chromatin dots in the nucleus.

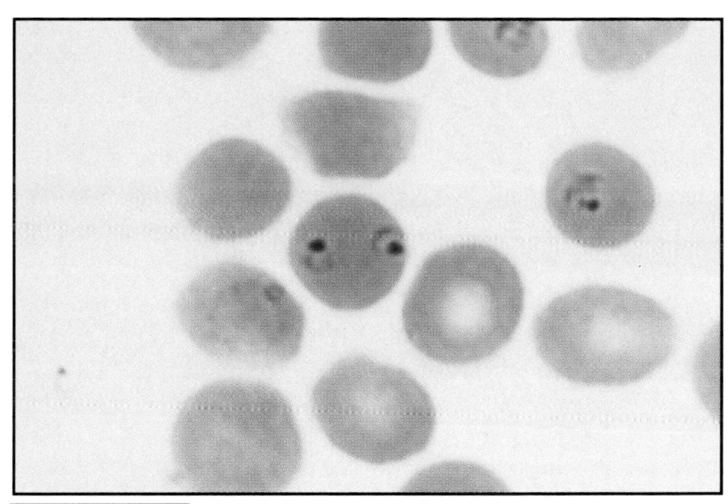

FIGURE 13-29 *Plasmodium falciparum* double infection of a red blood cell (X2640). This is commonly seen in *P. falciparum* infections.

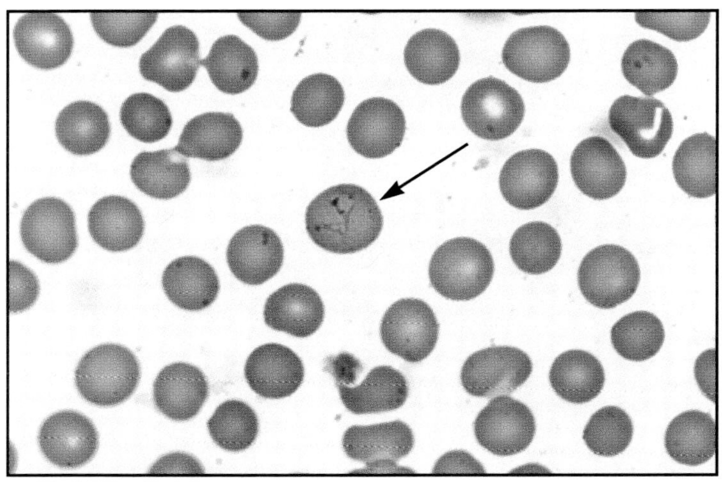

FIGURE 13-30 Erythrocyte infected with *Plasmodium vivax* (X1000). The parasite is in the ring stage, and the red cell exhibits characteristic Schüffner's dots in the cytoplasm. Schüffner's dots are also seen in red cells infected with *P. ovale*.

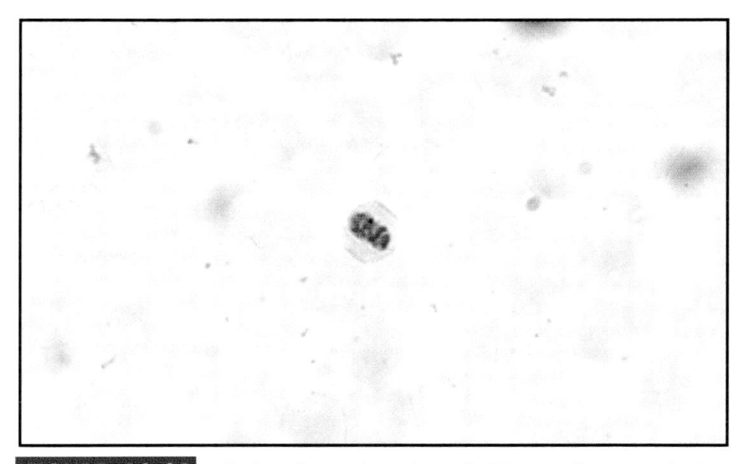

FIGURE 13-31 A band trophozoite of *Plasmodium malariae* (X1200).

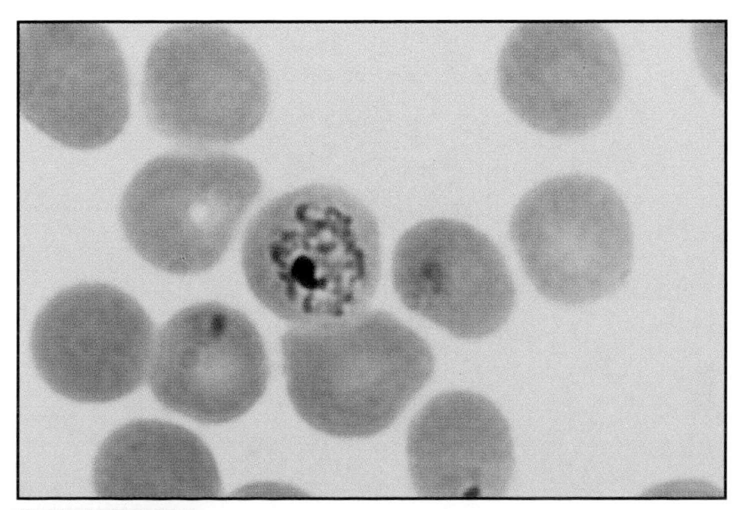

FIGURE 13-32 *Plasmodium falciparum* developing schizont in a red blood cell (X2640). These are usually not seen in peripheral blood smears since they reside in visceral capillaries.

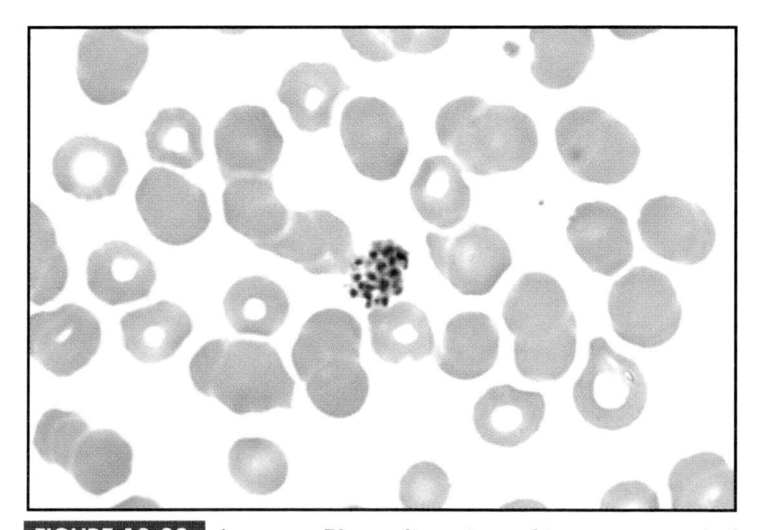

FIGURE 13-33 A mature *Plasmodium vivax* schizont composed of approximately 16 merozoites (X1200). More than 12 merozoites distinguishes *P. vivax* from *P. malariae* and *P. ovale*, which both typically have eight, but up to 12. *P. falciparum* may have up to 24 merozoites, but they are not typically seen in peripheral blood smears and so are not confused with *P. vivax*.

asleep. It is during this latter phase that the parasites reinfect the red cells, and the cycle repeats.

The sexual phase of the life cycle begins when certain merozoites enter erythrocytes and differentiate into male or female *gametocytes*. The sexual phase of the life cycle continues when gametocytes are ingested by a female *Anopheles* mosquito during a blood meal. Fertilization occurs and the zygote eventually develops into a cyst within the mosquito's gut wall. After many divisions, the cyst releases sporozoites, some of which enter the mosquito's salivary glands ready to be transmitted back to the human host.

Most malarial infections eventually are cleared, but not before the patient has developed anemia and has suffered permanent damage to the spleen and liver. The most severe infections involve *P. falciparum*. Erythrocytes infected by *P. falciparum* develop abnormal projections that cause them to adhere to the lining of small blood vessels. This can lead to obstruction of the vessels, thrombosis, or local ischemia which account for many of the fatal complications of this type of malaria — including liver, kidney and brain damage.

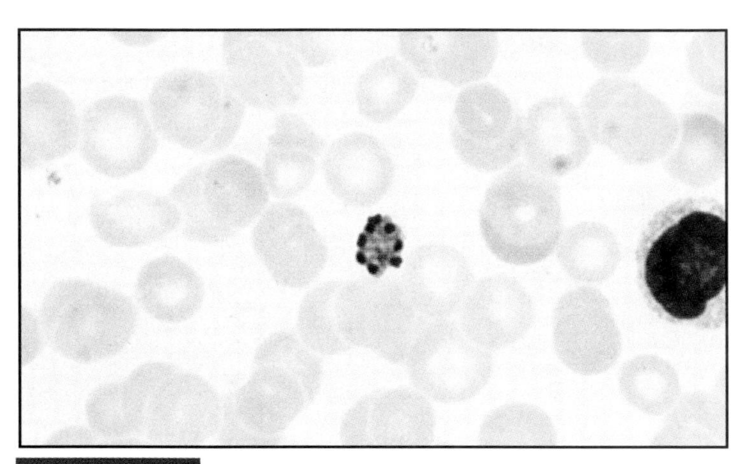

FIGURE 13-34 *Plasmodium malariae* schizont with 8 merozoites in a distinctive rosette arrangement (X1200).

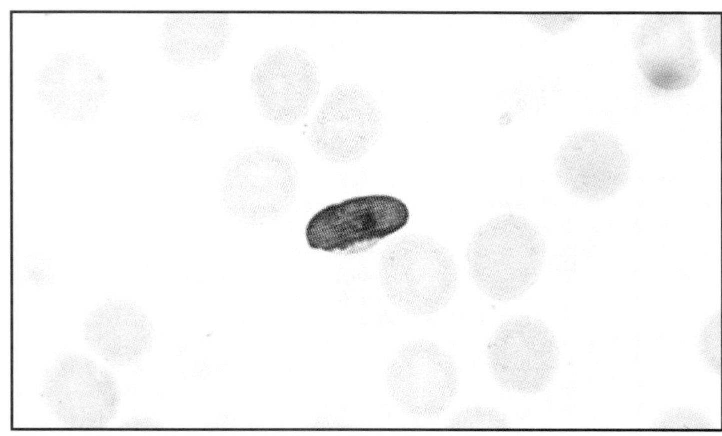

FIGURE 13-35 *Plasmodium falciparum* gametocyte in an erythrocyte (X1200). Differentiation between microgametocytes and megagametocytes is difficult in this species.

Toxoplasma gondii

Like other sporozoans, *Toxoplasma gondii* (Fig. 13-36) has sexual and asexual phases in its life cycle. The sexual phase occurs in the lining of cat intestines where *oocysts* are produced and shed in the feces. Each oocyst undergoes division and contains 8 sporozoites. If ingested by another cat, the sexual cycle may be repeated as the sporozoites produce gametocytes which in turn produce gametes. If ingested by another animal host (including humans) the oocyst germinates in the duodenum and releases the sporozoites. Sporozoites enter the blood and infect other tissues where they become trophozoites, which continue to divide and spread the infection to lymph nodes and other parts of the reticuloendothelial system. Trophozoites ingested by a cat eating an infected animal develop into gametocytes in the cat's intestines. Gametes are formed, fertilization produces an oocyst, and the life cycle is completed.

Infection via ingestion of the oocyst typically is not serious. The patient may notice fatigue or muscle aches. The more serious form of the disease involves infection of a fetus across the placenta from an infected mother. This type of infection may result in stillbirth, or liver damage and brain damage. AIDS patients may also suffer fatal complications from infection.

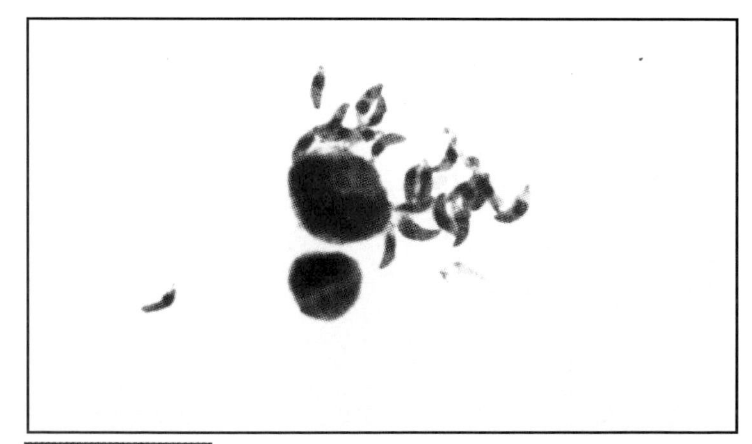

FIGURE 13-36 *Toxoplasma gondii* trophozoites (X1000). Notice the bow shaped cells with prominent nuclei.

Miscellaneous Protozoans Found in Clinical Specimens

Babesia microti

Babesia microti is a blood parasite found in the northeastern United States and other parts of the world. Mice (*Peromyscus*) are the main reservoir and the parasite is introduced into human hosts by a bite from an infected tick (*Ixodes dammini*). The parasites then infect RBCs and resemble young *Plasmodium* trophozoites (Fig. 13-37). Symptoms of *babesiosis*, which are not easily distinguished from other diseases, appear in nonsplenectomized individuals after about a week and include malaise, fever, and generalized aches and pains. Other *Babesia* species may be responsible for producing a fulminant hemolytic disease of immunosuppressed or splenectomized patients. Diagnosis is by detection of parasites in the blood.

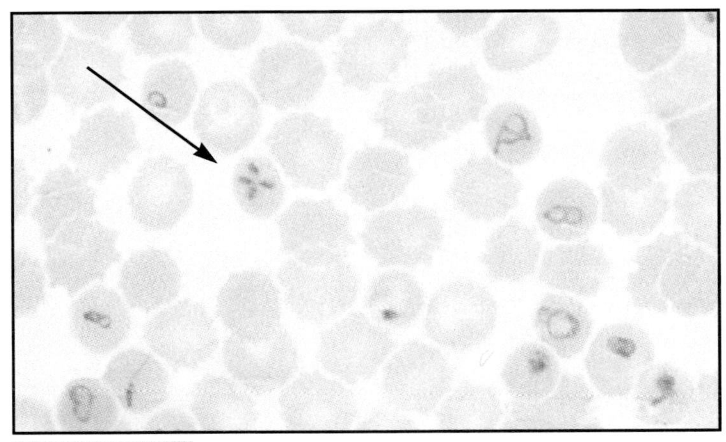

FIGURE 13-37 *Babesia microti* in human erythrocytes (X1000). The ring forms resemble *Plasmodium falciparum*, but are more variable in size, never pigmented, and occassionally form a cross-like tetrad (arrow).

Cryptosporidium parvum

Cryptosporidium parvum is a parasite of intestinal microvilli and causes *cryptosporidiosis*. Infection of immunocompromised patients results in a long-term disease characterized by profuse, watery diarrhea. Infective oocysts containing sporozoites (Fig. 13-38) are passed in the feces, so transmission is by fecal-oral contact. Infection may also occur through contact with infected animals. Diagnosis is made by finding oocysts in feces.

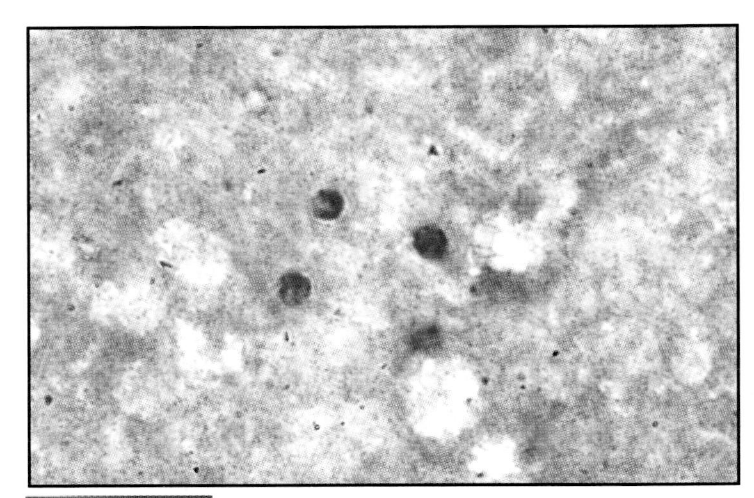

FIGURE 13-38 *Cryptosporidium parvum* oocysts from a human fecal sample (X1000, modified acid-fast stain). Oocysts contain sporozoites (not visible) and are the infective stage. They are typically about 5μm in size.

HELMINTH PARASITES

A study of helminths is appropriate to the microbiology lab because clinical specimens may contain microscopic evidence of helminth infection. The three major groups of parasitic worms encountered in lab situations are the trematodes (flukes), the cestodes (tapeworms) and the nematodes (round worms). Life cycles of the parasitic worms are often complex, sometimes involving several hosts, and are beyond the scope of this book. Emphasis here is on a brief background and clinically important diagnostic features of each worm.

Trematode Parasites Found in Clinical Specimens

Clonorchis sinensis

Clonorchis sinensis is the Oriental liver fluke and causes *clonorchiasis*, a liver disease. It is a common parasite of people living in Japan, Korea, Vietnam, China and Taiwan and is becoming more common in the United States with the influx of Southeast Asian immigrants. Infection typically occurs when undercooked infected fish is ingested. The adults migrate to the liver bile ducts and begin laying eggs in approximately one month. Degree of damage to the bile duct epithelium and surrounding liver tissue is due to the number of worms and the duration of infection. Diagnosis is made by identifying the characteristic eggs in feces (Fig. 13-39).

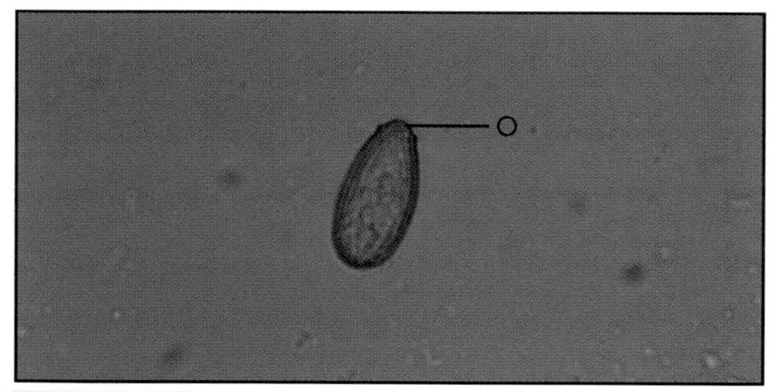

FIGURE 13-39 *Clonorchis sinensis* egg in a fecal specimen (X1000, D'Antoni's iodine stain). The eggs are thick shelled and between 27 and 35 µm long. There is a distinctive operculum (O) (positioned to give the appearance of shoulders) and often a knob at the aboperular end (not visible in this specimen).

Fasciola hepatica

Fasciola hepatica is a liver fluke commonly associated with domestic sheep and cattle. Human infection results when juvenile worms attached to aquatic vegetation are ingested. Penetration of the intestine and subsequently the liver leads the juveniles to the bile ducts where they develop into adults. Migration by the juveniles damages the liver; adults damage the bile ducts, gall bladder and liver, resulting in cirrhosis, jaundice or in severe cases, abscesses. Diagnosis is made by identification of eggs in feces (Fig. 13-40).

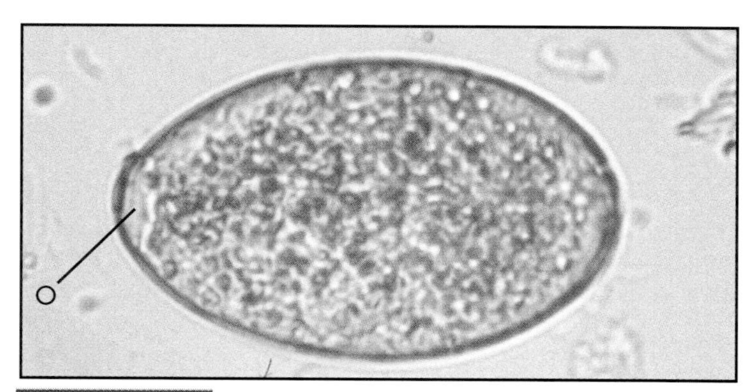

FIGURE 13-40 *Fasciola hepatica* egg in a fecal specimen (X1000, D'Antoni's iodine stain). The large eggs (130 to 150 µm long by 60 to 90 µm wide) are unembryonated in fecal samples and have an inconspicuous operculum (O). These eggs are difficult to distinguish from *Fasciolopsis buski*.

Fasciolopsis buski

Fasciolopsis buski is common in the Orient where it infects humans and pigs. Its life cycle is similar to *Fasciola hepatica,* but differs in that its site of infection is the small intestine, not the liver. Consequences of infection include inflammation of the intestinal wall and obstruction of the gut if the worms are numerous. Chronic infections lead to ulceration, bleeding, diarrhea, and abscesses of the intestinal wall. Metabolites from the worm may sensitize the host, which may cause death. Diagnosis is made by identification of the eggs in fecal samples (Fig. 13-41).

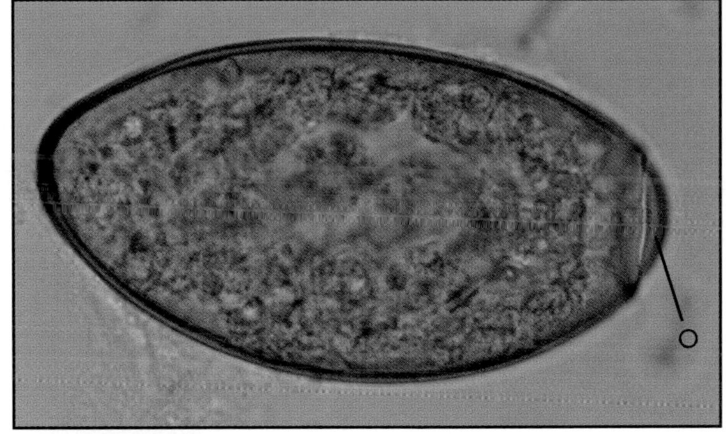

FIGURE 13-41 *Fasciolopsis buski* egg in a fecal specimen (X1000, D'Antoni's iodine stain). These eggs are very similar to those of *Fasciola hepatica.* The operculum (O) is visible at the right.

Paragonimus westermani

Paragonimus westermani is a lung fluke and one of several species to cause *paragonimiasis,* a disease mostly found in Asia, Africa and South America. *P. westermani* is primarily a parasite of carnivores, but humans (omnivores) may get infected when eating undercooked crabs or crayfish infected with the cysts. Ingested juveniles excyst in the duodenum and travel to the abdominal wall. After several days, they resume their journey and find their way to the bronchioles where the adults mature. Eggs (Fig. 13-42) are released in approximately two to three months and are diagnostic of infection. They may be recovered in sputum, lung fluids or feces. Consequences of lung infection are a local inflammatory response followed by possible ulceration. Symptoms include cough with discolored or bloody sputum and difficulty breathing. These cases are rarely fatal, but may last a couple of decades. Occasionally, the wandering juveniles end up in other tissues, such as the brain or spinal cord, which can cause paralysis or death.

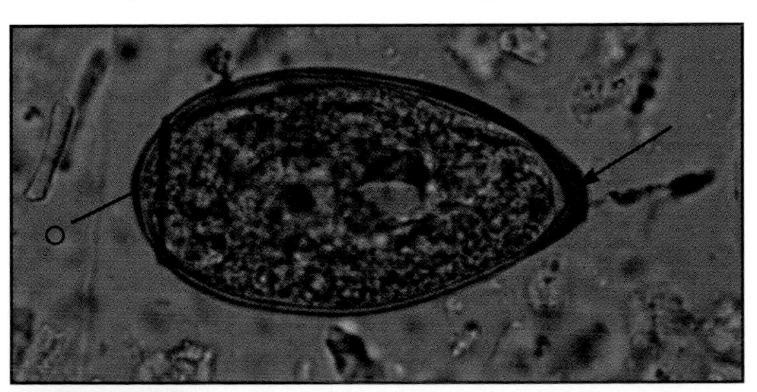

FIGURE 13-42 *Paragonimus westermani* egg in a fecal specimen (X1000, D'Anotni's iodine stain). *Paragonimus westermani* eggs are ovoid and range in size from 80 to 120 µm long by 45 to 70 µm wide. They have an operculum (O) and the shell is especially thick at the abopercular end (arrow). They are unembryonated when seen in feces.

Schistosoma haematobium

Schistosoma haematobium is an African and Middle Eastern blood fluke that causes *urinary schistosomiasis.* Infection occurs via contact with fecally contaminated water containing juveniles of the species. The juveniles penetrate the skin, enter circulation and continue development in the liver. After about three weeks in the liver, the adult worms colonize the veins associated with the urinary bladder and begin to lay eggs (Fig. 13-43). Some eggs pass through the walls of the veins and then the bladder to be passed out in urine, but a majority of them become trapped in the wall and initiate a build-up of fibrous tissue as well as an immune response. Symptoms of the disease include hematuria and painful urination. There is also a high probability of developing bladder cancer. If there is a high parasite load and the infection is chronic, other parts of the genitourinary system may become involved. Diagnosis is by finding eggs in urine or feces.

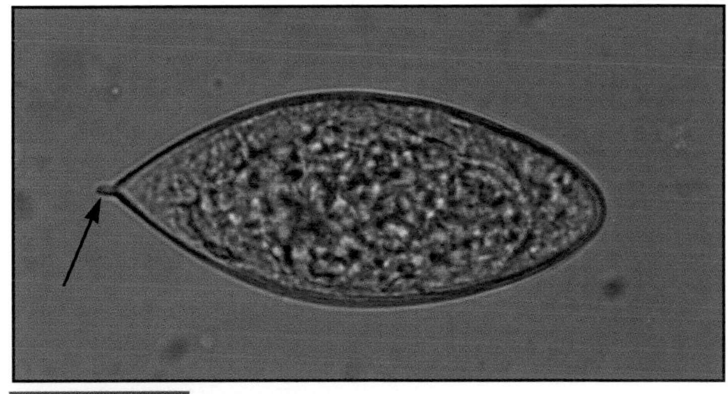

FIGURE 13-43 *Schistosoma haematobium* egg in a urine specimen (X1000). *Schistosoma haematobium* eggs are large (112 to 170 µm long by 40 to 70 µm wide), thin-shelled and lack an operculum. There is a distinctive terminal spine (arrow). Each egg contains a larva called a *miracidium.*

Schistosoma japonicum

Schistosoma japonicum is a Southeast Asian blood fluke. It has a life cycle similar to *S. haematobium*, but the adults reside in veins of the small intestine. Adults produce eggs (Fig. 13-44) that penetrate the intestine and pass out with the feces. Presence of eggs in the feces indicates infection. Some patients are asymptomatic, whereas others have bloody diarrhea, abdominal pain and lethargy. In some cases, eggs reach the brain and the infection may be fatal.

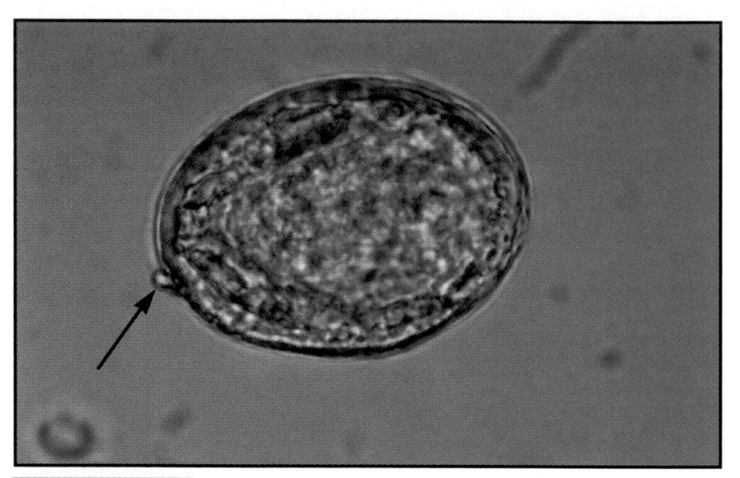

FIGURE 13-44 *Schistosoma japonicum* egg in a fecal specimen (X1000). *Schistosoma japonicum* eggs are thin-shelled and lack an operculum. They range in size from 70 to 100 µm long by 55 to 65 µm wide. There may be a small spine (arrow) visible near one end. Each egg contains a larva called a miracidium.

Schistosoma mansoni

Schistosoma mansoni is found in Brazil, some Caribbean islands, Africa and parts of the Middle East. It has a life cycle similar to *S. haematobium*, but the adults reside in the veins of the hepatic portal system. Eggs (Figs. 13-45 and 13-46) from the adults penetrate the intestinal wall and are passed out with the feces. Presence of eggs in the feces indicates infection. Symptoms are similar to those produced by *S. japonicum* infection.

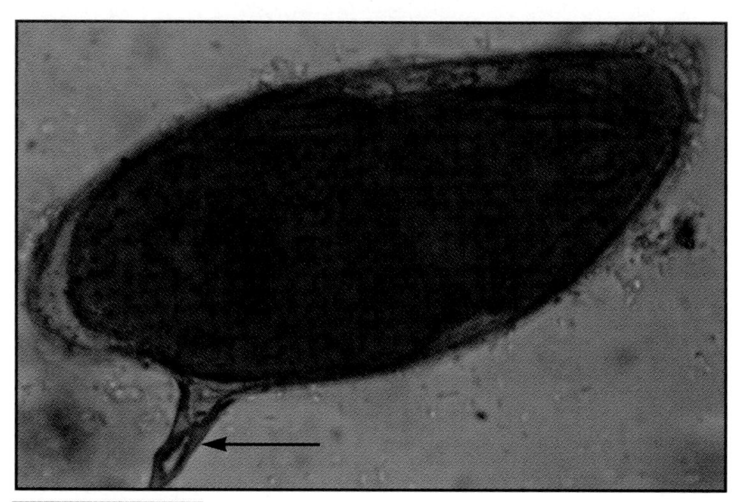

FIGURE 13-45 *Schistosoma mansoni* egg in a fecal specimen (X1000, D'Antoni's iodine stain). *Schistosoma mansoni* eggs are large (114 to 175 µm long by 45 to 70 µm wide) and contain a larva called a miracidium. They are thin-shelled, lack an operculum, and have a distinctive lateral spine (arrow).

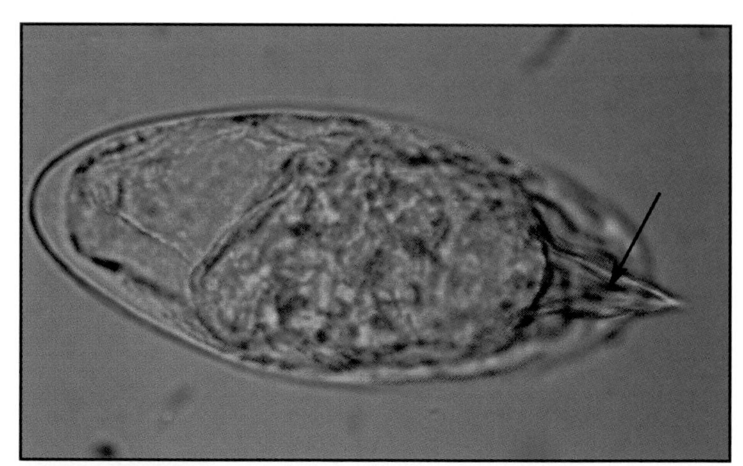

FIGURE 13-46 *Schistosoma mansoni* egg in a fecal specimen (X1000). The lateral spine may be oriented in such a way as to be difficult to see. In this specimen, the spine (arrow) is above the egg and the egg may be misidentified as *S. haematobium*.

Cestode Parasites Found in Clinical Specimens

Diphyllobothrium latum

Diphyllobothrium latum is the broad fish tapeworm. It is found in Northern Europe and the Great Lakes and west coast regions of North America. The closely related *D. ursi* is responsible for infection in northeastern North America. *D. latum* juveniles infect the muscle of fish-eating carnivores, which pass the infection on when they are eaten raw or are undercooked. Adults develop in the carnivore's intestine and begin egg production between one and two weeks later. Infection may result in no symptoms or mild symptoms, such as diarrhea, nausea, abdominal pain and weakness. In heavy infections, mechanical blockage of the intestine may occur. Rarely, infection results in pernicious anemia due to the worm's uptake of vitamin B_{12}. Diagnosis is commonly made by finding the eggs (Fig. 13-47) or more rarely proglottids (Fig. 13-48) in feces.

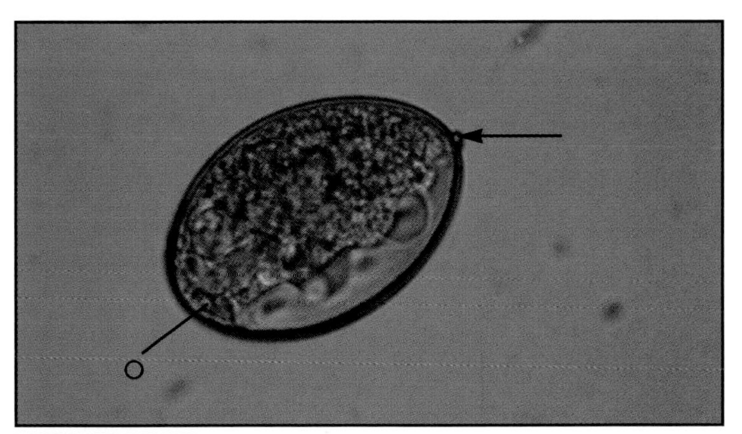

FIGURE 13-47 *Diphyllobothrium latum* egg in a fecal specimen (X1000, D'Antoni's iodine stain). The eggs are unembryonated when passed in feces and have an operculum (O). A knob (arrow) is often present at the abopercular end. Their dimensions are 58 to 75 µm long by 44 to 50 µm wide.

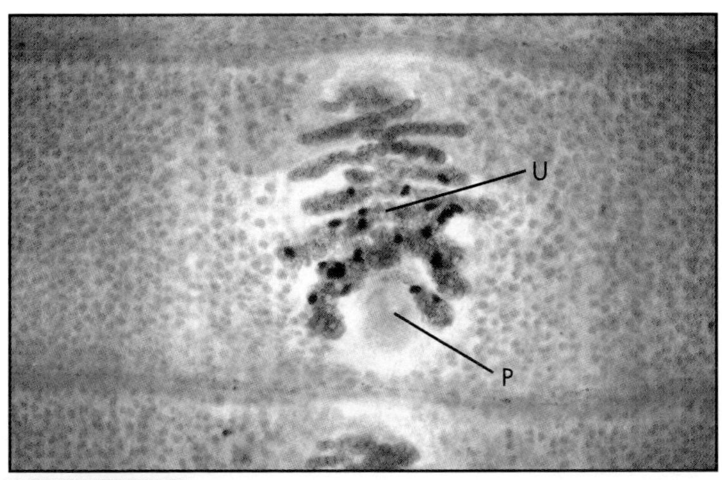

FIGURE 13-48 *Diphyllobothrium latum* proglottid (X32). Proglottids are typically wider than long and contain a rosette-shaped uterus (U) opening to a midventral pore (P).

Dipylidium caninum

Dipylidium caninum is a common parasite of dogs and cats. Human infection usually occurs in children. The adult worms reside in dog or cat intestines and release proglottids (Fig. 13-49) containing egg packets that migrate out of the anus. When these dry, they look like rice grains. Larval fleas may eat the eggs and become infected. If the dog, cat or child ingests one of these fleas, the life cycle is completed in the new host. Infection may be asymptomatic or produce mild abdominal discomfort, loss of apetite and indigestion. Diagnosis is made by identifying the egg packets (Fig. 13-50).

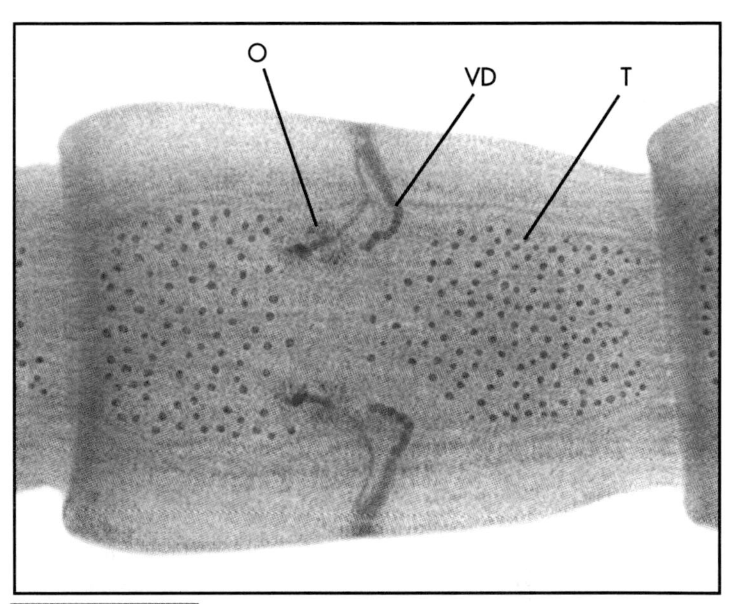

FIGURE 13-49 *Dipylidium caninum* maturing proglottid (X64). Visible in the proglottid are the testes (T), vasa deferentia (VD), and ovaries (O). The reproductive openings on each side of the proglottid give this worm its common name — the "double-pored tapeworm."

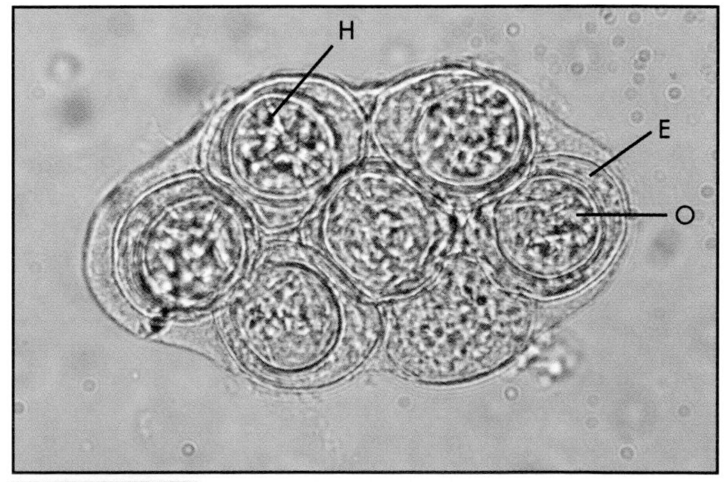

FIGURE 13-50 *Dipylidium caninum* egg packet (X1000). Each *Dipylidium caninum* egg packet is composed of 5 to 15 eggs (E) each with an oncosphere (O). The oncospheres contain six hooklets (H).

Echinococcus granulosus

The definitive host of *Echinococcus granulosus* is a carnivore, but the life cycle requires an intermediate host, usually an herbivorous mammal. Humans involved in raising domesticated herbivores (*e.g.*, sheep with their associated dogs) are most susceptible as intermediate hosts and develop *hydatid disease*. Ingestion of a juvenile *E. granulosus* leads to development of a *hydatid cyst* in the lung, liver, or other organ, a process that may take many years. The cyst (Fig. 13-51) has a thick wall and develops many protoscolices within (Fig. 13-52). The protoscolices, if ingested, are infective to the definitive host. Symptoms depend on the location and size of the hydatid cyst which interferes with normal organ function. Due to sensitization by the parasite's antigens, release of fluid from the cyst can result in anaphylactic shock of the host. Diagnosis is made by detection of the cyst by ultrasound or X-ray.

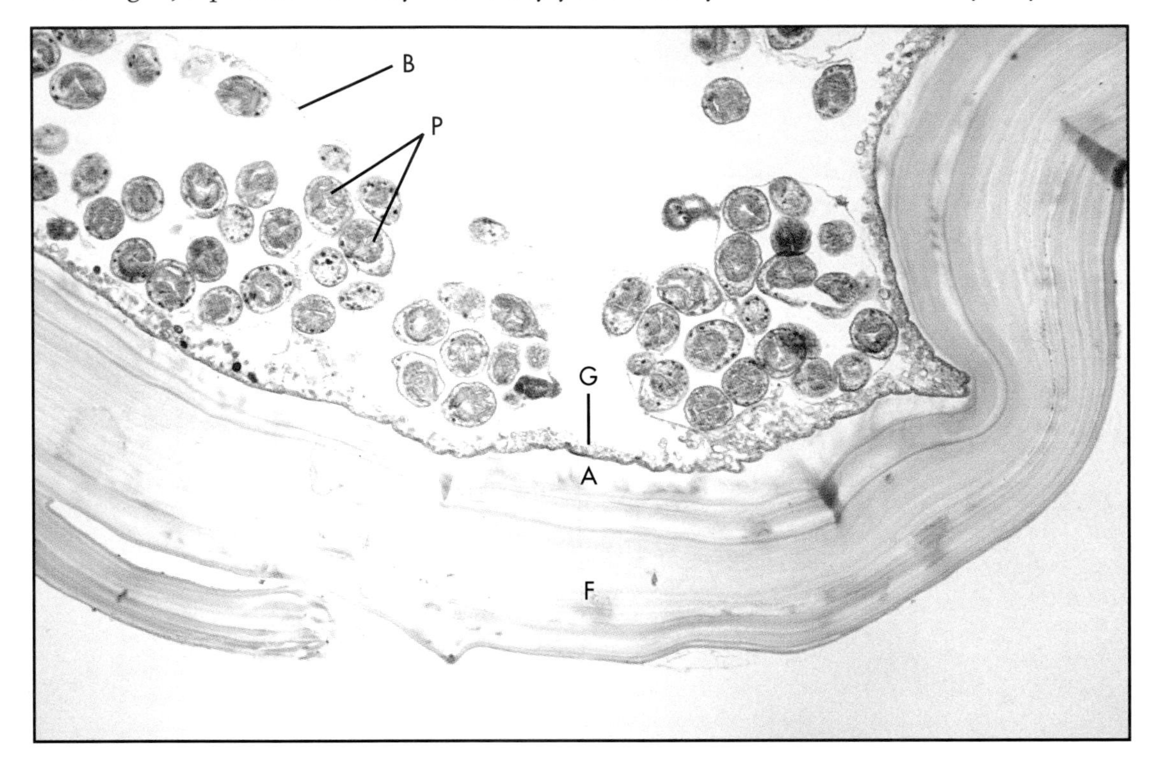

FIGURE 13-51 *Echinococcus granulosus* cyst in lung tissue (sec., X96). The cyst wall consists of a fibrous layer of host tissue (F), an acellular layer (A), and a germinal epithelium (G) that gives rise to stalked brood capsules (B). Brood capsules produce many *E. granulosus* protoscolices (P).

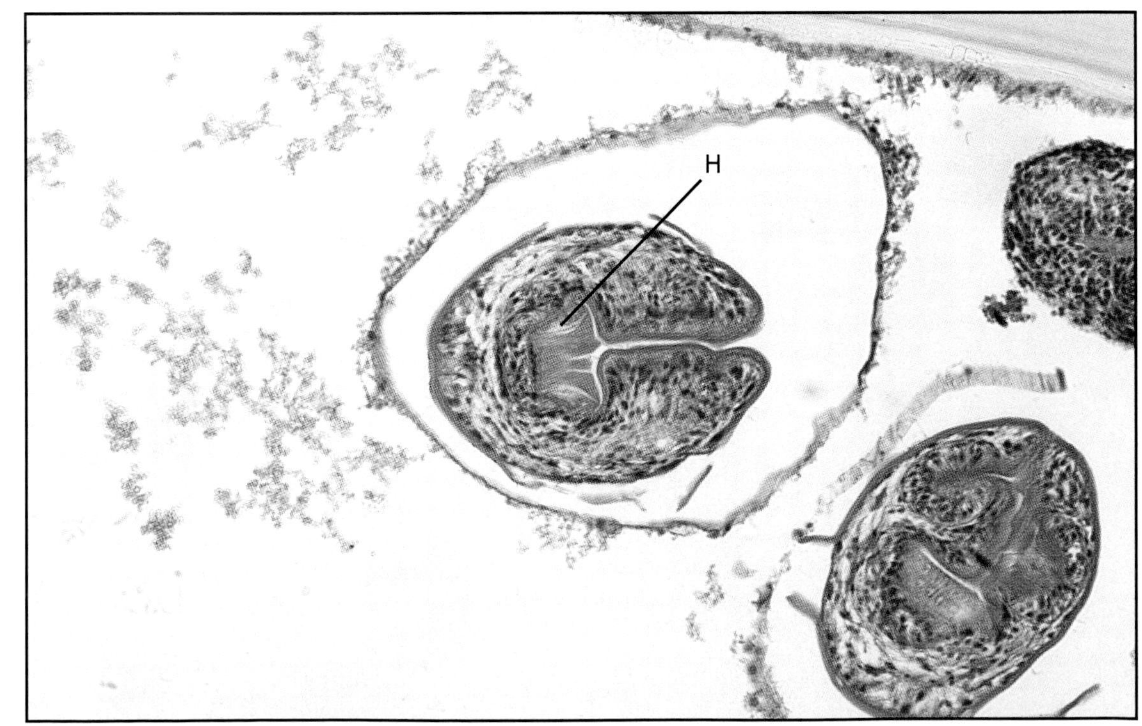

FIGURE 13-52 *Echinococcus granulosus* protoscolex (l.s., X320). The protoscolex contains an invaginated scolex with hooks (H). Upon ingestion by the host, the protoscolex evaginates and produces an infectious scolex that attaches to the intestinal wall, matures and produces eggs.

Hymenolepis diminuta

Hymenolepis diminuta is a tapeworm that uses beetles as the intermediate host and rats as the definitive host. Humans may also be infected via ingestion of beetles associated with stored grains. The adult worms develop in and attach to the intestinal mucosa, with one host harboring numerous worms. They release eggs (Fig. 13-53) which exit with the feces and are useful in diagnosis. Proglottids are rarely found in feces and are not of diagnostic use. Many infections are asymptomatic, but some result in mild abdominal discomfort and digestive upset.

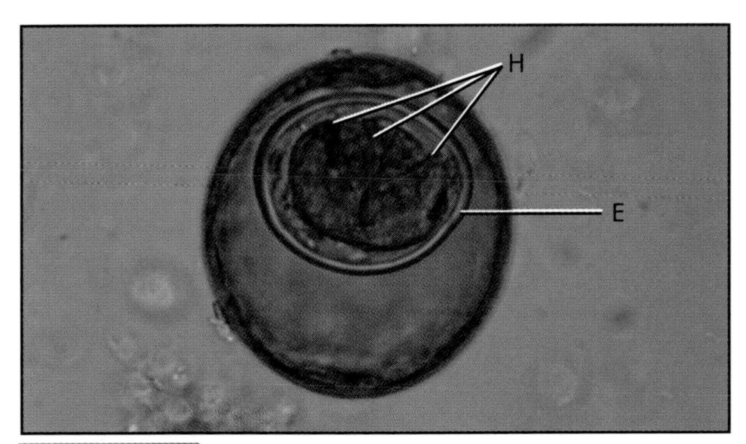

FIGURE 13-53 *Hymenolepis diminuta* egg in feces (X1000, D'Antoni's iodine stain). The spherical, thick shelled eggs of *Hymenolepis diminuta* range in size from 70 to 85 μm long by 60 to 80 μm wide. The embryo (E) is centrally positioned separate from the wall. Three pairs of hooks (H) are also present.

Hymenolepis (Vampirolepis) nana

Hymenolepis nana is the dwarf tapeworm and is the most common cestode parasite of humans in the world. When eggs (Fig. 13-54) are ingested, the *oncospheres* develop into juveniles in the lymphatics of intestinal villi. These juveniles are then released into the lumen within a week and attach to the mucosa to mature into adults. Infection may involve hundreds of worms, yet symptoms are usually mild: diarrhea, nausea, loss of appetite or abdominal pain. Eggs may reinfect the same host or pass out with the feces to infect a new host. Eggs in the feces are used for identification, but proglottids are not as they are rarely passed.

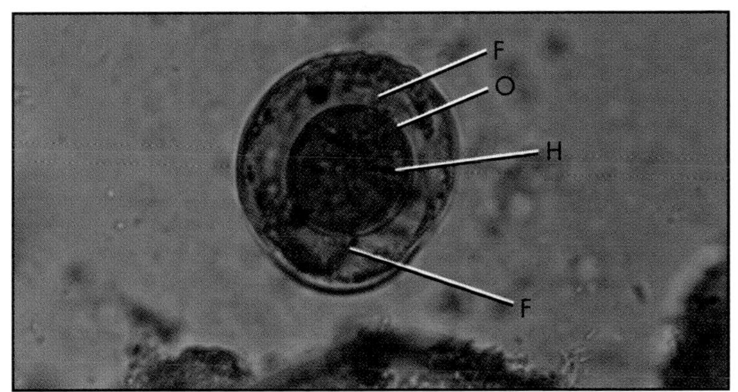

FIGURE 13-54 *Hymenolepis nana* egg in feces (X1000, D'Antoni's iodine stain). The *Hymenolepis nana* egg resembles the egg of *H. diminuta*, but it is smaller (30 to 47 μm in diameter) and has a thinner shell. The oncosphere (O) is separated from the shell and contains six hooks (H). Another distinguishing feature is the presence of between four and eight filaments (F) arising from either end of the oncosphere.

Taenia spp.

Two taeniid worms are important human pathogens. These are *Taenia saginata* (*Taeniarhynchus saginatus*) — the beef tapeworm — and *Taenia solium* — the pork tapeworm.

T. saginata infects humans who eat undercooked beef containing juvenile worms. In the presence of bile salts, the juvenile develops into an adult and begins producing gravid proglottids within a few weeks. Symptoms of infection are usually mild nausea, diarrhea, abdominal pain and headache. Diagnosis to species is impossible with only the eggs (Fig 13-55); specific identification requires a scolex or gravid proglottid (Fig. 13-56).

The *T. solium* life cycle is similar to *T. saginata*, but the host is pork, not beef, so human infection occurs when undercooked pork is eaten. If eggs are ingested, a juvenile form called a *cysticercus* develops. Cysticerci may be found in any tissue, especially subcutaneous connective tissues, eyes, brain, heart, liver, lungs and coelom. Symptoms of cysticercosis depend on the tissue infected, but mostly they are not severe. However, death of a cysticercus can produce a rapidly fatal inflammatory response. As with *T. saginata*, diagnosis to species is impossible with only the eggs; specific identification requires a scolex or gravid proglottid (Figs. 13-57 and 13-58).

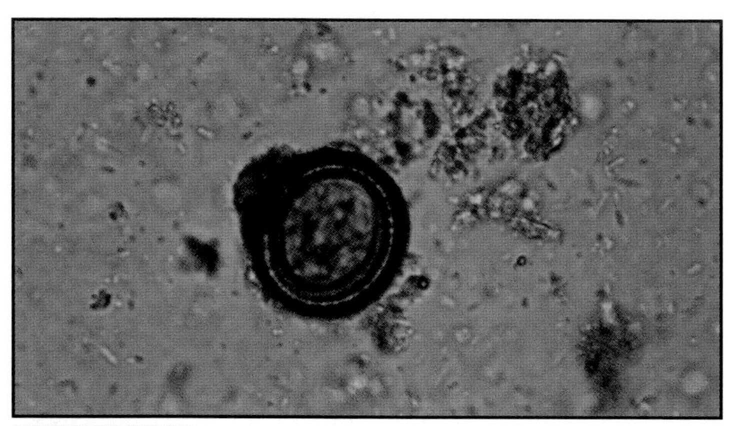

FIGURE 13-55 *Taenia spp.* egg in feces (X1000). Taeniid eggs are distinctive looking enough to identify to genus, but not distinctive enough to speciate. Eggs are spherical and approximately 40 µm in diameter with a striated shell. The oncosphere contains six hooks.

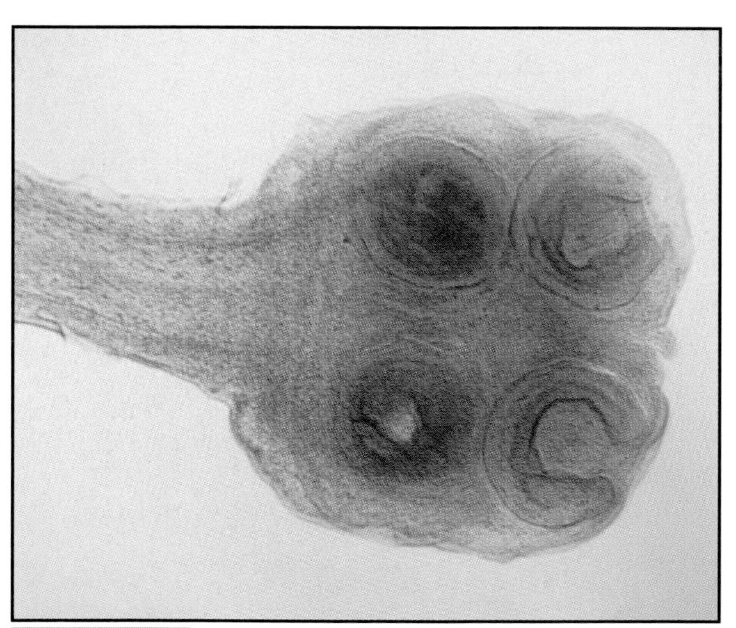

FIGURE 13-56 *Taenia saginata* scolex (X160). The *Taenia saginatus* scolex has four suckers and no hooks.

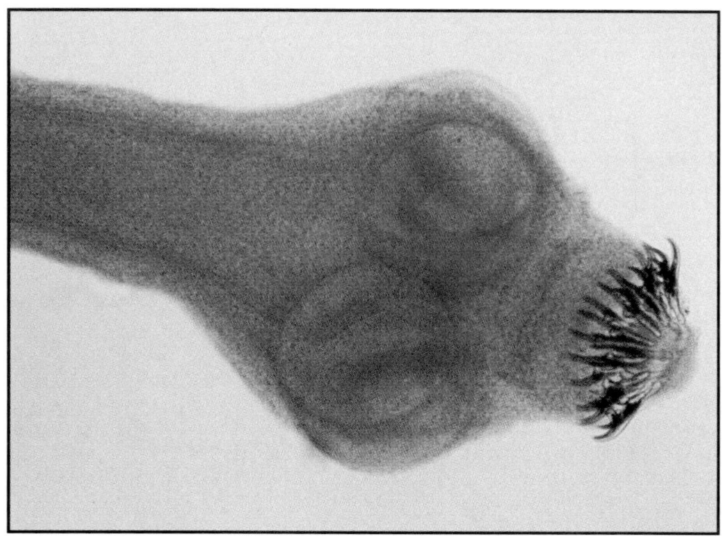

FIGURE 13-57 *Taenia solium* scolex (X64). The *Taenia solium* scolex has two rings of hooks and four suckers.

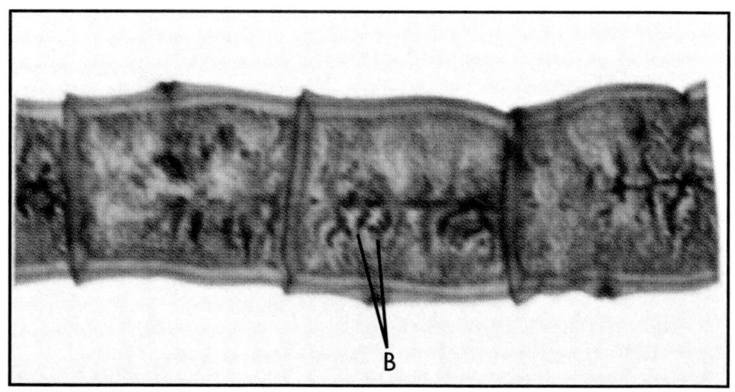

FIGURE 13-58 *Taenia solium* proglottid (X10). The uterus of *Taenia solium* proglottids consists of a central portion with 7 to 13 lateral branches (B). These are easily distinguished from *T. saginata* proglottids, which have between 15 and 20 lateral branches.

Nematode Parasites Found in Clinical Specimens

Ascaris lumbricoides

Ascaris lumbricoides is a large nematode — females may reach a length of 49 cm! Human infection occurs when eggs in fecally contaminated soil or food are ingested. Juveniles emerge in the intestine, penetrate its wall, then migrate to the lungs and other tissues. After a period of development in the lungs, the juveniles move up the respiratory tree to the esophagus and are swallowed again. Adults then reside in the small intestine and produce eggs (Figs. 13-59 and 13-60). Infection may result in inflammation in organs other than the lungs where juvenile worms settled incorrectly. *Ascaris* pneumonia occurs in heavy infections due to lung damage caused by the juveniles. If secondary bacterial infections occur, the pneumonia can be fatal. Blockage of the intestines and malnutrition also are possible in heavy infections. Lastly, under certain conditions, worms can wander to other body locations and cause damage or blockage. Identification of an *Ascaris* infection is made by observing the eggs in feces.

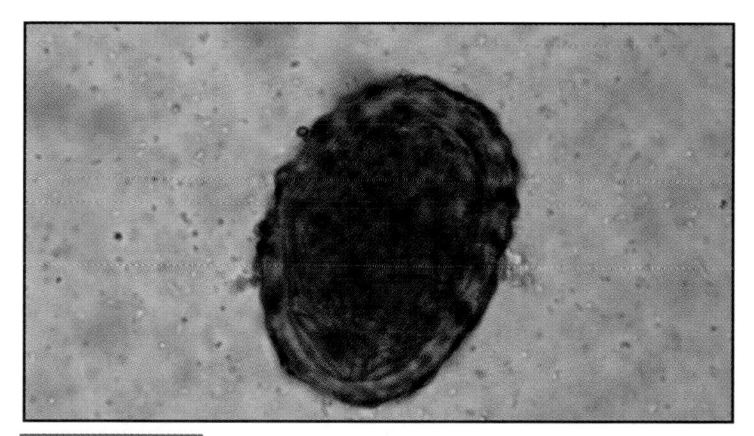

FIGURE 13-59 Fertile *Ascaris lumbricoides* egg in feces (X1000, D'Antoni's iodine stain). Fertile *Ascaris lumbricoides* eggs are 55 to 75 µm long and 35 to 50 µm wide and are embryontated. Their surface is covered by small bumps called mammillations.

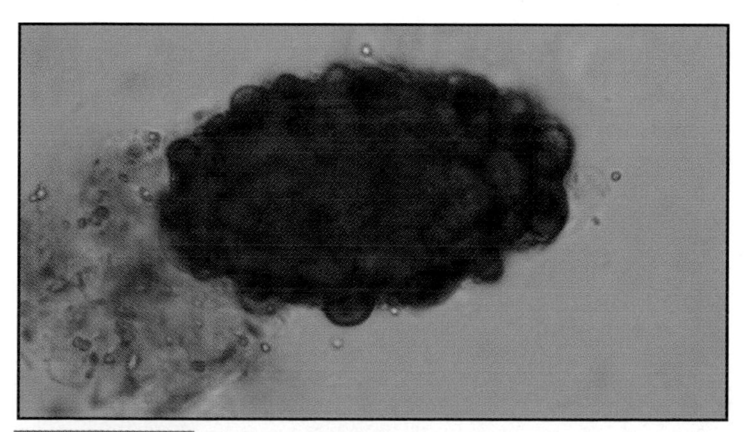

FIGURE 13-60 Infertile *Ascaris lumbricoides* egg in feces (X1000, D'Antoni's iodine stain). Infertile eggs are longer (up to 90 µm) than fertile eggs. There is no embryo inside.

Capillaria hepatica

Capillaria hepatica is mostly a rodent parasite, but human infections do occur. Infection results when food or soil contaminated with eggs is ingested. Juveniles emerge in the small intestine and migrate to the liver where development into adults occurs. Eggs (Fig. 13-61) are deposited in the liver, but cannot develop there. For further development, the infected host's liver must be eaten by a predator. The eggs pass through the predator's gut and are deposited in the soil with the feces. The main symptom of infection is hepatitis with eosinophilia, but other symptoms of liver disfunction may be present. Identification is made by liver biopsy or postmortem examinations.

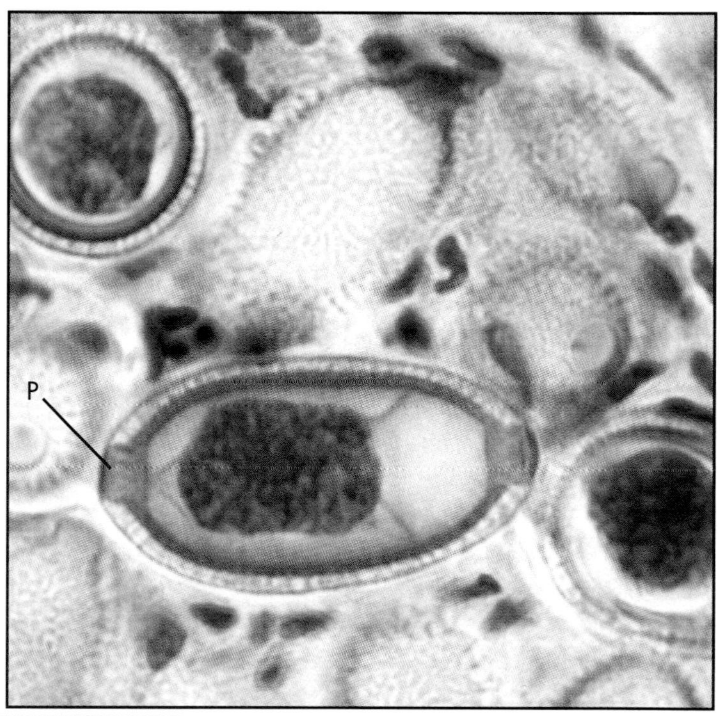

FIGURE 13-61 *Capillaria hepatica* eggs in the liver (sec., X960). *Capillaria hepatica* eggs are 51 to 67 µm long by 30 to 35 µm wide and have "plugs" (P) at either end. These are only passed in the feces if an infected liver has been eaten, but still must be distinguished from the similar eggs of *Trichuris trichiura* which has much more prominent plugs at each end.

Enterobius vermicularis

Enterobius vermicularis is the human pinworm. It is found worldwide and is especially prevalent among people in institutions (such as orphanages and mental hospitals) because the conditions favor the fecal-oral transmission of the parasite. Bedding, clothing and the fingers (from scratching) become contaminated and may be involved in transmission. Poor sanitary habits of children make them especially prone to infecting others. Transmission may also involve eggs (Fig. 13-62) being carried on air currents and then inhaled by a susceptible host. After ingestion, the eggs hatch in the duodenum and mature in the large intestine where the adults reside. Adult females emerge from the anus at night to lay between 4,600 and 16,000 eggs in the perianal region. About one-third of pinworm infections are asymptomatic. The two-thirds that are symptomatic usually do not produce serious symptoms. Diagnosis is made by identifying the eggs associated with the patient. Since the eggs are laid externally, they are rarely found in feces and a procedure to collect eggs from the perianal region is used.

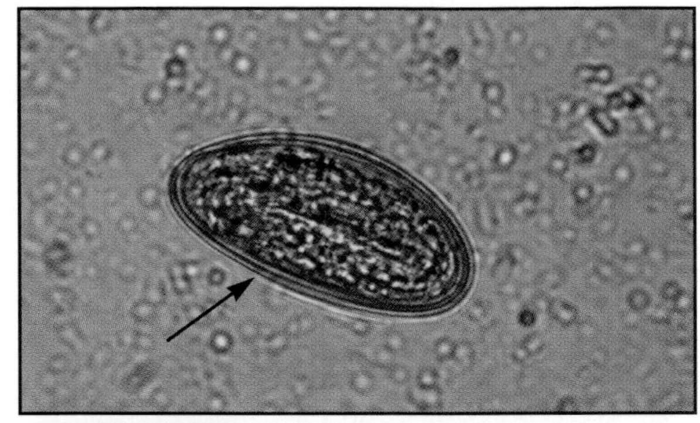

FIGURE 13-62 *Enterobius vermicularis* egg (X1000, D'Antoni's iodine stain). The eggs of *Enterobius vermicularis* are 50 to 60 µm long and 20 to 40 µm wide with one side flattened (arrow). They are usually embryonated in typical preparations.

Hookworms *(Ancylostoma duodenale* and *Necator americanus)*

Ancylostoma duodenale and *Necator americanus* have very similar morphologies and life cycles, and the eggs are indistinguishable, so they are considered together here. Infection occurs when juveniles penetrate the skin, enter the blood and travel to the lungs. They penetrate the respiratory membrane and are carried up and out of the lungs by ciliary action to the pharynx, where they are swallowed. When they reach the small intestine, they attach and mature into adults that feed on blood and tissues of the host. Eggs (Fig. 13-63) are passed in the feces and are diagnostic of infection. The severity of hookworm disease symptoms is related to the parasite load, and most infections are aymptomatic. As a rule, severe symptoms of bloody diarrhea and iron deficiency anemia are only seen in acute heavy or chronic infections. Diagnosis is made by identifying eggs in feces or by examining adult worms (Fig. 13-64) for characteristic mouth features.

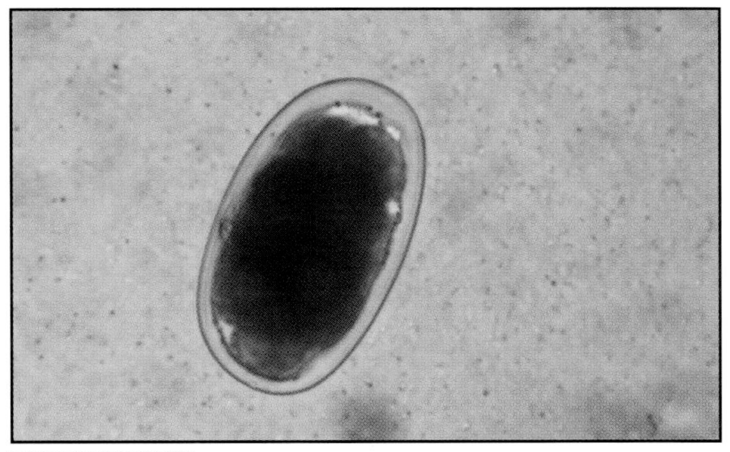

FIGURE 13-63 Hookworm egg in feces (X1000, D'Antoni's iodine stain). Hookworm eggs are 55 to 75 µm long and 36 to 40 µm wide. They have a thin shell and contain a developing embryo (seen here at about the 16 cell stage) that is separated from the shell when seen in fecal samples.

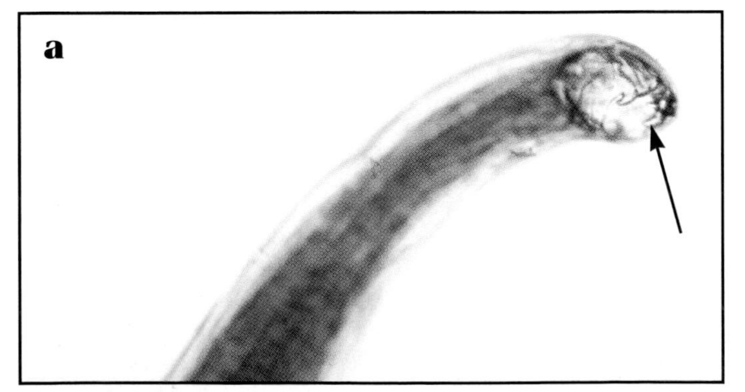

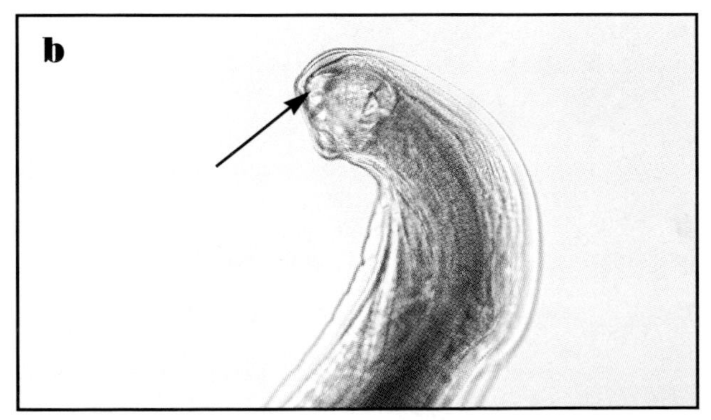

FIGURE 13-64 Heads of adult hookworms. *(a) Ancylostoma duodenale* head (X64) showing the pairs of chitinous teeth (arrow). *(b) Necator americanus* head (X160) showing the cutting plates (arrow).

Onchocerca volvulus

Onchocerca volvulus is found in Africa, Mexico, and parts of South and Central America. It causes *onchocerciasis* ("river blindness") and is transmitted through bites of infected black flies (*Simulium* spp.). Juveniles enter the tissues and develop into adults in about a year. Adults then reside in subcutaneous tissues and become surrounded by a collagenous capsule (Fig. 13-65). Microfilariae (Fig. 13-66) develop and may be picked up by a black fly when feeding to complete the cycle. Damage due to the adult worm is negligible, at worst forming a nodule. The microfilariae are more troublesome. Dead microfilariae may cause dermatitis followed by a thickening, cracking and depigmentaiton of the skin. Living microfilariae may infect the eyes, die, and stimulate an immune response. *Sclerosing keratitis*, which results in blindness, is one consequence of chronic eye inflammation. Diagnosis is by demonstration of microfilariae in skin snips.

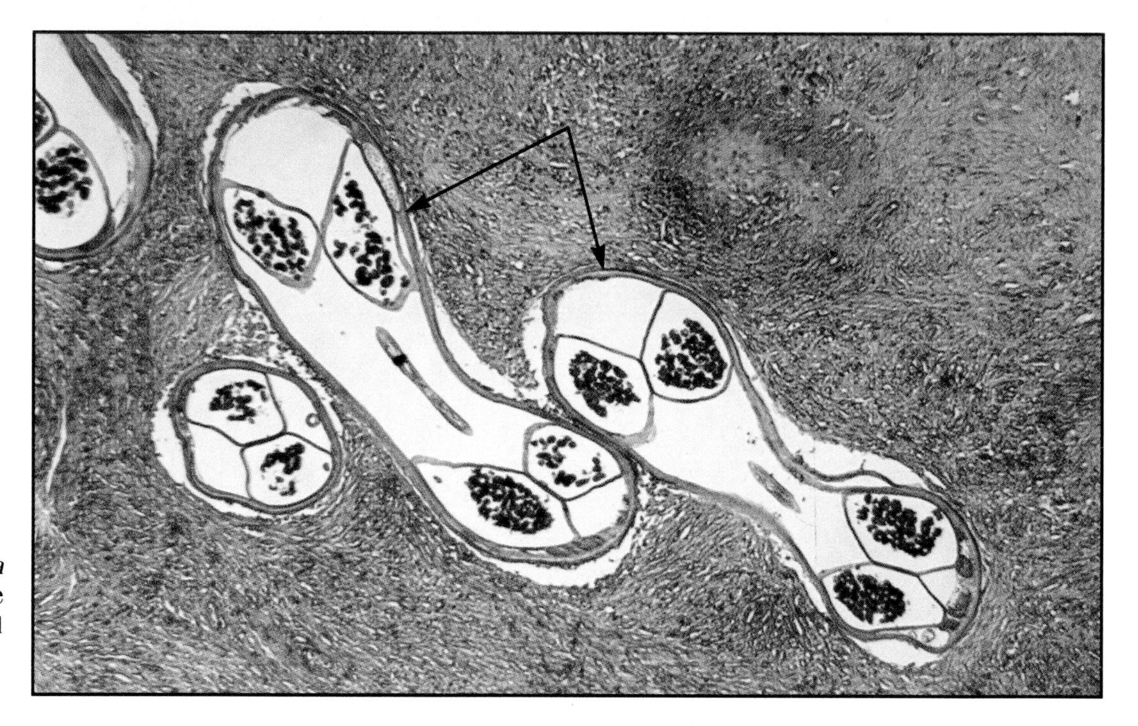

FIGURE 13-65 Adult *Onchocerca volvulus* in a nodule (sec., X40). The adult worm (arrows) is highly coiled within a fibrous nodule beneath the skin.

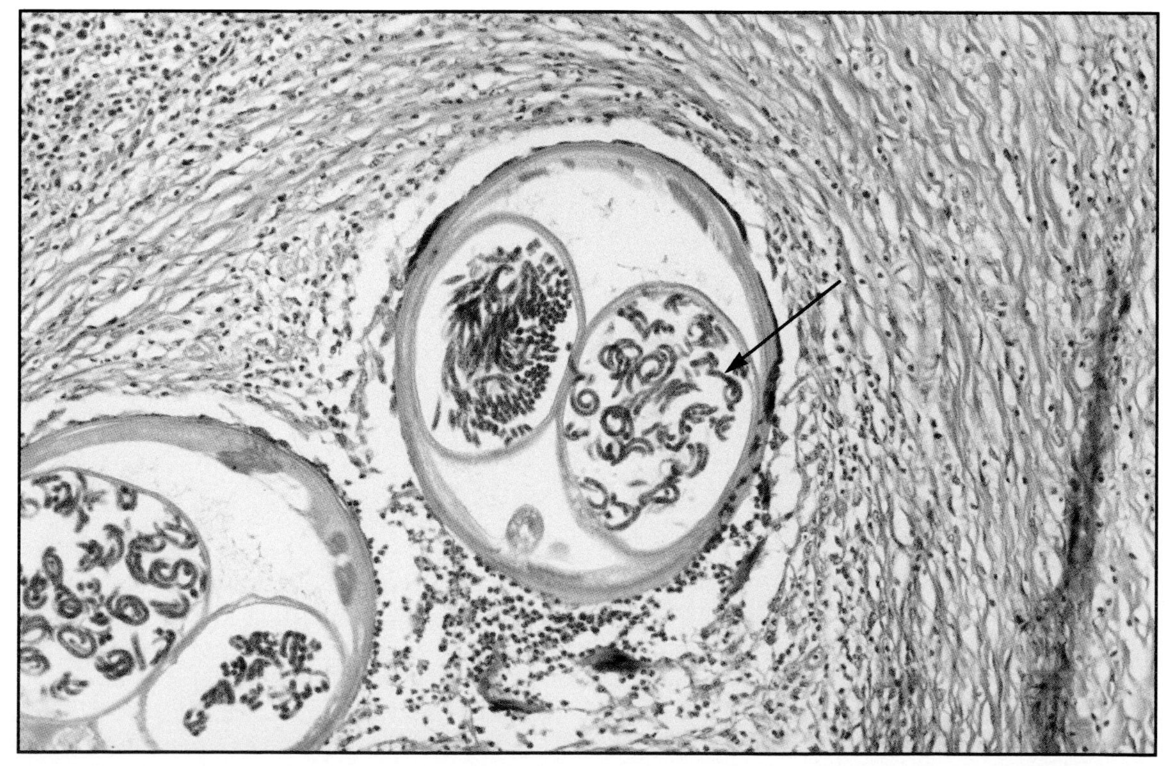

FIGURE 13-66 Section through an adult *Onchocerca volvulus* showing developing microfilariae (arrow) (X160).

Stongyloides stercoralis

Strongyloides stercoralis infection occurs by penetration of the skin by infective juveniles from fecally contaminated soil. The juveniles then migrate to the lungs and develop into parthenogenetic females that migrate to the pharynx, are swallowed, and then burrow into the intestinal mucosa. Each day they release a few dozen eggs that develop into juveniles (Fig. 13-67) before they are passed in the feces. These juveniles may become infective or may follow a developmental path that produces free-living adults. These adults eventually produce more infective juveniles and the cycle is completed. Symptoms of infection may be itching or secondary bacterial infection at the site of entry by the infective juveniles, a cough and burning of the chest during the pulmonary phase, and abdominal pain and perhaps septicemia during the intestinal phase. Diagnosis is by finding rhabditiform larvae in fresh fecal samples.

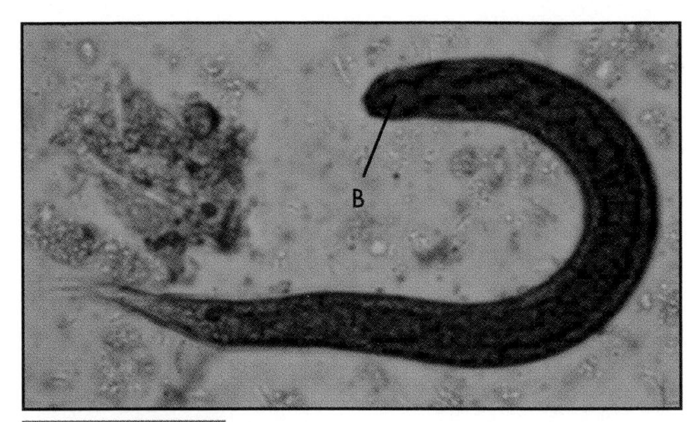

FIGURE 13-67 *Strongyloides stercoralis* rhabditiform larva in feces (X1000, D'Antoni's iodine stain). These larvae may be distinguished from hookworm larvae (which are rarely in feces) by their short bucal cavity (B).

Trichinella spiralis

Trichinella spiralis produces *trichinosis*, a disease of carnivorous mammals. Infection occurs when undercooked meat (*e.g.*, pork) containing infective juveniles in *nurse cells* is eaten. These juveniles emerge from their nurse cells and enter the intestinal mucosa. Between two and three days later, the juveniles have developed into mature adults that burrow within rows of the intestinal epithelial cells. Juveniles emerge from the females and are distributed throughout the body. When they are in skeletal muscle, they enter the muscle fibers and each induces the formation of a nurse cell (Figs. 13-68 and 13-69). In as little as four weeks, these juveniles become infective. Humans, unless they are eaten, are a dead-end host for this parasite. Symptoms of infection are many and varied because the juveniles migrate throughout the body. Some consequences of infection are pneumonia, meningitis, deafness, and nephritis. Death may occur due to heart, respiratory, or kidney failure, but most infections are subclinical. Diagnosis is by muscle biopsy.

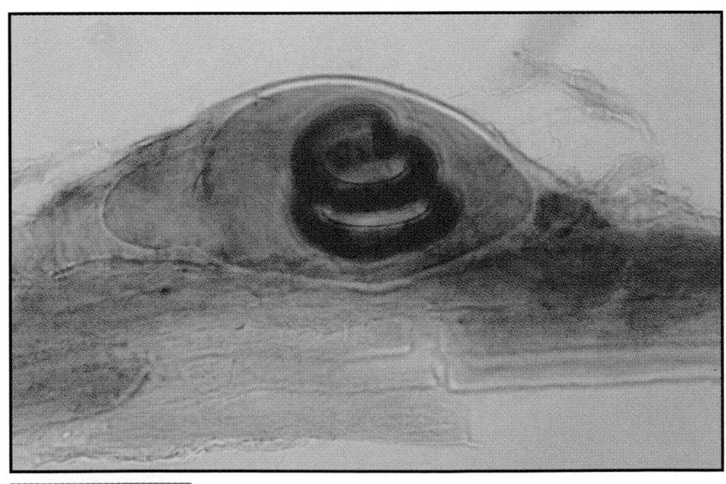

FIGURE 13-68 *Trichinella spiralis* larva in skeletal muscle (W.M., X260). The spiral juvenile and its nurse cell are visible in this preparation.

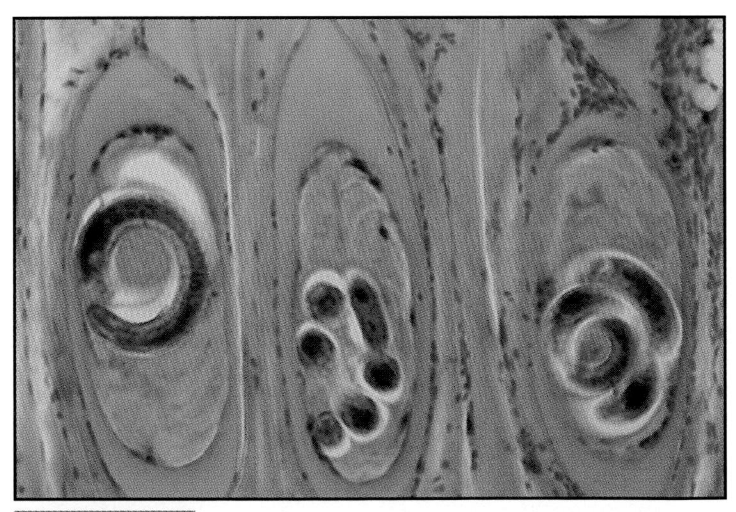

FIGURE 13-69 *Trichinella spiralis* larvae in skeletal muscle (sec., X260). Each larva has entered a different skeletal muscle cell and converted it into a nurse cell that sustains it with nourishment.

Trichuris trichiura

The whipworm *Trichuris trichiura* is a parasite of the large intestine. Infection occurs through ingestion of eggs in fecally contaminated soil or plants. The juveniles then emerge and penetrate the mucosa of the large intestine. As they grow, their posterior projects into the lumen while the anterior remains buried in the mucosa and feeds on cell contents and blood. The adult females release up to 20,000 eggs a day (Fig. 13-70) which pass out with the feces and are diagnostic of infection. Most infections are asymptomatic. With heavy worm burdens (more than 100), dysentery, anemia, and slowed growth and cognitive development are common.

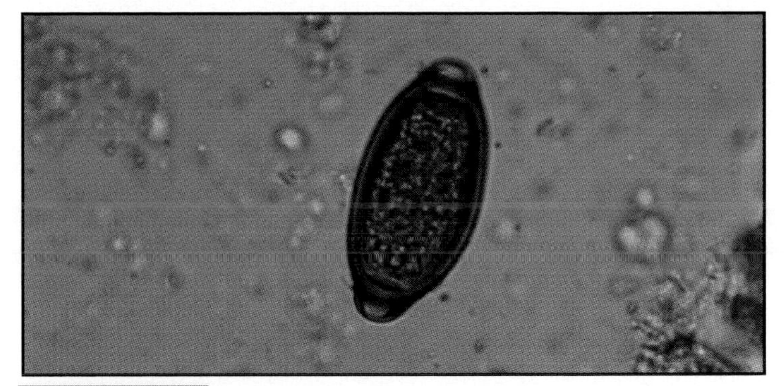

FIGURE 13-70 *Trichuris trichiura* egg in feces (X1000, D'Antoni's iodine stain). The barrel shaped eggs of *Trichuris trichiura* have distinctive plugs at either end. They are 50 to 55 µm in length by 22 to 24 µm wide.

Wuchereria bancrofti

Wuchereria bancrofti is a filarial worm that causes *lymphatic filariasis*. Infection occurs from the bite of a mosquito harboring infective juveniles. Upon injection into the host, the worms migrate into the large lymphatic vessels of the lower body and mature. Adults are found in coiled bunches and the females release microfilariae (Fig. 13-71) by the thousands. The microfilariae enter the blood and circulate there, often with a daily periodicity — most abundant at night when the mosquito vector is active and hidden away in lung capillaries during the day when it is not. Some infections are asymptomatic, whereas others result in acute inflammation of lymphatics associated with fever, chills, tenderness and toxemia. In the most serious cases, obstruction of lymphatic vessels occurs and results in *elephantiasis*, a disease caused by accumulation of lymph fluid in the tissues, an accumulation of fibrous connective tissue and a thickening of the skin. Diagnosis of infection is made by identifying microfilariae in blood smears.

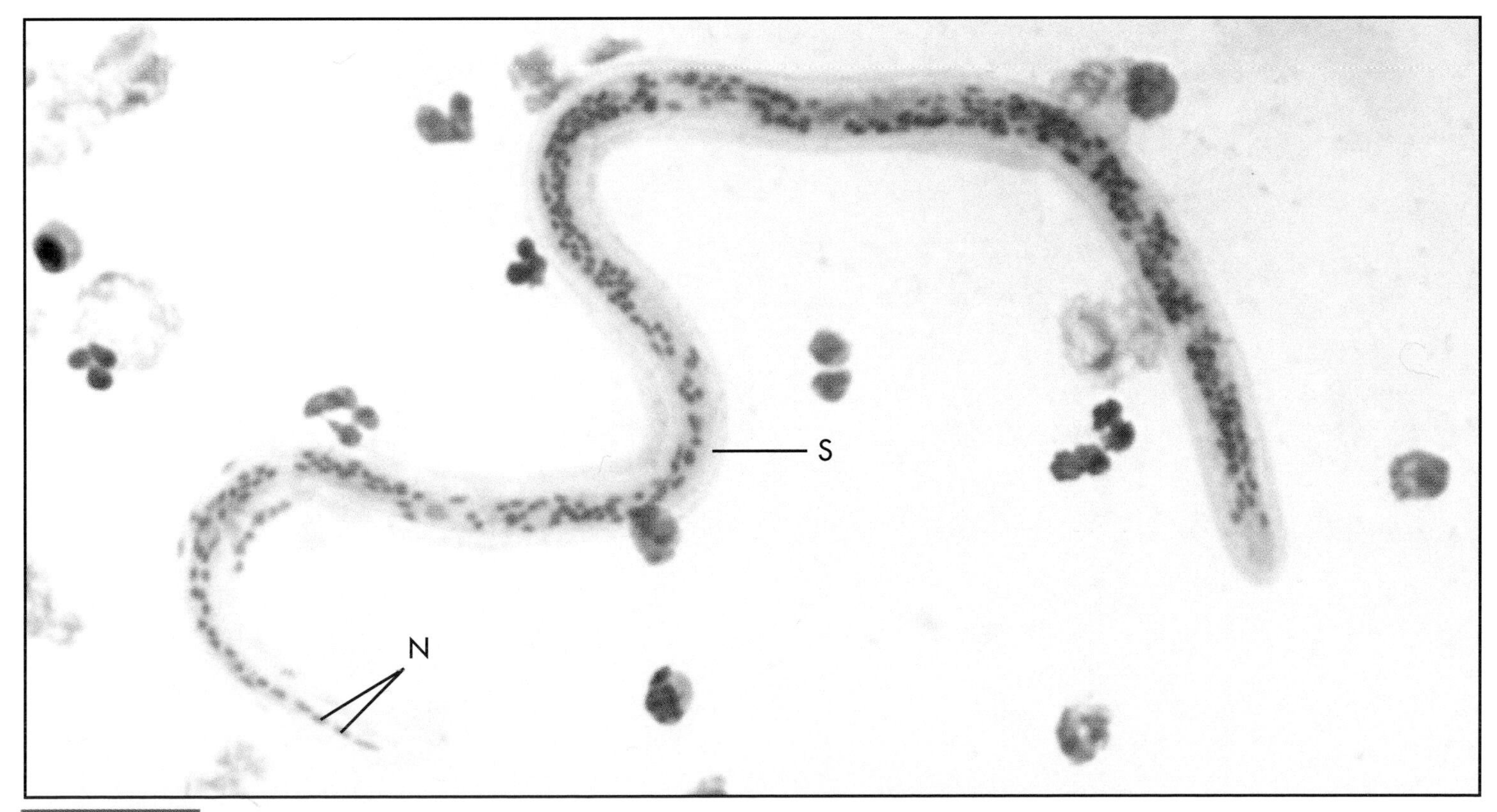

FIGURE 13-71 *Wuchereria bancrofti* microfilaria in a blood sample (X960). The microfilariae of *Wuchereria bancrofti* can be distinguished from others in the blood by the sheath (S) and the single column of nuclei (N) not extending to the tip of the tail.

Glycolysis

APPENDIX

The glycolytic pathway (components are shown in red type) is used in energy metabolism. Each glucose oxidized in glycolysis yields two pyruvic acids, 2 $NADH+H^+$, and a net of 2 ATPs by substrate phosphorylation. The $NADH+H^+$ may be oxidized in an electron transport chain or a fermentation pathway, depending on the organism and the environmental conditions. The former yields ATP, the latter usually does not.

Though its intermediates are carbohydrates, many are entry points for amino acid, lipid, and nucleotide catabolism. Many intermediates are also a source of carbon skeletons for the synthesis of these other biochemicals. *Major details have been omitted from these other pathways. Single arrows may represent several reactions and other carbon compounds not illustrated may be required to complete a particular reaction.*

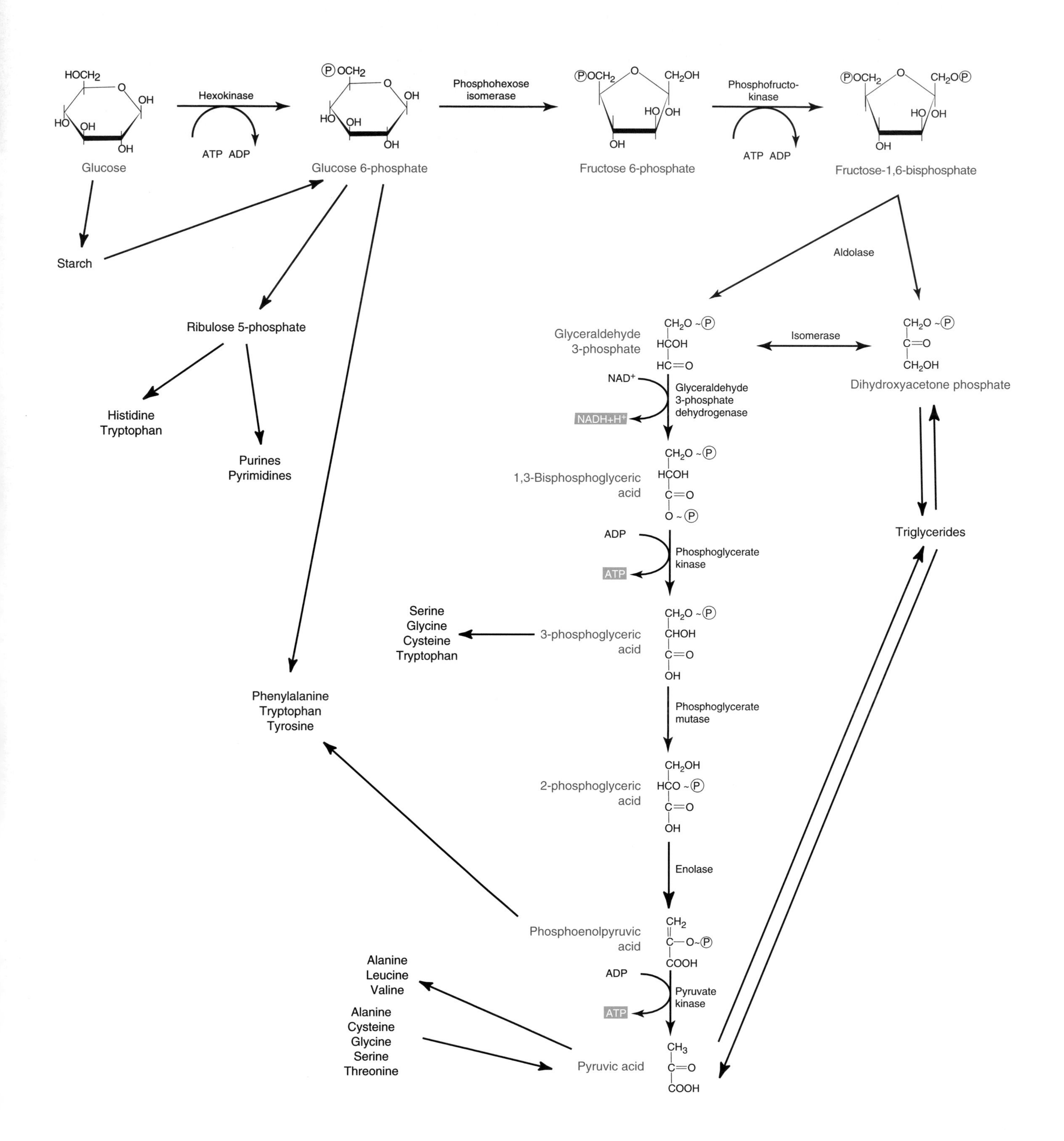

FIGURE A-1 Glycolysis and associated pathways. The components of glycolysis are shown in red.

APPENDIX B

Fermentation

The accompanying diagram illustrates some major fermentation pathways exhibited by microbes (though no single organism is capable of all of them — see Table B-1). Pyruvic acid (shown in the blue box) is typically the starting point for each. End products of fermentation are shown in red. Fermentation allows a cell living under anaerobic conditions to oxidize reduced coenzymes (such as $NADH+H^+$ and shown in blue) generated during glycolysis or other pathways. Some bacteria (aerotolerant anaerobes) rely solely on fermentation and do not use oxygen even if it is available.

Notice that fermentation end products typically fall into three categories: acid, gas, or an organic solvent (an alcohol or a ketone). The specific fermentation performed is the result of the enzymes present in a species and is often used as a basis of classification.

TABLE B-1 Major Fermentation Pathways

Fermentation	Major End Products	Representative Organisms
Alcoholic Fermentation	Ethanol and CO_2	*Saccharomyces cerevisiae*
Homofermentation	Lactic acid	*Streptococcus* and some *Lactobacillus*
Heterofermentation	Lactic acid, ethanol, and acetate	*Streptococcus, Leuconostoc* and *Lactobacillus*
Mixed Acid Fermentation	Acetic acid, formic acid, succinic acid, CO_2, H_2 and ethanol	*Escherichia, Salmonella, Klebsiella* and *Shigella*
2,3-Butanediol Fermentation	2,3-Butanediol	*Enterobacter, Serratia and Erwinia*
Butyric Acid/Butanol Fermentation	Butanol, butyric acid, acetone and isopropanol	*Clostridium, Butyrivibrio* and some *Bacillus*
Propionic Acid Fermentation	Propionic acid, acetic acid and CO_2	*Propionibacterium, Veillonella* and some *Clostridium*

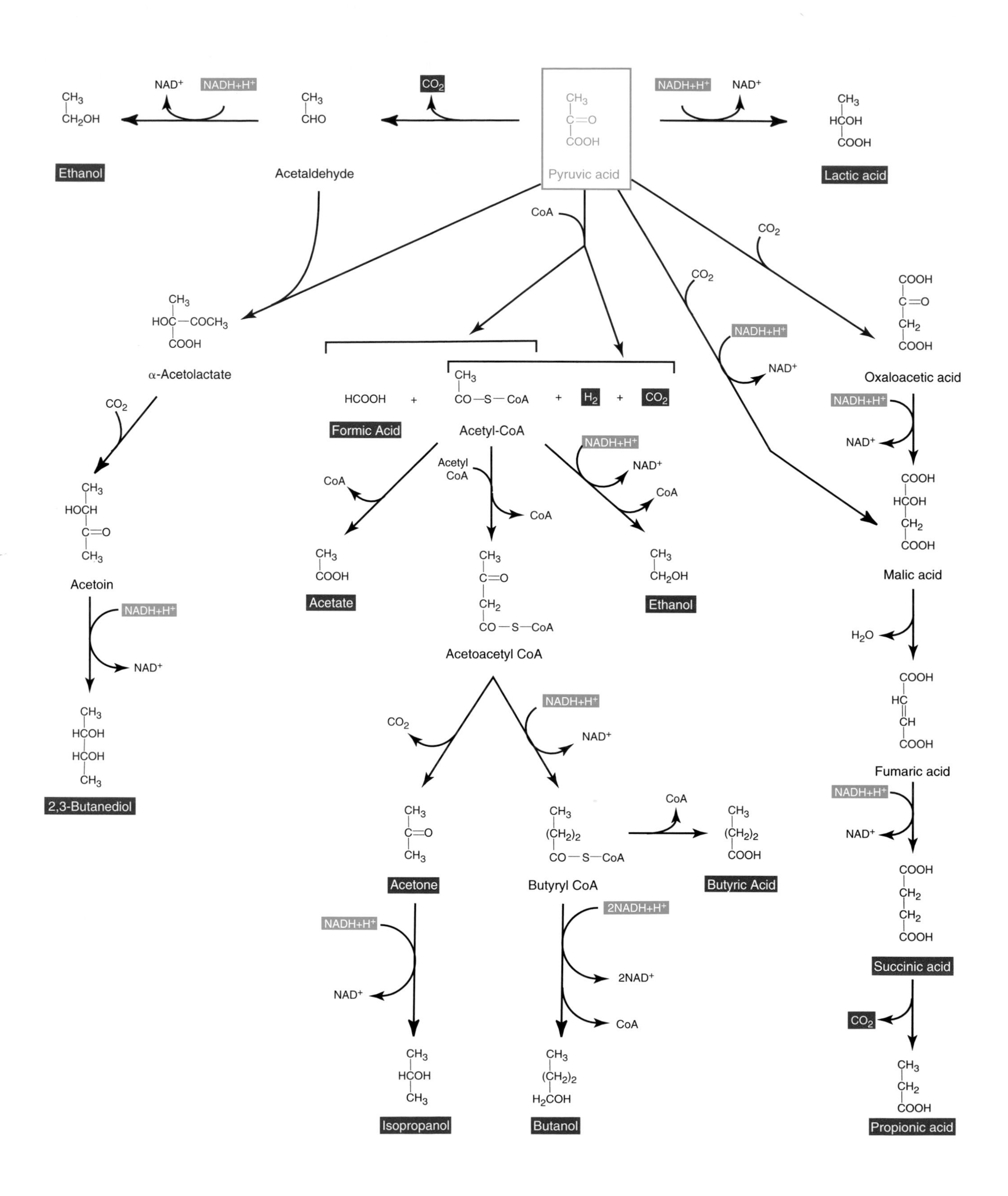

FIGURE B-1 Fermentation pathways. Fermentation products (shown in red) are usually an acid, gas and/or organic solvent.

APPENDIX C

Entry Step and the Krebs Cycle

The Krebs cycle is a major metabolic pathway used in energy production. Pyruvic acid produced in glycolysis or other pathways is first converted to acetyl-coenzyme A during the *entry step*. Acetyl-CoA enters the Krebs cycle through a condensation reaction with oxaloacetic acid. Products for each pyruvic acid that enters the cycle via the entry step are: 3 CO_2, 4 $NADH+H^+$, 1 $FADH_2$, and 1 GTP. The energy released from oxidation of reduced coenzymes ($NADH+H^+$ and $FADH_2$) in an electron transport chain is then used to make ATP.

Like glycolysis, many of the Krebs cycle's intermediates are entry points for amino acid, nucleotide and lipid catabolism, as well as a source of carbon skeletons for synthesis (anabolism) of the same compounds. These pathways are shown, but *details have been omitted. Single arrows may represent several reactions and other carbon compounds not illustrated may be required to complete a particular reaction.*

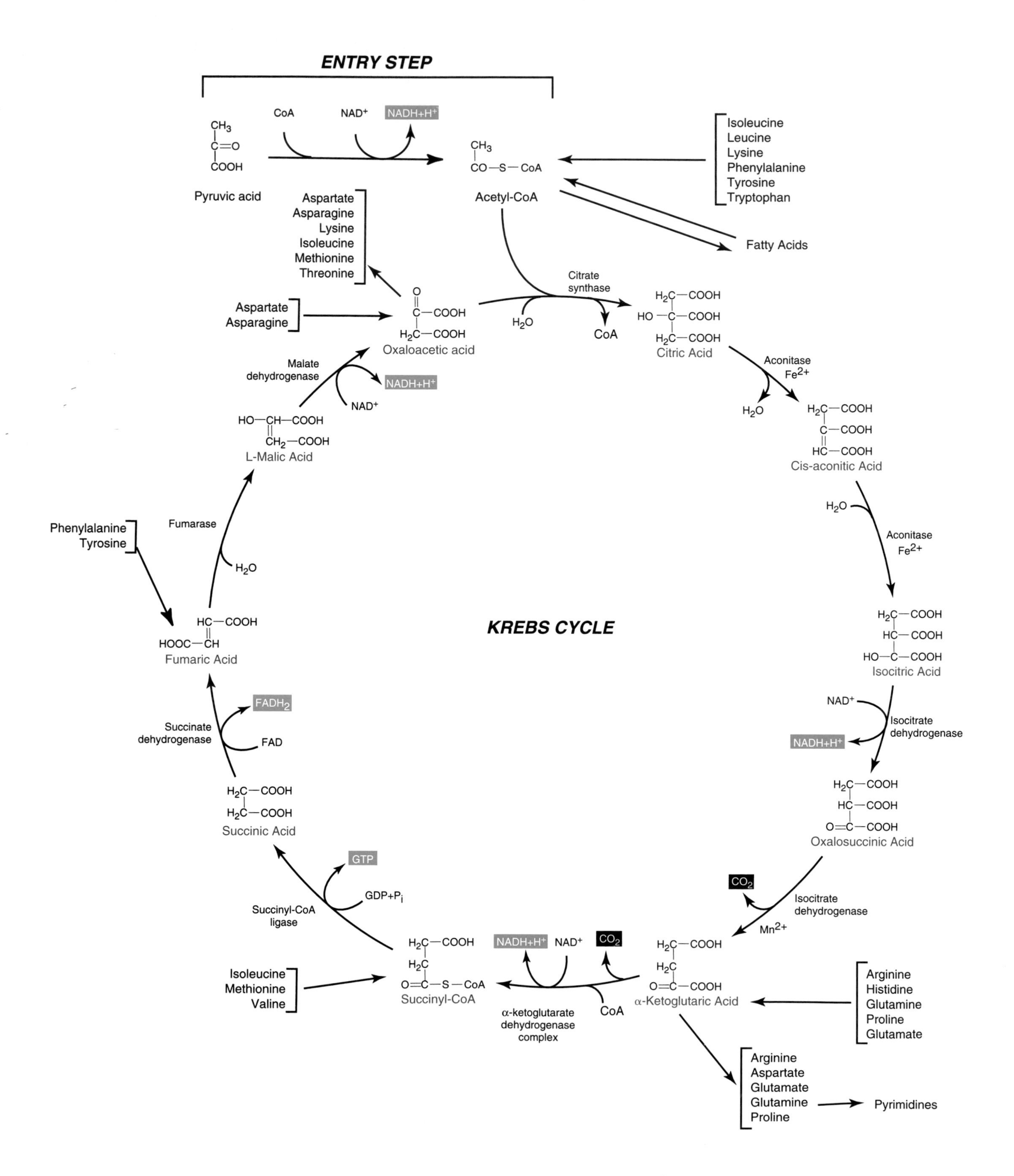

FIGURE C-1 The Entry Step and Krebs Cycle. Associated anabolic and catabolic pathways are also shown.

Selected References and Further Reading

APPENDIX

Baron, Ellen Jo, Lance R. Peterson and Sydney M. Finegold. *Bailey & Scott's Diagnostic Microbiology, 9th Ed.* Mosby–Year Book, Inc., 1994.

Battone, Edward (Editor). *Schneierson's Atlas of Diagnostic Microbiology.* Abbott Laboratories, 1984.

Braude,, Abraham I., Charles E. Davis and Joshua Fierer (Editors). *Infectious Diseases and Medical Microbiology.* W.B. Saunders Co., 1986).

Brooks, Geo. F., Janet S. Butel and L. Nicholas Ornston. *Jawetz, Melnick & Adelberg's Medical Microbiology, 20th Ed.* Appleton & Lange, 1995.

Brown, BarbaraBarabara A. *Hematology - Principles and Procedures, 6th Ed.* Lea and Febiger, 1993.

Gerhardt, Philipp (Editor-in-Chief). *Manual of Methods for General Bacteriology.* American Society for Microbiology, 1981.

Gerhardt, Philipp (Editor-in-Chief). *Methods for General and Molecular Bacteriology.* American Society for Microbiology, 1994.

Gillespie, Stephen. *Medical Microbiology Illustrated.* Butterworth-Heineman Ltd., 1994.

Holt, John G. (Editor). *Bergey's Manual of Determinative Bacteriology, 9th Ed.* Williams and Wilkins, 1994.

Howard, Barbara J. *Clinical and Pathogenic Microbiology, 2nd Ed.* Mosby-Yearbook Inc., 1994.

Junqueira, L. Carlos, Jose Carneiro and Robert O. Kelley. *Basic Histology, 8th Ed.* Appleton & Lange, 1995.

Koneman, Elmer Ww., *et. al. Color Atlas and Textbook of Diagnostic Microbiology, 5th Ed.* J. B. Lippincott company, 1997.

Krieg, Noel R. and John G. Holt (Editor-in-Chief). *Bergey's Manual of Systematic Bacteriology, Volume 1.* Williams & Wilkins, 1984.

Laskin, Allen I. and Hubert A. Lechevalier. *CRC Handbook of Microbiology, Volume V, 2nd Ed.* Chemical Rubber Company, 1984.

Lehninger, Albert L., David L. Nelson, and Michael M. Cox. *Principles of Biochemistry, 2nd. Ed.* Worth Publishers, 1993.

Lennette, Edwin H. (Editor-in-Chief). *Manual of Clinical Microbiology, 4th Ed.* American Society for Microbiology, 1985.

Lewin, Benjamin. *Genes IV.* Oxford University Press, New York. 1990.

MacFaddinMacFaddin, Jean F. *Biochemical Tests for Identification of Medical Bacteria, 2nd Ed.* Williams & Wilkins, 1980.

Mandlestam, Joel, Keneth McQuillen and Ian Dawes. *Biochemistry of Bacterial Growth, 3rd Ed.* Blackwell Scientific Publications, 1982.

Moat, Albert G. and John W. Foster. *Microbial Physiology, 3rd Ed.* Wiley-Liss, Inc., 1995.

Murray, Patrick R., (Editor-in-Chief). *Manual of Clinical Microbiology, 6th Ed.* American Society for Microbiology, 1995.

Murray, Robert K., Daryl K. Granner, Peter A. Mayes and Victor W. Rodwell. *Harper's Biochemistry, 23rd Ed.* Appleton & Lange, 1993.

Murray, Patrick R., (Editor-in-Chief). *Manual of Clinical Microbiology, 6th Ed.* American Society for Microbiology, 1995.

Roberts, Larry S., and John Janovy, Jr. *Foundations of Parasitology, 5th Ed.*, Wm. C. Brown Publishers, Dubuque, IA. 1996.

Roitt, Ivan, Jonathan Brostoff and David Male. *Immunology, 3rd Ed.* Mosby–Year Book Europe, Limited, 1993.

Ryan, Kenneth J. (Editor). *Sherris Medical Microbiology - AnAnb Introduction to Infectious Diseases, 3rd Ed.* Appleton and Lange, 1994.

Schaechter, Moselio, Gerald Medoff and Barry I. Eisenstein. *Mechanisms of Microbial Disease, 2nd Ed.* Williams & Wilkins, 1993.

Sneath, Peter H. A., Nicholas S. Mair, M. Elisabeth Sharpe and John G. Holt (Editor-in-Chief). *Bergey's Manual of Systematic Bacteriology, Volume 2.* Williams & WilkinsWilkens, 1986.

Stine, Gerald J. *AIDS Update 1994-1995.* Prentice-Hall, Inc. 1995.

Stites, Daniel P., Abba I. Terr and Tristram G. Parslow. *Basic and Clinical Immunology, 8th. Ed.* Appleton & Lange.

Stokes, E. Joan and G. L. Ridgeway. *Clinical Microbiology, 7th Ed.* Edward Arnold (UK) Williams & Wilkins, 1988.

Villar, Rodrigo. Personal interview. 20 Mar. 1998.

Voet, Donald and Judith Voet. *Biochemistry, 3rd Ed.* John Wiley & Sons. 1995.

Index